R. W. Horobin

Histochemistry

Histochemistry

An Explanatory Outline of Histochemistry and Biophysical Staining

Richard W. Horobin

With 77 Figures and 55 Tables

Gustav Fischer
Stuttgart
New York

Butterworths
London · Boston · Sydney
Wellington · Durban · Toronto

Address of Author

Richard W. Horobin, Senior Lecturer in the Department of Human Biology and Anatomy,
The University of Sheffield, Sheffield S1O 2TN, England

CIP-Kurztitelaufnahme der Deutschen Bibliothek

Horobin, Richard W.:
Histochemistry : an explanatory outline of histochemistry and biophysical staining
Richard W. Horobin.
Stuttgart ; New York : Fischer;
London ; Boston ; Sydney ; Wellington ; Durban ; Toronto : Butterworths, 1982.
ISBN 3-437-10700-3 (Fischer)

Distribution rights for Continental Europe (incl. Soviet Union):
Gustav Fischer Verlag Stuttgart

Distribution rights for all other areas:
Butterworths London
This book is sold subject to the Standard Conditions of Sale of Net Books and may not be re-sold in the UK below the net price given by the Publisher in their current price list.

Wollgrasweg 49 · D-7000 Stuttgart 70 (Hohenheim)
Typeset and Printed by Mayr, Miesbach
Printed in Germany

ISBN 3-437-10700-3 (Gustav Fischer)
ISBN 0-407-00248-0 (Butterworths)

This book is dedicated to those invaluable persons
who thrust challenges upon me, and also to those who
gave me support: but most of all to those rare and
precious individuals who provide both caring and challenge

Preface

The histochemical staining of cells and tissues is currently a major tool applied across the whole range of biological and medical investigation, both in basic research and in practical applications such as clinical diagnosis. Where once staining prior to microscopy was limited to morphological studies, the histochemical approach is now a common component of investigations in such fields as physiology, biochemistry, and pharmacology. The blinkers of the traditional specialisations are eroding rapidly under the impact of broader, integrating biological ideas, and also of quantitative microscopical techniques.

The present book is directed towards several groups of readers within these wide subject areas. Primarily perhaps to bench-workers seeking an understanding of the scientific basis of staining procedures, and a guide for rational action when methods do not proceed as expected. The book is also relevant to those requiring a critical awareness of the type of information a histochemical investigation can generate, together with a knowledge of the limits and pitfalls of this methodology: of course many such investigators will not themselves be concerned with the manipulative details of the techniques. Finally, I hope to provide a stimulus for future studies, and some measure of knowledge for all those who find pleasure in the images arising from histochemistry and biological staining, and who wonder at the origins of such visual delights.

The over-arching objective of this book is therefore to contribute to the scientific foundation of the arts and crafts making-up the practice of histochemistry and biological staining. In keeping with this, an emphasis is given throughout the book to the common physico-chemical aspects of the technically diverse methodologies. Histochemistry of course has its own specialist literature, but to approach the prime objective it has proved necessary to draw ideas and information from physico-chemically and biochemically related fields, such as chromatography, pharmacology, photography, tanning, textile dyeing and so on. The cited literature reflects this, and offers bibliographic links with these related fields. Hopefully therefore, the reader will be aided in achieving a critical understanding of existing methods, and both trouble-shooting and technical innovation will be facilitated.

Certain characteristics of this book result from the overall objective. The explanatory nature means that no encyclopaedic coverage of techniques is attempted. Indeed practical methods are not described: this work is no substitute for a bench-book such as those of Lillie and Bancroft. The wide-ranging use of non-histochemical and non-biological literatures is another consequence. Another bias, which should receive mention here, is the emphasis on staining. When fixation is discussed, for example, the focus is on the influence of fixation on subsequent staining: the important influences of fixation on morphology being mentioned only in passing.

Some comment on the lay-out of material in the book is pertinent. Certain chapters are of a general character. In particular Chapter 1 provides an outline of the objectives of biological staining, of the physical principles underlying selective staining, and of ways of checking the significance of staining results. Chapter 11 also contains general essays: on the usefulness of staining theories, and on the problems arising from reagent impurities for example. However most of the book is written around a skeleton made up of groups of practical procedures: such as fixation, staining with dyestuffs, metal impregnation, and selective extraction as a test of significance. Hence these chapters are practically orientated, although they refer to and are largely organised around the general theoretical ideas first sketched in Chapter 1.

To end with a point of nomenclature: the reader will note the persistant use of the phrase "biological staining". This signifies a concern with the whole spectrum of techniques which introduce colour or electron scattering properties into biological material, to be followed by microscopical observation. Certain of these procedures are intended to demonstrate specific chemical groupings, or specific compounds. It is the use of such methods in particular which some authors describe as "histochemistry". Many other workers however regard the terms "biological staining" and "histochemistry" as synonyms. The former, narrower, usage has indeed a certain practical value. From the generalising, physico-chemical perspective adopted in this book however such a distinction is blurred, even confusing.

List of Contents

1 An overview of biological staining

Much of this book is a mass of detail. Furthermore the Chapters are largely organised around particular areas of technique, such as methods involving labelled antibodies or metal ions, or ways of fixing or blocking. Yet underlying the multitudes of facts and unifying the diverse procedures there is a set of basic ideas. These may be skeletally expressed in the form of questions:

1. Why do we want to stain anything? What information can be obtained using biological staining methods?
2. How do we carry out biological staining procedures? What are the objectives of the various manipulations?
3. How do the biological staining procedures work? Namely:
   a) What is present to be stained? This question requires discussion of topics such as fixation and other loss-limiting procedures.
   b) Why does anything stain? What for instance are the sources of stain-tissue affinity? Are there alternative modes of staining, not involving such affinities?
   c) As such affinities exist, why doesn't everything stain? What are the sources of selective affinity, and of differential rates of staining and destaining?
   d) Given that affinities may exist between stain and tissue, why does the tissue remain stained after its removal from the staining solution?
4. What is the significance of staining *or* of non-staining? What tests of significance, either conceptual or manipulative, are available in biological staining?

We will consider each question separately.

1.1 What information does biological staining provide?

The essence of the information provided is: what/where. Typically we simultaneously obtain data regarding what is present and also where in the organism/tissue/cell/organelle it is found. Such information is obtained by first staining the biological material, then viewing the stained preparation using the light or electron microscope. There is considerable variation as to precisely what is stained, we may stain everything, more or less, or stain selectively.

When everything stains information may be obtained from observations of the shapes of the various entities. Such essentially morphological data is generated by systems such as the light microscopists classic H & E oversight stain, by the electron microscopists lead citrate and by the biochemists negative staining of macromolecules and virus particles.

Selective staining can be used to provide information of diverse kinds. Thus the entities colour-coded, or otherwise selectively visually labelled, may be defined biologically. Examples include tissue elements such as muscle and connective tissue fibres; micro-organisms such as bacteria, fungi and viruses; and sub-cellular structures such as plant cell-walls or the Golgi apparatus . . . the list could be extended indefinitely. When staining such entities the chemical nature of the substrate is informationally irrelevant, indeed may often not be known. Contrast this with methods which selectively stain defined chemical functions, say phospholipids, Fe(III) ions, or the enzyme lactate dehydrogenase.

It is instructive to pause here and compare such histochemical demonstrations with standard biochemical assays. The two procedures are complimentary. Thus a histochemical method for demonstrating lactate dehydrogenase will commonly yield precise information on the distribution of this enzyme through tissues and cells, but it is unlikely to yield quantifiable results, through see Section 11.5. Contrariwise a biochemical assay of lactate dehydrogenase will be expected to be quantitative, but to find just where in a tissue the enzyme is localised is more difficult. Hopefully investigators will choose the technique most suited to the problem, though in the past there has been an unfortunate reluctance to do so. Such inflexibility does seem to be waning, and it becomes more common to find studies combining the two approaches. An early and dramatic example may be found in an experimental study of the rejection of transplanted kidneys. Biochemical assays showed large changes in the levels of several enzymes (TYLER et al., 1964). A parallel study of the distribution of these enzymes using biological staining methods showed however that these changes were not due to alterations of the kidney cells but rather to massive invasion of the organ by lymphoid cells (JANIGAN et al., 1964). Amongst more recent examples is an investigation of the intracellular distribution of non-specific esterase (BOCKING et al., 1974). Homogenate studies indicated that more than 50% of the enzyme was particle-bound. However observation of the loss of 70–80% of the enzyme from sections into the incubation medium prompted further investigation. This showed that the original biochemical observations were artifactual and were, at least in part, due to large amounts of non-microsomal esterase becoming irreversibly attached to the microsomes during homogenisation. Another recent, and extensive, investigation of enzyme distribution, within the tissues and cells of rabbit parotid, demonstrates that histochemistry using the light and electron microscopes, homogenate studies, and quantitative morphology may be powerfully allied (COPE and WILLIAMS, 1977; PRATTEN et al., 1977).

1.2 How do we carry out biological staining? And why do we do what we do?

Our objectives are to study the structure and contents, sometimes even the function, of tissues and cells. Starting, typically, from living tissue we wish to demonstrate certain of the constituents by staining, and then to view both the tissue structure and staining using the microscope.

To achieve these ends we must take active steps to protect form and content of the native tissue from hydrolytic enzymes of micro-organisms and of autolytic lysosomes, and also from mechanical and chemical degradation induced by tissue processing itself.

Such processing involves not only the staining procedures but also conversion of tissues into layers thin enough to be viewed intelligibly down the microscope. This requires mechanical subdivision of tissues, with consequent risk of damage by shear forces and by evaporation from the increased surface area. The tissue will also be exposed to a range of solvents, typically both organic and aqueous, cold and hot, with consequent risk of extraction and modification of the contents, and also of structural damage by osmotic effects.

Measures to limit damage to structure and losses of content include fixation, freezing, and addition of protective polymers to staining solutions. Thin layers, suitable for microscopic viewing, are obtained by procedures such as smearing and sectioning. To facilitate this latter procedure we must either freeze the tissue or embed it in a matrix such as paraffin wax or a plastic resin. Though staining may be carried out on whole cells and tissues, or even on whole organisms, it is usually performed on the sections or smears. In light microscopy this commonly involves prior removal of any matrix.

1.3 What is present to be stained? Fixation and its alternatives

First consider what may happen if we stain biological material without a prior stabilising or loss-limiting process. Imagine a piece of tissue excised from an organism and stored in buffered isotonic saline. Various processes could be observed: gross extraction of many cell constituents into the buffer, changes in morphology, autolysis of hydrolase-rich cells, growth of micro-organisms. Alternatively imagine that such a tissue fragment, prior to detectable changes of the type listed above, is dehydrated and embedded in some matrix. The osmotic, mechanical and extractive influences of the solvents involved could again give rise to gross alterations and distortions of the living form. In either scenario the further processes of sectioning, removal of matrix and staining will commonly induce further losses and changes. General discussions of this topic may be found in Sections 3.1 and 3.6, and in Section 11.8. Accounts are also given of the particular situations arising with enzyme histochemistry (Sections 7.1 and 7.2) and with immunostaining (Section 8.9).

So how can we limit these losses and degradations of content and form? For descriptive convenience we will first consider procedures for limiting losses of content.

A few materials, such as cellulose, chitin and keratin, do prove insoluble in the solvents typically used in biological staining processes. However most substances in their native form can prove soluble at some stage or another: proteins, nucleic acids and polysacharides may be extracted into aqueous solutions, lipids into dehydrating solvents and non-polar embedding media. Even if not extracted such soluble materials may well move around inside the cell, confusing our efforts to discover the *in vivo* localisations.

Several quite different strategies have been attempted to limit such extractions and diffusions. Firstly we may *reduce contact between tisssue and potential extractants.* Tissues may be freeze-dried and embedded directly, avoiding contact with fixative and dehydrating solvents. Alternatively tissues may be frozen and sectioned in this form using a cryostat, even avoiding exposure to embedding medium. Another varient is the use of fixative and processing solutions in which those substrates of interest are insoluble. An example is the use of alcohol if one wishes to stain sites of calcium ions or of glycogen. Indeed vapour phase fixation has been recommended in this context. Which brings us to the second major strategy, namely the *formation of insoluble derivatives* from soluble tissue components by treatment *with a fixative agent.* Such insolubilisation is achieved by a variety of mechanisms. Proteins for instance, and probably nucleic acids also, are denatured and so insolubilised by fixative agents as diverse as heat, organic solvents, or reactive chemical compounds. Proteins can also be crosslinked by certain fixative agents, such as the bi-functional compound glutaraldehyde. Analogously, unsaturated lipids can be crosslinked by osmium tetroxide or dichromates. Formation of insoluble salts and complexes may also be exploited. Examples are the digitonin-cholesterol complex, and cetyl pyridinium-glycosaminoglycan salts. The mechanistic bases of fixation are discussed in Sections 3.4 and 3.5. The final strategic varient is the *trapping of soluble compounds* inside a membrane or an insoluble matrix. The latter possibility can arise from fixation. Denatured cellular protein for instance enmeshes many compounds, such as glycogen, which are chemically unaltered by the fixative. The former approach can utilize a semipermeable cellulose or celloidin membrane applied to the surface of each section. Such membranes physically restrict the egress of large molecules. A procedure rather similar in principle is to add high concentrations of polymers, such as polyvinyl alcohol or dextran, to the staining solutions. Such solutions may indeed be applied gelled, with acrylamide or agarose, rather than as a simple solution. Finally note that loss-limiting due to membranes may be inate. Thus enzymes of membrane-bound organelles such as lysosomes and mitochondria are lost much more slowly than those of the cytosol.

It is now pertinent to ask if such procedures, so far considered only in the light of limitation of losses of tissue contents, also stabilise form and structure: in so far as content and form are qualities which may be divorced. The various types of procedure differ markedly.

For instance fixation originated as a way of preserving a life-like form for study by light microscopy. Techniques such as the use of membranes or of solution additives can however hardly be expected to stabilise structures not involving macromolecules. Furthermore loss-limiting procedures can themselves induce changes in form. Fixation is the dramatic examplar, with each fixative giving rise to its own, often characteristic, morphology. This is so marked that only a few fixatives (formaldehyde, glutaraldehyde, and osmium tetroxide for instance) of the many considered suitable for light microscopy proved satisfactory for electron microscopy, with its higher resolving power. There are several reasons for this, as considered in Section 3.8. Thus fixatives such as alcohol cause such gross shattering of tissues and cells that ultrastructural detail is substantially obscured, see for example Fig. 14. Another factor is poor fixation of protein, which is likely to results in poor preservation of structural detail, protein being such a basic structural material. For electron microscopy lipid plays an equivalent role since many organelles are either membrane bound or *are* membranes. However failure to completely retain content or structure may sometimes be desirable. An example is the use of acetic acid-alcohol fixation prior to Feulgen-DNA measurements on frog red blood cells. This fixative is ineffectual at retaining protein, and this is very useful since otherwise the strong absorbance of haemoglobin would be optically inconvenient. Choosing an optimum fixation regime is still an art, the essence of which is compromise.

A forward-looking question is useful here, namely *what effects does fixation have on subsequent staining?* Again we initially focus on changes in chemical content, and for simplicity this account will be restricted to proteins. One difference between fixatives is that they retain different amounts of tissue proteins, which obviously will affect staining. Alcohol for instance retains proteins rather poorly, whereas glutaraldehyde is very effective. These are overall consequences and there may be differences between proteins, for instance picric acid may selectively retain basic proteins. For details see Section 3.6. Usually fixation results in the loss of native reactive groupings, for example primary amino groups are destroyed by aldehydes, and thiols by mercury salts. Such changes naturally involve the formation of new reactive groups, which is sometimes useful: see Section 3.7. The denaturation process which is inevitably part of protein fixation does for example make the chemical groups of a protein hydrophobic core more accessible to many reagents and dyes. At least two classes of fixative-induced chemical change demand specific mention. Firstly the overall electrical charge carried by a protein, since a change of charge will influence any staining procedure making use of ionic reagents, for instance acid and basic dyes. Fixatives such as organic solvents have little effect on overall electric charge. However after treatment with most other fixatives this property is commonly shifted, either becoming more negative (after aldehyde treatment for instance) or more positive (for example after dichromates or osmium tetroxide). This classic topic receives attention in Section 3.7. Another important change is seen with enzymes or with proteins which are to be stained by a labelled antibody, and emphasises a basic dilemma of the fixation strategy. Thus if we do not fix tissues then much protein will be lost and morphological detail obscured. However if we do fix, then enzymic and antigenic activity will diminish, since fixation of proteins involves their simultaneous denaturation. For a more detailed account of this dilemma see Section 7.2 and 8.9. Such a change of subtle nature rather than of gross content reminds us to consider other fixative-induced changes in the physical nature of cells and tissues. Fixatives physically alter the living tissue, for example major changes are produced in surface area and in the tendancy to swell or shrink. Such changes are differential, so whereas alcohol greatly increases the surface area and allows subsequent swelling, glutaraldehyde has little effect on the surface area and inhibits swelling. These changes have dramatic consequences for staining procedures dependent on diffusion rates of reagents and products through the tissues. A general survey of this topic may be found in Section 3.8. More specific accounts are to be found in Section 4.16 (effects on acid and basic dyeing), Sections 5.2 and 5.3 (effects on reactive staining), Section 8.8 (effects on immunostaining). Such fixative effects, physical and chemical, only occur after the fixative agent has contacted the structure concerned. So rates of penetration of fixation into tissue are important practical variables. The range of penetration rates is large, with alcohol being amongst the slowest and acetic acid amongst the fastest. An account of this topic is given in Section 3.9.

Some further comments on practicalities are now apposite. That there is no single ideal procedure for stabilising the life-like contents and form of cells and tissues is only too apparent. Hence compromises and double-barrelled procedures abound. For instance, cutting frozen sections is often preceded by a very brief period of fixation, or is combined with use of semi-permeable membranes or of additives to the staining solutions. Analogously, within the traditional fixation strategy, it is commonplace to use fixative sequences or mixtures. In this way we hope to control excesses of swelling or shrinking, for example by using alcohol-acetic acid mixtures; or to retain materials of widely differing chemical type, using glutaraldehyde-osmium tetroxide, to retain protein and lipids for example.

1.4 Why does anything become stained? An account of reagent-tissue affinities, and of other causes

There are several rather distinct reasons, namely:

1. The cells may eat the staining reagent. Reagents may be taken up from solution by the physiological processes of a live cell, usually termed vital staining.
2. The entity to be studied may not be stained, but its surroundings are, demonstrating its form in outline. This is called negative staining.
3. The more general case of stain-tissue affinities. The sources of such affinities are extremely varied, as discussed later. The stain may be the same substance as is present in the staining solution, as with most dyestuffs. Alternatively the stain may be a product derived from a reagent in solution by some reaction involving the tissue, as with the periodic acid-Schiff procedure or as in enzyme histochemistry.

Processes (1) and (2) of this list will be briefly discussed, after which we will focus on the sources and nature of stain-tissue affinity (3). It will be appreciated that more than one cause may underlie the binding of a given stain to the tissues.

Vital staining, though hoary with tradition, is still extremely useful in certain cases, for instance in assessing the viability of cell cultures or tracing neuronal processes. Since vital staining often involves pinocytosis and phagocytosis this staining procedure is sometimes used to study these activities, or the cells and organs where these processes are common. For a discussion of vital staining see Section 9.4.

Negative staining is applied over a remarkably wide linear scale. At one end of this scale molecular biologists study the shapes of macromolecules and of virus particles by evaporation of solutions containing, say, a virus together with an electron-scattering heavy-metal salt which provides the contrasting outline when viewed in the electron microscope. At the other end of the scale histologists trace the course of tissue channels, such as blood vessels and bone cannaliculi, by injecting pigmented gelatin or impregnating with silver salts.

We will now consider, in some detail, staining attributable to *stain-tissue affinity.* However before plunging into an account of the sources of such affinities consider the histochemical usage of the word affinity. There are two meanings extant, namely:

- as a synonym for stain-tissue attractive forces;
- as shorthand for saying that, under the conditions of use, a given staining agent prefers the tissue to the staining solution.

The second usage is the one which proves useful to us and can be put into more exact form. When we use affinity in this way to describe the tendancy of a reagent to remain in solution or to move into the tissue we are making two assumptions. One is that the staining process approaches or reaches equilibrium, and in practice this is not always so. The other assumption is that the magnitude of such affinities depend on *all* the factors influencing such an equilibrium, on changes in solvent and tissues, and not just on changes directly involving the stain. A more extended essay on this topic may be found in Section 4.3. That said we can move on to consider some of the possible contributions to these affinities.

As the first major contribution consider various *entropy factors.* At the commencement of a dyeing procedure, all the staining reagent is in the solution and none in the tissue. This is often a more ordered situation than when, at equilibrium, reagent is found both in the solution and in the tissue. Thus during the staining process the degree of order of the system decreases, that is the entropy of the system increases. As the laws of thermodynamics tell us such processes take place spontaneously, so staining can occur. An example of this is when the dye is merely dissolved in the tissue, as is probably the case with the staining of fat droplets by Sudan dyes. Such processes are described in Section 4.4, Model 1. Such positive entropy changes will not however always be sufficient to ensure staining. Specific dye-tissue attractions for instance may involve an increase in order, reducing the affinity-favouring positive entropy change. Another common entropy-contribution to biological staining is *hydrophobic bonding* occuring with aqueous solutions. This is considered at some length in Section 4.4, Model 3. Here the positive change in entropy contributing to stain-tissue affinity arises from alterations in the structure of the water of the solvent phase during the staining process. These alterations occur because aggregates of water molecules are stabilised by hydrophobic chemical groupings such as hydrocarbon side chains and rings on reagent, product and tissue molecules. When the hydrophobic portions of reagent or product move into contact with those of the tissue some water aggregates transform into non-associated random form. A decrease in the degree of order of the solvent of this kind, corresponding to an increase in entropy, favours staining, Hydrophobic bonding may well be an important contribution to stain-tissue affinity in any staining system involving aqueous solutions, and where reagent and product carry hydrophobic groups. Examples are organic dyestuffs and reactive stains, and many enzyme histochemical substrates and final reaction products. For an account of the estimation of the hydrophobic/hydrophilic character of staining reagents see Section 11.14.

Now consider contributions to affinity resulting from negative changes in enthalpy, commonly arising from the action of attractive forces. Attractions usually discussed are between reagent or products and the tissue, but here we first discuss *stain-stain attractions* as sources of stain-tissue affinity. Such attractions will contribute to affinity when they occur only or principally in the tissues and do not occur in the solution. A well known phenomenon where dye-dye attraction aids affinity is metachromatic dyeing. Certain ionic dyes stain different tissue sites different colours, and it is considered that these spectral differences reflect site-dependent dye-dye attractions. This dye aggregation, since it must form multiply changed species, occurs most readily where there are high concentrations of counter ions. With basic (cationic) dyes metachromatic staining takes place in the strongly sulphated mucosubstances of cartilage matrix and mast cell granules. Metachromasia is further discussed in Section 4.2 and also in Section 4.4, Model 5. In addition to metachromasia there are many techniques exploiting product-product attractions, resulting in the formation of crystals of stain of low mobility and solubility. Examples of such methods include metal impregnation, in which metallic crystals of silver and gold are formed (Chapter 6); enzyme histochemistry, where organic or inorganic pigments may be deposited (Section 7.6); and reactive methods, such as Perl's method for the demonstration of ferric ions, in which the crystals produced are of an inorganic coordination complex.

Still discussing enthalpic contributions to affinities we will now turn to *stain-tissue attractions,* which are sometimes erroneously considered as the only sources of stain-tissue affinity.

The most widespread of such attractions are *van der Waals forces*, which act between all stains and tissue substrates to some degree, as considered more extensively in Section 4.4, Model 2 and Section 7.7. The term «van der Waals forces» is an abbreviation for the various polar attractions, monopole-dipole, dipole-dipole, dipole-induced dipole, dispersion forces and so on. These are all short range forces, which increase in strength with increases of magnitude of the dipole moments and polarisabilities of reagents and substrate molecules. Various implications for biological staining arise. When using aqueous solutions, contributions to affinity from van der Waals attractions will be most significant for poorly hydrated substrates, since the hydrated tissue materials will be separated from stain molecules by a sheath of solvent. Also molecules with extensive systems of delocalised electrons (e. g. those possessing large conjugated systems) are polarisable and often possess large dipoles. Overall we can anticipate van der Waals attractions to be important when the substrates are non-polar, proteins such as elastin for example, or contain numerous aromatic rings as do nucleic acids; and also when the reagents are large aromatic compounds such as poly-azo dyes or bis-tetrazolium salts, or are halogenated compounds as are some dyes and enzyme substrates. Note that some but not all of these structural features are also requirements for effective hydrophobic bonding. Finally consider the corollary, that van der Waals attractions are likely to be less important between small ionic dyes or other polar reagents and highly ionic substrates. Procedures for the quantitative assessment of the degree of conjugation and the hydrophobic/hydrophilic character of staining reagents are described in Section 11.14.

Probably the most widely described reagent-tissue interactions are *Coulombic attractions* (or salt links or electrostatic bonds) arising between ionic stains and tissue counterions. For instance between the coloured cations of basic dyes and tissue structures rich in anions, such as phosphated DNA and RNA, and carboxylated and sulphated mucosubstances. Viewing Coulombic influences solely in this light is however rather misleading since both solution and tissue are, overall, electrically neutral. Acid and basic dyes, and the proteins and other tissue components, are salts. It is useful to regard staining with ionic reagents as an entropy-driven ion-exchange process with acid or basic dye ions from the solution being exchanged for, say, chloride or sodium ions from the tissues. This ion-exchange viewpoint offers explanations for such complexities as the marked influence on staining of inorganic salts, and the degree of tissue swelling. Finally note that such Coulombic effects arise if either the reagent *or* the product are ionic. In such cases, even non-ionic substrates acquire charge as staining procedes, consider staining non-ionic glycogen by either the periodic acid-Schiff (two ionic reagents) or by the Best's carmine (carminic acid is anionic) procedures. Such effects are widely distributed. Hence subsequent discussions mention acid and basic dyes (Section 4.4, Model 4), reactive stains (Section 5.2), metal impregnation (Section 6.2), enzyme histochemistry (Section 7.7), immunostaining (Section 8.7), and colloidal iron and other rather ill-defined procedures (Section 9.2).

Both the examples of stain-tissue attractions so far discussed, Coulombic and van der Waals attractions, are non-di~~rected. Now we~~ will turn to describing forms of directed attractions, that is to bonding.

Stain-tissue *hydrogen bonding* is sometimes suggested as contributing to affinity. Note that hydrogen bonding is a polar bond occurring when a hydrogen atom is situated between two strongly electronegative atoms such as nitrogen or oxygen, but is only covalently bonded to one of them. Amino, carbonyl and hydroxyl moieties are amongst the common functional groups which can take part in hydrogen bonding. It is significant that water, being effectively composed of hydroxyl groups, will be extensively hydrogen bonded. Such bonding will link water to water, so forming the aggregated water referred to earlier, and also to solute and tissue molecules. Consequently hydrogen bonding is unlikely to be an important source of stain-tissue affinity when water is the solvent, since even though stain-tissue hydrogen bonds may well be formed their formation will require prior breaking of water-tissue and water-stain hydrogen bonds. Thus little nett gain will ensue. In non-aqueous solvents however stain-tissue hydrogen bonding could be more significant. This has been proposed for the Best's carmine stain for glycogen, which uses a substantially methanolic solvent. More details are found in Section 4.4, Model 2. We now turn to the archetypal directed link, the covalent bond. Such contributions to stain-tissue affinities are formed between a wide variety of atoms. Staining methods in which non-metallic atoms become so linked are typically regarded by histochemists as mere organic chemistry, for example the periodic acid-Schiff and the Feulgen nucleal procedures. Such methods are sometimes described as *chromogenic or reactive staining*, of which an account is available in Section 5.2. However when one of the linked atoms is a metal, histochemical nomenclature is unclear. Consider a dye or other organic ligand whose colour and staining properties are enhanced or altered by the presence of suitable metals ions. Such a phenomenon is commonly referred to as *mordanting* and any observed enhancement of affinity is often attributed to formation of tissue-metal and metal-dye covalent bonds. Such effects can take place, and have been convincingly documented for certain wool-dyeing procedures. However the techniques described in the biological staining literature as mordanting involve organic ligands such as Gallocyanin and Hematein, with metal ions such as Al^{+++}, Fe^{+++} and Cr^{+++}, and the tissue sites proposed are typically nucleic acid phosphate groups. Some doubt has been cast on the role of the mordanting mechanism in such cases, as described in Section 4.4, Model 2. Some staining reactions which are similar in chemical principle and which do seem to take place in tissues, such as the chelation of Ca^{++} by dyes such as Alizarin Red S and the reaction of tissue thiols with Mercury Orange, are *not* termed mordanting by histochemists.

There may well be further negative-enthalpy contributions to tissue-stain affinities, charge-transfer complexes have been suggested for instance. However as the present account is illustrative not encyclopaedic it is time to call a halt.

1.5 Why doesn't everything always stain?

Having discussed reasons why staining can occur, we can ask how is *selective* staining achieved? This is a significant question, since it is selective colouration which is usually required in biological staining. Several factors can be distinguished. Thus:

- certain tissue sites will have higher affinities for a given stain than do other sites. In practice such differential affinities will be confused by the fact that . . .
- even if all tissue sites had the same affinity for a given stain, the concentration of such sites will vary from one tissue structure to another;
- furthermore even if there were a uniform distribution of tissue sites possessing a similar affinity for a given stain, there may be differential rates of penetration of

reagents into the tissues, or of reaction inside the tissues, or of loss of reagents and products from the tissues.

- Finally even if there is no differential uptake or retention of stain, the colour of the stain at different sites may not be uniform; there may be metachromasia or polychromasia.

Understanding any particular technique will usually require a consideration of more than one of these factors.

As an example of selective staining controlled by *differential affinities* consider the colouration of fat-droplets by Sudan-type dyes. Here the majority of tissue components fail to stain since they are hydrated and hence have only a trivial affinity for the hydrophobic, non-ionic Sudan dyes. Traditional stains for elastic fibres provide another example. These stains are large ionic dyes, usually applied from non-aqueous dyebaths of such acidity or alkalinity that the only important contributions to stain-tissue affinity are from van der Waals attractions. Under these conditions the few structures of high affinity include elastic fibres, composed as they are of the unusually hydrophobic protein elastin. A more complete discussion of this phenomenon may be found in subsequent Chapters, as seen with dyeing methods (Section 4.9), with metal impregnation (Section 6.2), with immunostaining (Sections 8.2 and 8.7), and a variety of other procedures (Chapter 9). When the major source of affinity is a stain-tissue covalent bond, then the selectivity of staining will reflect that of the chemical reaction giving rise to the bond. For instance the -Hg-Cl moiety of Mercury Orange reacts with thiol groups and little else. Analogously the acid hydrolysis-Schiff reagent sequence of the Feulgen nucleal procedure will produce red derivatives at few sites other than those containing the deoxyribosyl residue of DNA. Selective reactivity of this kind is dealt with at greater length in Section 5.3.

Some selective staining methods involving differential affinities are easier to understand if we focus on the anti-staining characteristics of the non-staining tissue sites. For example a small cationic dye will stain those structures with overall negative charges, DNA-rich chromatin, RNA-rich ribosomes and so on. However under acidic conditions such dyes will not stain the proteins, since these macromolecules will be positively charged. If the pH of the dyebath is raised sufficiently the proteins become anionic, then they will stain also. Thus the selectivity of the nuclear staining depends on the charges carried by reagent and substrate, and hence on pH. Another Coulombic effect hinges around the amount of electrolyte in the solution, inorganic salt as well as ionic dyestuffs. As predicted by theoretical models such as the Donnan equilibrium, and as often seen in practice, addition of inorganic salts to a dyebath can decrease the staining of those substrates carrying the same charge as the dye ion, whilst increasing the staining of substrates with the same charge. An extended account of the influence of salt effects and other coulombically mediated processes on the staining selectivity achieved with acid and basic dyes is provided by Section 4.9, Affinity Model 4. Naturally there is nothing magical about dyestuffs, and analogous effects are seen with other ionic staining reagents. Examples include metal ions (Section 6.4), labelled antibodies (Section 8.7), reagents for the demonstration of enzymes (Section 7.7), and also reagents such as colloidal iron (Section 9.2).

Many selective staining patterns in fact reflect *differential concentrations* of staining sites. A fairly simple example is seen with the staining by the Eosin component of the Hematoxylin and Eosin oversight stain. Under usual conditions of use all protein chains will be expected to carry overall positive charges. However the most intense staining, especially with procedures using longer staining times, is of physically dense structures such as cell cytoplasms rich in mitochondria and secretion granules, hair shafts and superficial keratin, and red blood cells. This topic is considered in Section 4.8.

Reactive stains for proteins, such as the Procion and Remazol dyes, show similar patterns of colouration. Though the sources of stain-tissue affinity differ, the bases of selectivity are the same.

We now take a large conceptual leap, away from affinities and towards rates of diffusion and reaction. In many staining procedures such rates, acting differentially in different tissue structures, are the dominant factor in establishing the staining patterns. To such an extent indeed that staining processes allowed to approach equilibrium may often yield an unwanted pattern of colouration. As a preliminary we may ask *what characteristics of reagent or product on the one hand, and of tissues on other, control the rates of diffusion into and out from the tissues?*

One relevant tissue characteristic is physical density. Mitochondria, red blood cells and secretion granules for instance are dense, and slow to be penetrated by staining reagents. However once stained they are also slow to loose stain. A factor leading to low densities, and so promoting rapid reagent diffusion, is ease of tissue swelling. Swelling is often enhanced by limited macromolecular cross-links, presence of numerous polar functional groups and use of aqueous solvents. All these factors promote hydration and so swelling. Thus protein-rich materials such as goblet cell mucin and collagen fibres contain hydratable polysaccharides and are only weakly cross-linked, take up and also lose reagents with great rapidity when using aqueous staining solutions. Conversely structures such as elastic fibres and hair shafts are comprised of heavily cross-linked proteins, are poorly swollen in water, and are only slowly penetrated by staining reagents; though stain once taken up is slow to diffuse out again.

A tissue feature releated to swelling, and not in practice readily distinguishable, is gross surface area. For instance both goblet cell mucin and collagen fibres are ultrastructurally of high surface area, being composed of spherical vesicles and fibrils respectively. Such issues are the subject matter of Sections 4.11 to 4.16.

Yet another relevant tissue feature is electric charge. If a tissue substrate carries the same charge as a staining reagent then the reagent's rate of entry will be slowed compared to that of an uncharged or oppositely charged substance. So a substrate of high charge density, such as cartilage matrix, very rapidly acquires a surface stain of a cationic dye such as Alcian Blue. However if the layer is sufficiently pronounced, and the surface layer of Alcian Blue on cartilage matrix can be crystalline, the staining of the deeper layers of the tissue will be grossly slowed.

We have by this stage been obliged to consider both tissues and reagent characteristics. A more general statement of the effects of tissue-stain affinities on staining rates can in fact be made: the higher the affinities of reagents or products with the tissue the slower their rates of diffusion through that tissue. Consequently reagents and products of small size will tend to be fast diffusing, as indeed they will be even in free-solution, and *vice versa* of course.

Consider finally that many tissue characteristics depend on the mode of fixation. For example diffusion rates tend to rise as the fixative is changed from Zenker's fluid to formalin, and to rise again on changing from formalin to Carnoy's fluid. Available information is summarized in Section 4.16. Analogous effects seen during immunostaining are described in Section 8.8.

Making use of this background data we can now consider staining patterns controlled by *differential rates of entry of reagents* into different tissue elements. Such systems may involve only a single reagent, for example the dyes used to selectively stain goblet cell mucins. These dyes are large, so slow diffusing. Their cationic character facilitates staining of the anionic glycosaminoglycan-rich mucins. Though these dyes have affinities for other anionic tissue sites, such as nuclear chromatin, short periods of staining only allow the rapidly penetrated mucins, and not the slower staining materials, to become coloured. This and similar procedures are discussed in Section 4.18, Rate Model 1.

Other entry-controlled methods involve pairs of dyes, one dye large and slow diffusing, the other small and fast. After short staining periods the larger dye of such a pair will be restricted to the faster staining of the tissue components with which it has affinity, whilst the small dye will have access to fast and slow staining structures. Thus if the dye ions are anionic and are applied at low pH when the fixed tissue proteins will be cationic, then the larger dye will stain collagen fibres selectively with the small dye providing a general background staining. Very many such methods exist, for example van Gieson's stain in which the larger dye is Acid Fuchsin and the smaller dye picric acid. An analogous example involving cationic dye ions is the Methyl Green-Pyronin stain for distinguishing DNA-rich and RNA-rich sites. The smaller red dye, Pyronin, coloures the slowly penetrated ribosome-rich cytoplasms and nucleoli, as well as the chromatin. The larger dye, Methyl Green, is restricted to the nuclear chromatin. There are many variants of such methods, such as using more than two dyes or using dyes sequentially, both of which processes are considered in Section 4.18, Rate Model 4. Other varients add to staining solutions colourless species with affinities for the tissue, as described in Section 4.19.

The same rate factors may be utilized in reverse, when selective staining is achieved by the exploitation of *differential rates of loss of reagent or product* from different tissue elements. At its simplest this involves washing out unreacted reagents from the tissue. Examples are the periodic acid-Schiff procedure for polysaccharides or the Mercury Orange method for thiols. Less trivial is the procedure traditionally termed differentiation. This typically entails initial non-selective staining of many tissue elements with a large, and hence slow diffusing, dye. The stained tissue is then washed with dye-free solvent, so that those structures from which dye diffuses slowest retain dye to become selectively coloured. Examples include the staining of the striations of voluntary muscle with Heidenhains iron-haematoxylin, of myelin sheaths with Luxol Fast Blue, and of Gram-positive bacteria with a Crystal Violet-iodine complex. Other examples are found in Section 4.18, Model 2. Such controlled-loss procedures are not limited to dyeing methods. The basis of the many silver impregnation techniques is an initial uptake of silver ions by the tissue in a rather non-selective way. The silver ions are then converted into crystals of silver metal by treatment with a reducing agent. Factors influencing the final distribution of silver crystals include the use of a reducing solution of such reactivity that only the relatively dense structures retain silver long enough for it to be reduced; and use of post-reduction washes with cyanide ions, which can complex with and so dissolve metallic silver, with this occurring most slowly in denser structures. This aspect of silver staining is discussed in Section 6.4.

When staining involves conversion of a reagent into a product then selectivity may be controlled by the *differential rates of reaction* at different sites, for details of which see Sections 5.2 and 5.3. Examples include the periodic acid oxidation step of the periodic acid-Schiff procedure, during which step 1,2-diol groups react faster than other substrates. However if the time of oxidation is sufficiently extended then «non-specific background staining» will occur. Many other examples are provided by enzyme histochemistry.

As a final source of selectivity consider *polychromatic staining*. In some instances there is no selectivity of the varieties discussed above, but a single staining reagent taken up or reacting at different tissue sites generates different colours. Metachromasia, mentioned earlier, is an example. A single dye ion such as Toluidine Blue may stain erythrocytes greenish-blue, cell nuclei azure, and mast cell granules purplish-red. Again, if a tissue section stained with the dye Eosin is viewed under ultra-violet illumination only that dye in the elastic fibres fluoresces, although dye is bound plentifully throughout the tissues. Finally consider those reactive staining procedures in which a given reagent reacts non-specifically with several functional groups, yielding products of different colours with different groups. One example is aminobromanthraquinone sulphonic acid, which yields orange derivatives with thiols and blue derivatives with alcohols and amines; another is seen in the reaction of certain fluorones with nucleic-acids, yielding blue-black derivatives from DNA, and yellow-red derivatives from RNA.

1.6 Why does the tissue remain stained after its removal from the staining solution?

Consider a procedure where uptake of stain depends on stain-tissue affinities. Why are such stains not subsequently lost into washing solutions, or dehydrating or processing solvents, or into the mountant? In short because post-staining reagents are carefully chosen such that the stain has a very low affinity for them, or only diffuses into them extremely slowly. This may be achieved in various ways. For instance by generation of a stain which is barely soluble in any of the solvents to which it will be exposed. Examples are crystals of silver or gold, or of metal sulphides, or metal co-ordination complexes such as Prussian Blue or Hatchetts Brown, or ionic dyes insolubilised by a large counterion.

Alternatively particular attention may be given to the nature of processing solvents and mountants. Thus if the stain is non-polar, as are fat stains and the formazans and azo-dyes generated as final reaction products in enzyme histochemistry, we use polar mountants such as gelatin or polyvinyl alcohol or glycerol dissolved in water. Conversely when the stain is polar, as with ionic dyes, the processing solvents and mountants are organic and hydrophobic.

1.7 How do we judge the significance of staining? Or indeed, of failure to stain?

There are various checks of significance which we may apply if staining does take place, and also a number of sceptical questions we should ask whether or not we see staining. There are also various useful general standpoints and cautions.

One of the checks of significance discussed at length in Section 11.2 is *selective extraction.* This method is based upon the following logic. If «substrate X» is believed to be responsible for the staining seen, then extraction of the tissue before staining with a medium designed to selectively remove substrate X should prevent staining. Hence if prior extraction prevents staining, substrate X is assumed present. Such procedures may involve solvent extraction, using organic solvents to remove lipids for instance; or chemical hydrolysis, as with the removal of DNA by perchloric acid, following which treatment the depolymerised fragments are readily lost into the solvent; or selectively may be achieved by use of an appropriate hydrolytic enzyme, amylase for the extraction of glycogen for example. A basic weakness is that extraction may not be sufficiently selective. Thus a solvent intended to dissolve only hydrophobic lipids may, if contaminated by a trace of water, remove polar lipids also. Non-selectivity can also occur when extraction procedures adventitiously remove tissue components on such a scale as to allow markedly increased rates of reagent diffusion. This can result in alterations of rate controlled staining patterns involving reagents quite unrelated to stains for substrate X. Such gross extractions are reduced by tissue fixation. However since fixation, almost by definition, alters native chemical form and properties, the problems of finding really selective extractants are then even more acute.

Another widely used control technique, discussed in Section 11.3, is *histochemical blockade.* This has as its operating principle the notion that if a particular tissue chemical structure X is required for staining to occur, then suitable chemical modification («blocking») of structure X should prevent staining. Hence if prior blocking does prevent staining, structure X may well be present. The blocking procedures vary from straight-forward organic chemistry, such as the conversion of carboxylic acids into esters or of aldehydes into hydrazones, to inhibition of enzymes, the blocking of antibodies with suitable antigens and *vice versa.* Blockade has proved a valuable histochemical tool, but naturally has its limitations and snags. These include two mentioned above, namely the possibility of grossly altering tissue porosity, in the present instance by reagents such as strong acids and bases, and also the inevitability of physical and chemical changes if fixation is used. Incidentally some procedures long described as blockades, such as the hydrochloric acid-aniline method for preventing staining of DNA, turn out on closer study to be selective extractions. Other characteristic snags of histochemical blockades include discovering that the chemistry of a blockade process is in fact less selective than had been assumed, or finding that the blockade may be reversed by a reagent or solvent subsequently used in staining. A final complication lies in the question: how complete must a blockade be before the process is useful? The answer depends on the mode of staining that is subsequently used. Consider for instance two ways of staining primary amine groups of proteins, one using an acid (that is anionic) dye, the other using a reactive stain such as a Procion dye. The former process depends on the protein maintaining an overall positive charge. Consequently, as proteins typically contain large numbers of potentially anionic carboxyl groups, the conversion of say half of the amines to non-cationic derivatives could block staining. However when staining with the Procion dye intense colouration may still be achieved when only a few percent of the amino groups remain intact.

A wide range of *other control procedures* are used at times, though perhaps not often enough. Sometimes ommissions constitute controls. For instance when we omit steps from a multi-step procedure, say periodic acid from the periodic acid-Schiff procedure; or substrate from the incubation medium in an enzyme histochemical method; or analogously when we use a non-specific serum in place of the supposedly specific one in labelled-antibody staining. Such methods are considered in more detail in Section 7.7, Section 8.7, and Section 11.4. Yet another approach to control methods in biological staining is the use of test objects and of model systems, as described in Section 11.12.

Many workers keep on hand sections containing known examples of various common structures or substances, to check out staining reagents and procedures when setting out to stain new tissues or when changing to new batches of stain. Model systems offer some of the same advantages. Such model systems may involve the use of a piece of cellophane to model polysaccharides for the periodic acid-Schiff process, or more elaborate encapsulations of cell homogenate fractions in acrylamide gel films. The intrinsic problem with model systems is not knowing whether they do indeed reflect the relevant tissue property. Not always, it seems.

Still discussing occasions when staining does occur, we may consider some sceptical questions. The first is: *Does the staining intensity reflect the concentration of the tissue substrate?* Often not, for many reasons. For instance tissue sections vary in thickness, both from one section to the next and also within a single section. Amongst more complex reasons is the general lack of reaction stoichiometry in biological staining. Though mere proportionality may be sufficient to allow some measure of quantitation we will then require some standard. This may be internal, diploid-haploid cell nuclei for instance if assessing DNA, or alternatively model systems may be used as external standards. Assessment of relative concentrations does not require a standard. Though too many difficulties exist in this area to be summarised succinctly, one can emphasise that the objectivity achievable with a microspectrophotometer is not synonomous with quantitation. These questions are discussed at length in Section 11.5.

Another important question may be expressed as follows: *Is the stain where the substrate was?* An initial sub-question involves considering whether the tissue was altered morphologically between life and staining. In the absence of fixation it is commonplace for morphological changes to occur at all stages up to and including staining. Organelles however do exhibit large variations in fragility and mobility. Fixation usually halts such changes, though the fixation process itself commonly alters morphology. This varies from the gross to the subtle, and all can mislead. A second aspect of the original question entails wondering if the stain is where the substrate is, disregarding the substrates own wanderings. Sometimes it just can't be, as with those metal impregnation and enzyme histochemical methods which produce stain in the form of crystals larger than the structures stained. Sometimes the stain may migrate through the tissue, to be trapped at some site of high affinity distant from the substrate. The localisation issue is considered in Section 11.6, and also, for the particular instances of enzyme histochemistry and immunostaining, in Sections 7.7 and 8.9.

We now turn to sceptical questions useful when staining does *not* occur. Firstly we may ask: *Have there been losses of substrate?* This may or may not relate to the previously discussed issue of morphological damage. Be that as it may, substrate losses can occur into working solvents at any stage from fixation to staining. Methods for limiting these losses were sketched earlier. Such losses can be of essence as well as substance, especially when studying structurally labile compounds, enzymes and antibodies for example. These problems receive general attention in Section 11.8. The special complications arising with enzyme histochemistry and immunostaining are given attention in Sections 7.1, 7.7 and Section 8.9 respectively.

Now for a second question: *Was there staining, which has been lost before viewing?* Stain can diffuse away from the site of the substrate. We have already considered the possibility that such stain may be finally absorbed elsewhere in the tissue. Other possibilities include such complete dispersion that we fail to perceive that staining ever took place («pale non-specific background staining»), or the stain may diffuse right out of the tissue. Other losses of stain are due to chemical and photochemical fading. Metal sulphides may oxidise in atmospheric oxygen to sulphates of weaker colour, and many dyes can be photolysed by the ultra-violet radiation of sunlight to form colourless derivatives. Alternatively colours may be inconveniently altered rather than totally lost. For example reversion of metachromatic staining to orthochromatic colouration on transferring tissue from the aqueous staining solution to non-aqueous dehydrating and mounting media is pronounced and commonplace. Such issues are spelled out in Section 11.10.

There are, finally, useful questions applicable to staining *and* non-staining situations. The first is: *Do we actually know how the staining process works?* Often we do not. After all books like this one are uncommon, at least in English. There are a few specialist monographs (BAKER, 1958; GURR, 1965), and also books such as the on-going classic of PEARSE (1968, 1972) which do provide extensive background discussions. Most biological staining texts however are essentially practical bench-books. Consequently trouble-shooting, and assessing the significance of staining, is very rule-of-thumb in many or most laboratories. Views concerning theory and it's value in biological staining and histochemistry may be found in Section 11.12. The questions kept until last are in principle less necessary than the others, though in practice prove only too necessary, namely: *Are staining reagents pure? And does it matter if they are not?* To summarise, reagents are often impure, and this is often significant. Dyestuffs in particular are commonly mixtures, containing many coloured and colourless impurities. Indeed it is not unknown for the dye named on the bottle's label to be absent from the bottle's contents. This situation is probably due to the fact that dyes are mostly manufactured to colour textiles and not sections of tissue. Furthermore dyes are not the easiest of substances to analyse and purify. This is exacerbated by the biological users being accustomed to low-cost products. In the past enzyme substrates also had a bad name, though this situation seems to have improved, at least with the reputable suppliers. Technical progress also continues to be made with the well known problems of purifying labelled-antibody preparations. Consequences of these various impurities vary from gross lack of specificity to total lack of staining, and may be trivial or disastrous for our biological understanding. The most general and pernicious problem however is that the widespread uncertainty removes confidence, and indeed it is true that one does not know whether a procedure found in the literature used pure reagents or not. A dyestuff-oriented discussion of this, often overlooked, problem area is provided by Section 11.13. Analogous issues arise with reagents for enzyme histochemistry (Section 7.7) and labelled antibody staining (Section 8.5).

2 The preparation of material to be stained

This is not a technical handbook, so the following account merely sketches some of the various routes from live tissues and cells to stained preparations awaiting microscopic study. Procedures in these transformations are usually directed towards assisting the simultaneous microscopic observation of morphology and of some aspect of a tissues constitution, chemical or biological.

Such a combined approach results in several requirements. We must ensure that the materials present *in vivo*, which we wish to stain, are in fact still present during the staining process. Similarly tissue and cellular, even molecular, structures must remain sufficiently intact throughout tissue processing and staining. These requirements are limited by the lability of many biological materials. In part this lability is biochemical, consider denaturation, autolysis, and attack by micro-organisms for instance. Mechanical damage due to shear and osmotic forces can also take place. Furthermore most biological materials are soluble in either water or in the common organic solvents. In addition, obtaining intelligible images in the microscope requires optical contrast together with the necessary degree of selectivity. It also demands preparation of the biological material in the form of thin layers, in which the entities to be studied are not too much smaller than the thickness of the layer. In practice a thickness of 1–20 μm for light microscopy and 0.05 μm for electron microscopy proves satisfactory. Consequently, prior to microscopic observation the structure and content of biological materials must be stabilised or at least preserved, a sufficiently thin layer of material obtained, and staining carried out.

Tissue stabilisation is typically achieved by treatment with fixative agents, though alternative procedures which limit loss of content exist. Thin layers, whether of tissues or of molecules, are usually obtained by smearing or spreading suspensions or solutions, or by slicing up solid blocks or lumps. If this latter sectioning procedure is adopted the soft texture of many live cells and tissues presents a difficulty, since for microtomy a certain rigidity is necessary. This may be achieved either by freezing wet tissues, or by embedding tissues in a liquid which is subsequently transformed into a solid matrix around and within cells. When such a matrix-forming liquid is not miscible with water the biological material must first be dehydrated, by solvent extraction or by exposure to a vacuum. The staining process itself is often carried out on matrix-free material before or, more usually, after sectioning or smearing. This may necessitate removal of the matrix, though an important exception is the staining of plastic sections for electron microscopy. In this instance the matrix is retained to provide mechanical support for the specimen in the electron beam.

A few of the variety of ways in which the necessary manipulative steps are combined in practice are shown in Fig. 1.

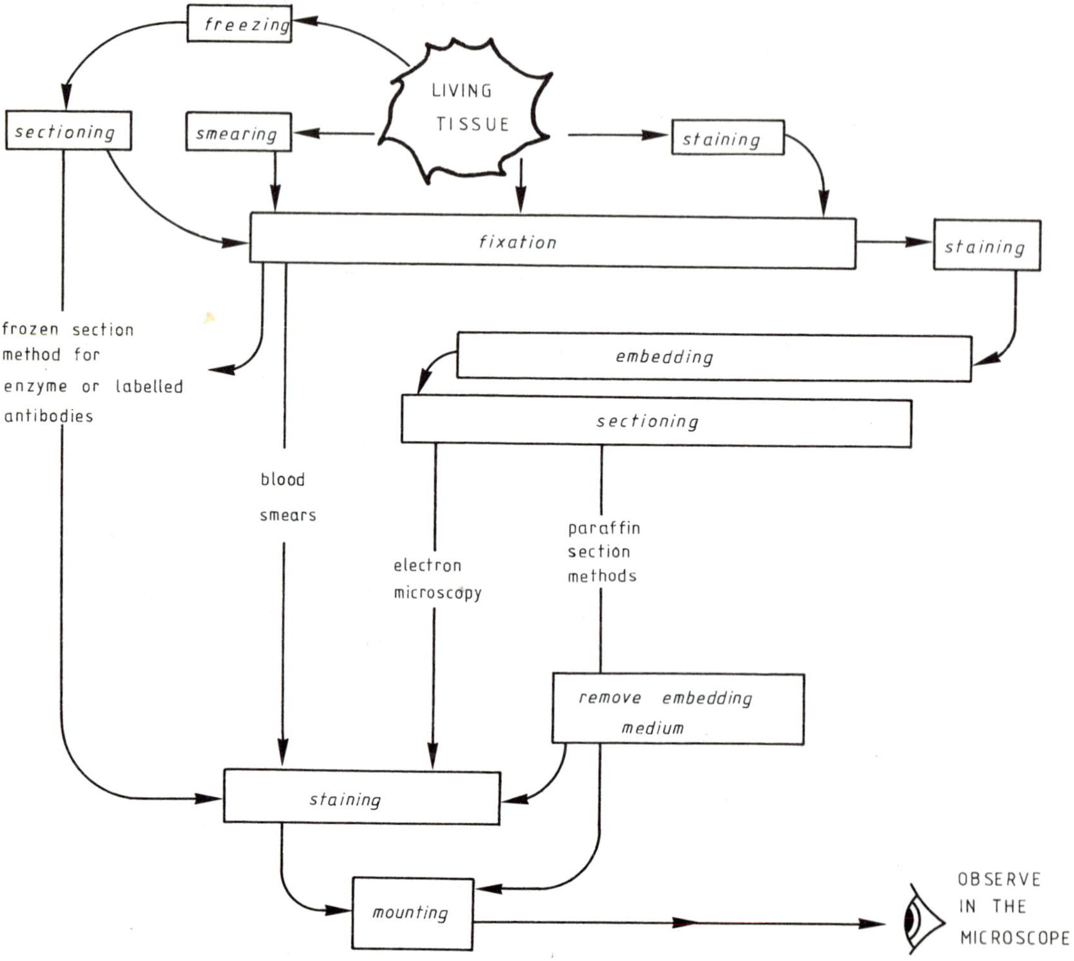

Fig. 1. Examples of some common routes from living tissues to stained material awaiting microscopic examination

3 Fixation and other Loss-limiting Procedures

3.1 Why we need to limit losses: the practical options

The necessity of protecting the structures and contents of most biological materials from enzymic hydrolysis, from osmotic and shear forces, and from solvent extraction during staining and prior processing steps was described in Section 1.2. To avoid repetition this Section offers in Fig. 2 a graphic visual demonstration of the need to protect tissues, taking the form of electron micrographs of protected and unprotected, specifically fixed and unfixed, tissues. As a further and numerical demonstration consider the losses of material from unfixed tissues or tissue constituents into water.

Sections of rat liver cut on a cryostat can, within a few minutes, lose up to three quarters of their nitrogenous material into aqueous incubation media (Jones, 1966). Large pieces of rat liver can lose a third of their dry weight if placed in water, though formaldehyde fixation reduces this loss by a half (Merriam, 1958). Isolated hepatocyte nuclei can lose a fifth of their dry weight into water within a few hours, a loss almost completely prevented by prior fixation in formaldehyde (Abramczuk, 1972).

It is noteworthy that different organelles vary in fragility, and membrane bound structures such as lysosomes and mitochondria act to retain and protect their enzyme content.

What procedures are available for limiting such losses? What is fixation for instance? Again, such questions were briefly introduced in Section 1.2. The possibilities for achieving the linked aims of stabilising structures and limiting losses of content were categorised as follows. Firstly by reducing contacts between tissue and potential extractants, by means of freezing, vapour fixation, careful choice of solvents and so on. Secondly by formation of insoluble derivatives from tissue components by treatment with fixative agents, which will be discussed at length in Sections 3.3 to 3.9. Finally by trapping soluble tissue components. This may be within tissue elements rendered insoluble by fixation, or within naturally impermeable structures. Alternatively, as will be described in Section 3.2, trapping may be achieved by coating the tissue with semi-permeable membranes or by adding large amounts of polymer to the staining solutions.

3.2 Limitation of losses without chemical modification of the tissue

This Section brings together methods whose common feature is that, unlike fixation, they do not inevitably involve chemical modification of the tissues to form insoluble

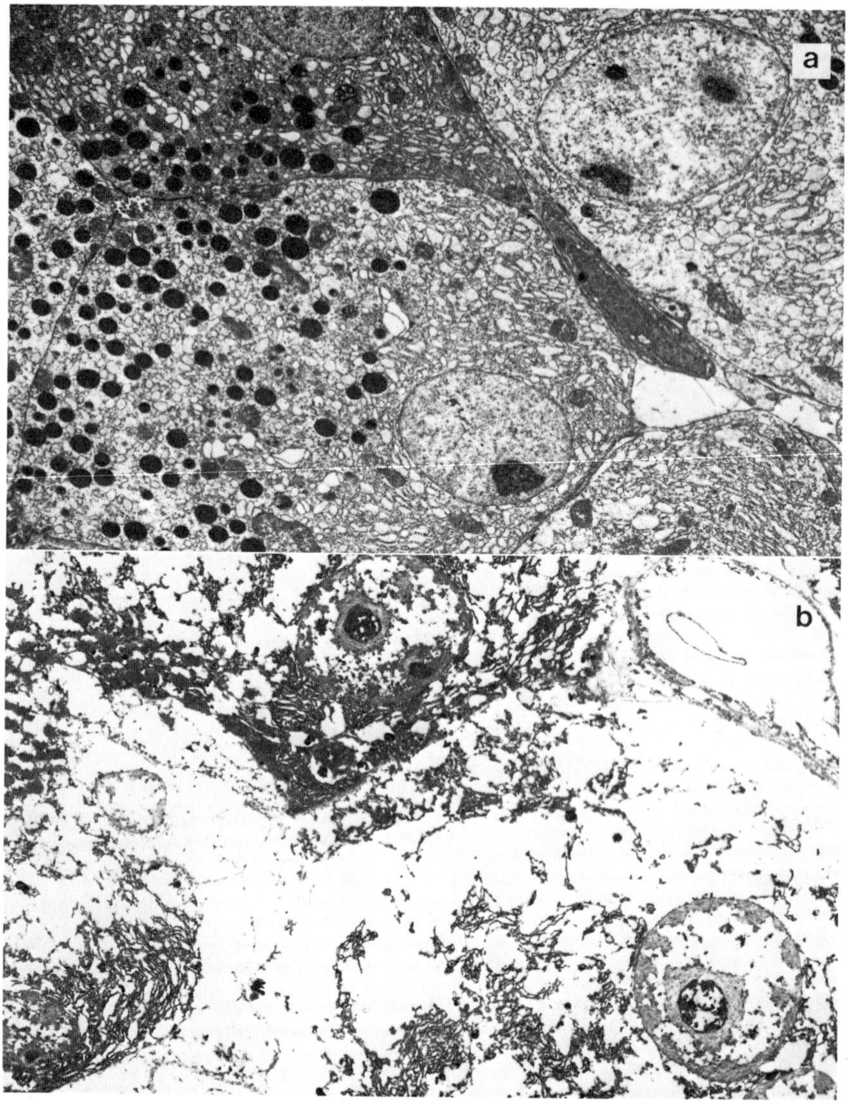

Fig. 2. Electron micrographs of rat pancreatic acinar cells, (a) prepared after preliminary fixation by the glutaraldehyde-osmium tetroxide procedure, and (b) prepared without prior fixation ×2400

derivatives. However manipulations more complex than mere avoidance of extractants are involved.

The initial strategy of this type is to store tissue at very low temperatures and to cut cryostat sections which are then stained directly. This certainly avoids losses prior to the staining step, though during this step such preparations may lose large amounts of substance. Furthermore structural damage induced by the freeze-thaw cycle, with its potential for ice-crystal formation, may be appreciable. Such effects can be reduced by careful technical manipulation, even to the extent of allowing the cutting of ultra-thin frozen sections (e. g. Varriano-Marston et al., 1977). Nevertheless for everyday use cryomicrotomy is best suited to membrane-bound structures such as mitochondria, with water soluble enzymes or co-enzymes being more difficult to demonstrate (Jones, 1972).

Fascinatingly a certain amount of structural damage is sometimes an asset, as when staining with labelled-antibodies, see Section 8.8 and 8.9. In general however such structural damage is unwanted, and indeed many workers combine cryotomy with brief periods of fixation. There have been other, usually more elaborate, attempts to get best of all worlds. For instance combining freeze-drying with embedding in a hydrophilic resin such as glycol methacrylate. Such embedded tissue can be sectioned and stained without removal of the polymer, and many tissue components are retained and can be stained (Mitrenga et al., 1974).

Which leads us to another, increasingly popular, alternative to fixation. Namely coating of the surface of a cryostat section with a thin layer of substance permeable to small reagent molecules but not to tissue macromolecules. Such layers have been applied in various staining systems, by Lake (1970), Meijer (1972) and others. The strategy is discussed further in Section 7.2. Though the limits are by no means known as yet, this procedure is not likely to preserve structure very effectively. Loss of small molecules is inevitable. A varient, seemingly rather similar in physical-chemical basis, is the addition of substances such as polyethylene glycols, polyvinyl alcohol or pyrrolidone to the staining solutions. Such procedures probably suffer from the same limitations as the use of semi-permeable membranes, but can on occasion reduce losses of macromolecules at least to negligible amounts. Such protective additives are compatible with several enzyme histochemical procedures (Altman, 1971), and also to some staining methods involving dyes or reactive reagents (Scott, 1974).

Finally we will consider preservation procedures whose status on the chemical-change dimension is sometimes confused. Both organic solvents and heat have by some been regarded as acting without chemical alteration of the tissues. It is true that many solvents, unlike other fixative reagents, are not likely to form covalent bonds with the tissues. Hence they were termed «non-additive fixatives» by Baker (1958) in his influential book. However this does not imply that solvents are non-reactive, some will produce hydrolysis of labile bonds and all are likely to denature nucleic acids and proteins in very definite reactions. Analogously heat, which may be applied via boiling saline (Lillie and Fullmer, 1976, p. 67) or as microwave radiation (Bernard, 1974; Petrer and Schardein, 1977), has sometimes been thought of as strangely etherial: as if boiling an egg was not a chemical process, for an elegant discussion of which see Moore (1963, p. 300). Denaturation can indeed occur during the embedding in hot paraffin wax, even if it did not happen earlier (Bruchhausen et al. 1969).

3.3 Fixatives: substances used to chemically modify native tissues

Organic solvents such as acetone and alcohols, electrolytes such as mercuric chloride and cetylpyridinium chloride, inorganic compounds such as osmium tetroxide and chromyl chloride, and a variety of reactive organic compounds such as aldehydes, olefins and triazines have all been used as fixative agents. Only a few of the more commonly used substances will be discussed here. Though in practice mixtures of fixative agents are often used, here only the actions of isolated primary fixative agents will be described. Comment on characteristics of fixative mixtures is given later, in Section 3.8. Readers wishing to sample the incredible variety of substances used for fixation (all apparently successfully!) will find a fascinating list given by HOPWOOD (1969, p. 332).

Since several accounts of probable and possible properties and reactions of fixatives are available (BAKER, 1958, p. 19–51; HOPWOOD, 1969; PEARSE, 1968, p. 70–105) only

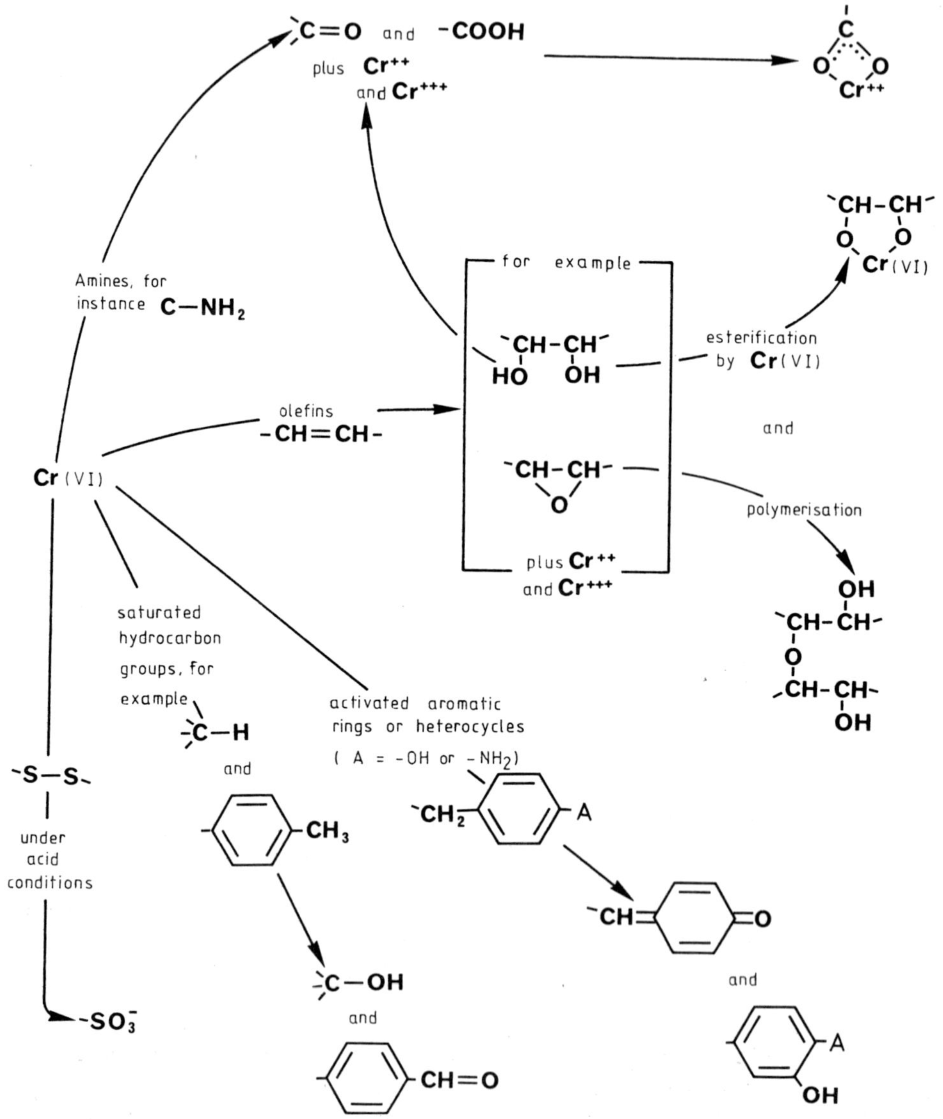

Fig. 3. Some reactions of Cr (VI) ions, as present in chromate and dichromate salts. Many of these reactions are strongly pH dependant

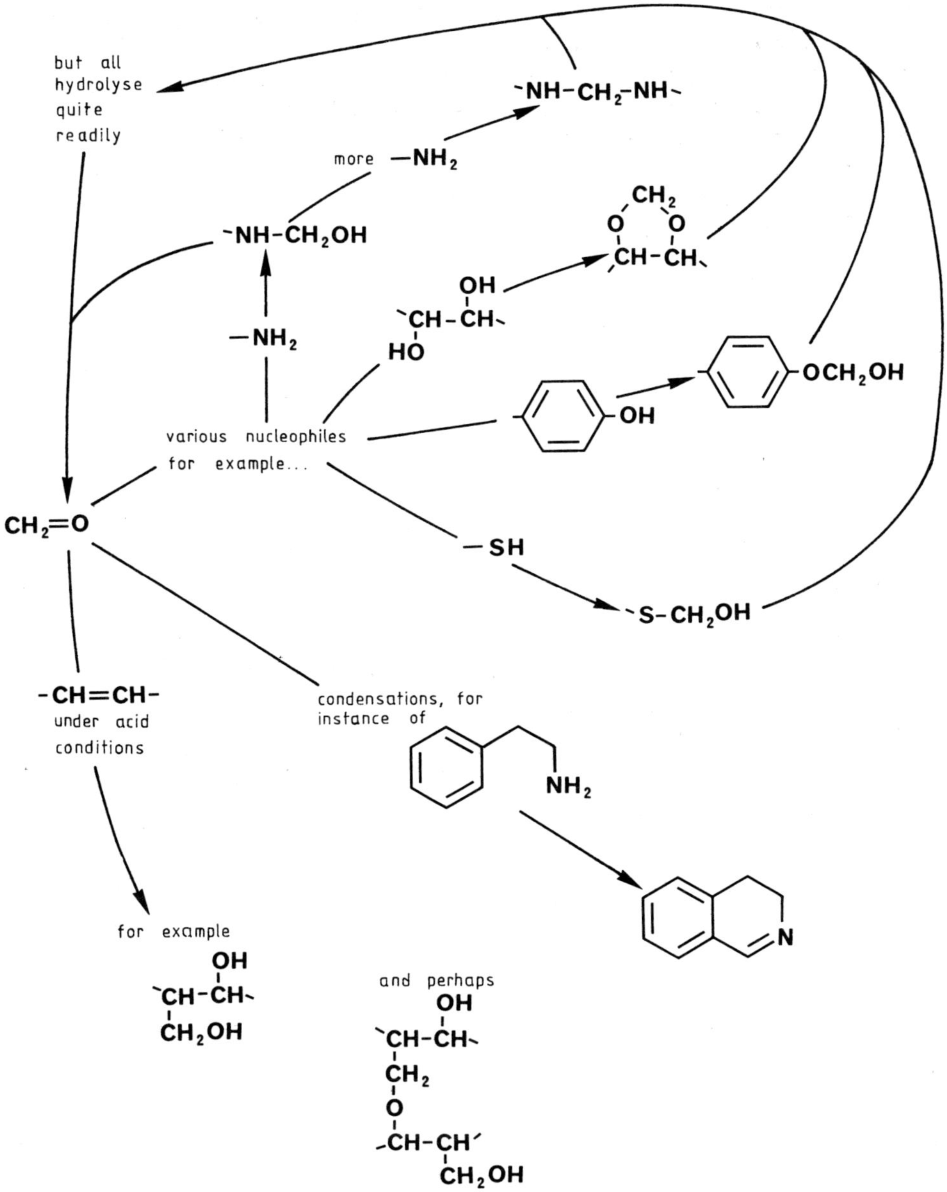

Fig. 4. Some reactions of formaldehyde. Most of these reactions occur whether the formaldehyde is present as the gas or in aqueous solution

an outline of the reactions of several common fixatives with organic compounds will be given here. This summary takes the form of a series of charts, Figs. 3 to 6, and two Tables, 1 and 2. The charts were constructed with the aid of the sources cited above plus some other general texts, in particular RODD's «Chemistry of the carbon compounds» (1962), WALKER's «Formaldehyde» (1964), and the biennial chapters on metal-protein

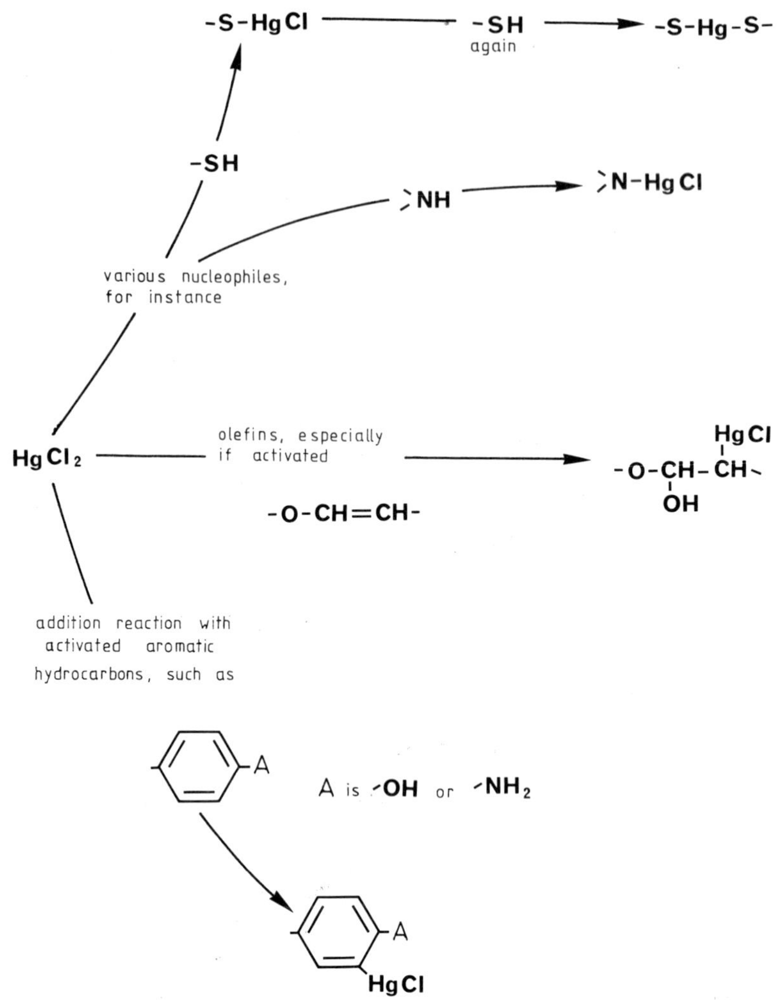

Fig. 5. Some reactions of mercuric chloride and similar salts. Note that mercuric salts of differing degrees of ionisation differ in reactivity

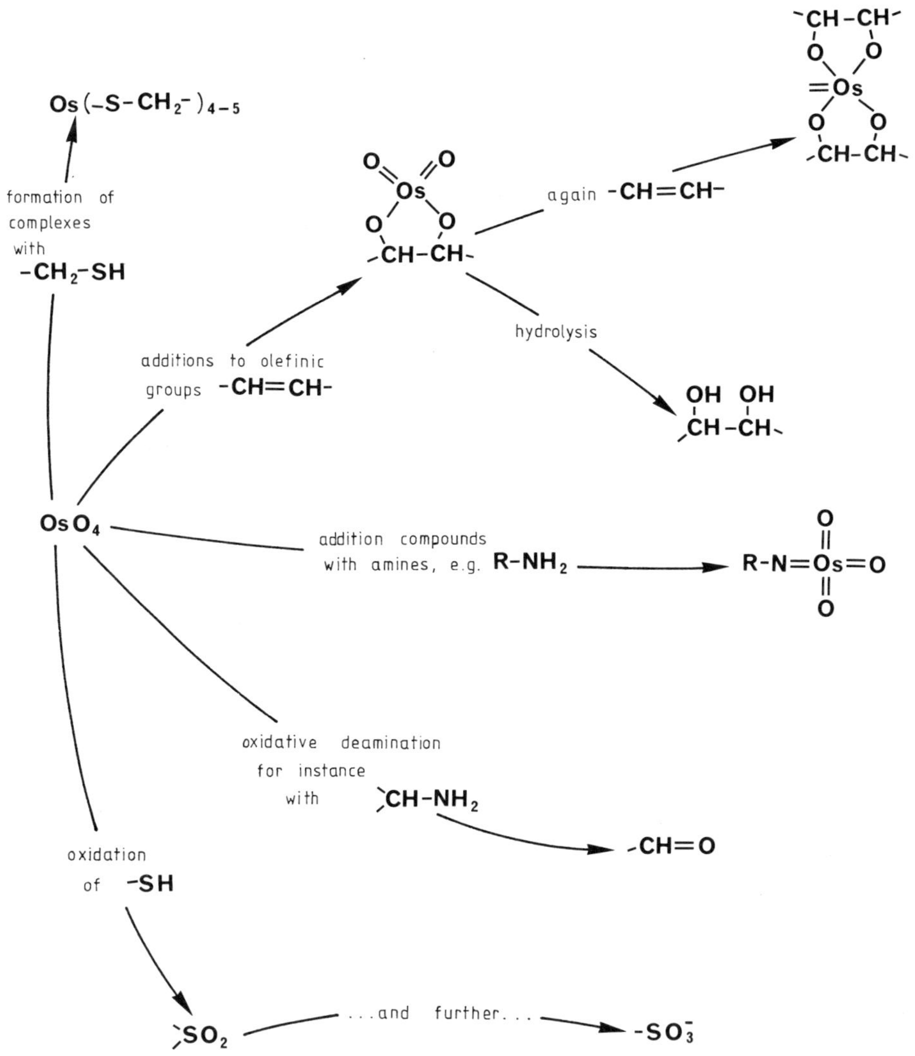

Fig. 6. Some reactions of osmium tetroxide. Most of these reactions occur whether the reagent is present as a vapour or dissolved in some solvent

interactions in «Amino-acids, Peptides and Proteins, Specialist Periodical Reports» published by the Chemical Society (London).

3.4 How is the fixation of tissues achieved?

Essentially by the insolubilisation of those all-prevading biopolymers, proteins. In addition, since much intracellular detail is made up of or built around lipid-rich membranes, the insolubilisation of lipids is a requirement for electron microscopy.

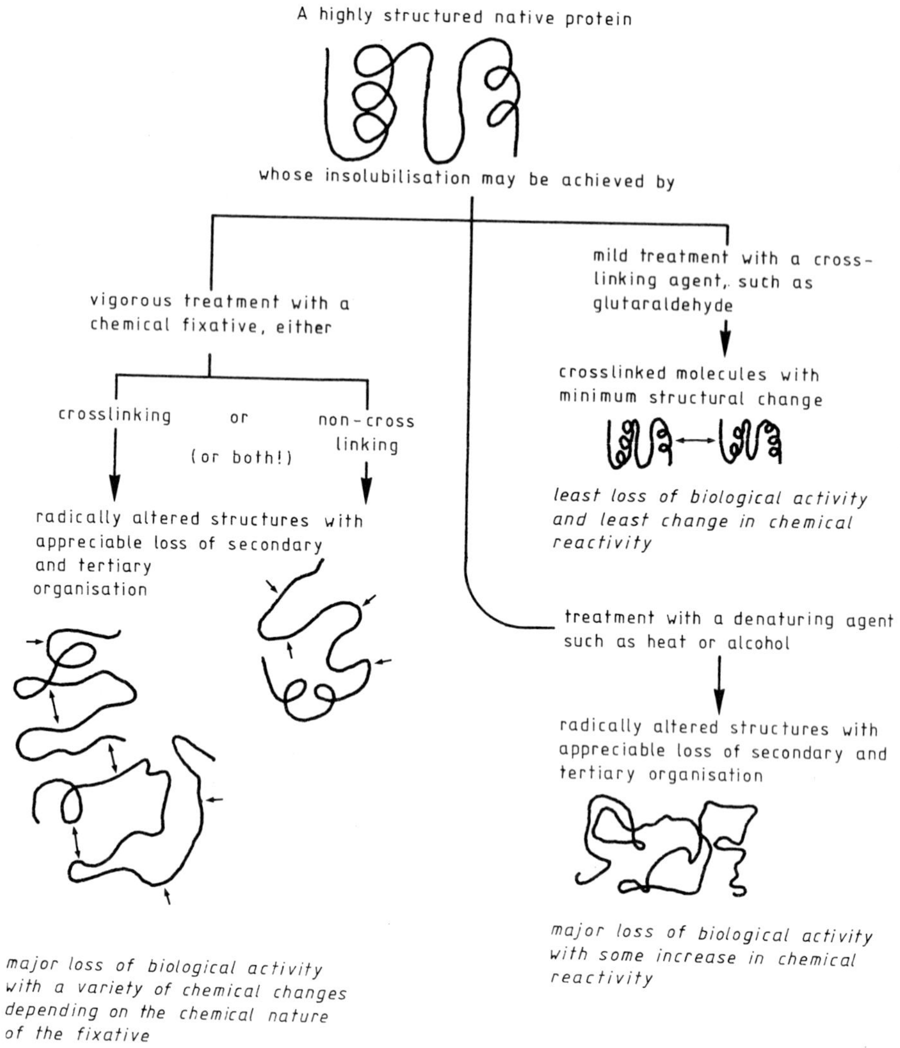

Fig. 7. Structural changes in proteins associated with fixation. Points of gross chemical modification are indicated by arrowing

Initially the *insolubilisation of proteins* by fixatives will be considered. This may be achieved in at least two distinct ways. By crosslinking separate protein molecules into giant aggregates, and also by disrupting the secondary and tertiary structures of the proteins to produce insoluble derivatives. These two processes, with certain consequences, are shown diagrammatically in Fig. 7. Many authors writing on fixation regard crosslinking as the quintessence of the fixative process. Consider for example PEARSE (1968) who says «...the essential feature is thus... the formation of crosslinks...», MILLONIG and MARINOZZI (1968) who say «...chemical fixation involves the establishment of bonds between the molecular consistents of tissues...», HOPWOOD (1969a) who

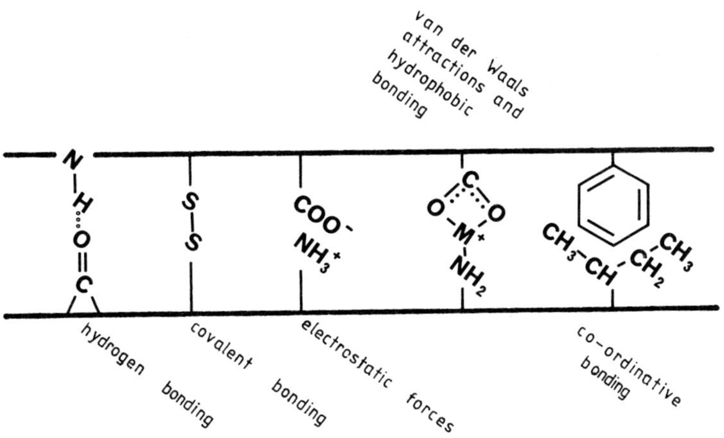

Fig. 8. Examples of links between protein chains which help to maintain the secondary and tertiary structures. The chains are represented by the heavy lines

says «Fixation may also be described... as the formation of a macromolecular network between the various cellular constituents...» and also of course WOLMAN's influential pictorial representation of crosslinks (1955, his Fig. 3). However the contention here is that the change in secondary and tertiary structure of the proteins and not cross-linking is better regarded as the characteristic molecular event of the fixation process, as was suggested by SINGER (1952), and indeed by GABE (1976) who stated that «...the essential problem of histological fixation is to denature proteins...». For this reason, and also since cross-linking is conceptually straightforward even if chemically complex, the second type of fixation involving of protein structure will now be discussed in detail. A brief account of the crosslinking of proteins will then follow.

As an introduction consider how the secondary and tertiary structures of proteins are maintained. As illustrated in Fig. 8 there are a wide variety of processes involved: covalently bonded bridges, electrostatic interactions, hydrogen bonding, co-ordination of metal ions, van der Waals attractions, and the water-structure effect usually termed

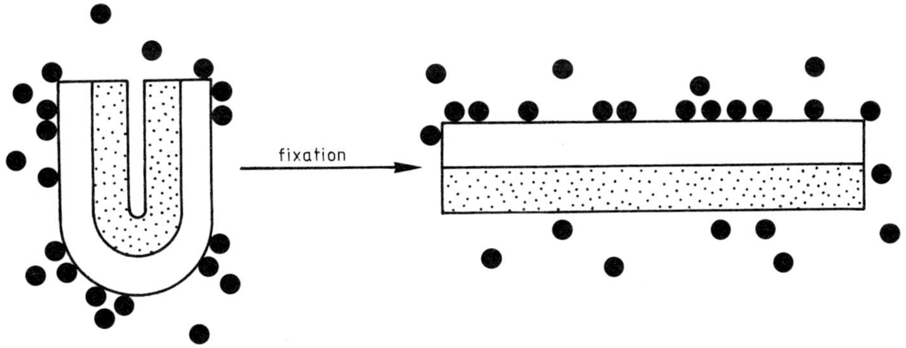

Fig. 9. Exposure of the lipophilic amino acid residues in a protein's core (stippled region) caused by loss of secondary and tertiary structure. The region of polar amino acid residues unshaded and water molecules are represented as bold circles

hydrophobic bonding. For brief accounts for the latter four topics see Supplements 5, 6, 3 and 7 respectively. However though the breaking of a disulphide bridge or hydrogen bond, or loss of a co-ordinated metal ion, or alteration in electrostatic attractions can indeed result in major structural changes, none of these factors constitutes the essential stabilising factor of native proteins. Such a role may however be assigned to hydrophobic bonding. To explain this consider proteins which are soluble under physiological conditions. When the molecules are in the conformation in which they occur in their natural surroundings most of the polar (that is hydrophilic and in particular ionisable) amino acid residues are situated on the surface of the molecules in contact with the aqueous solvent. On the other hand the non-polar or lipophilic amino acid residues are found predominantly inside the protein's core. Now why does the

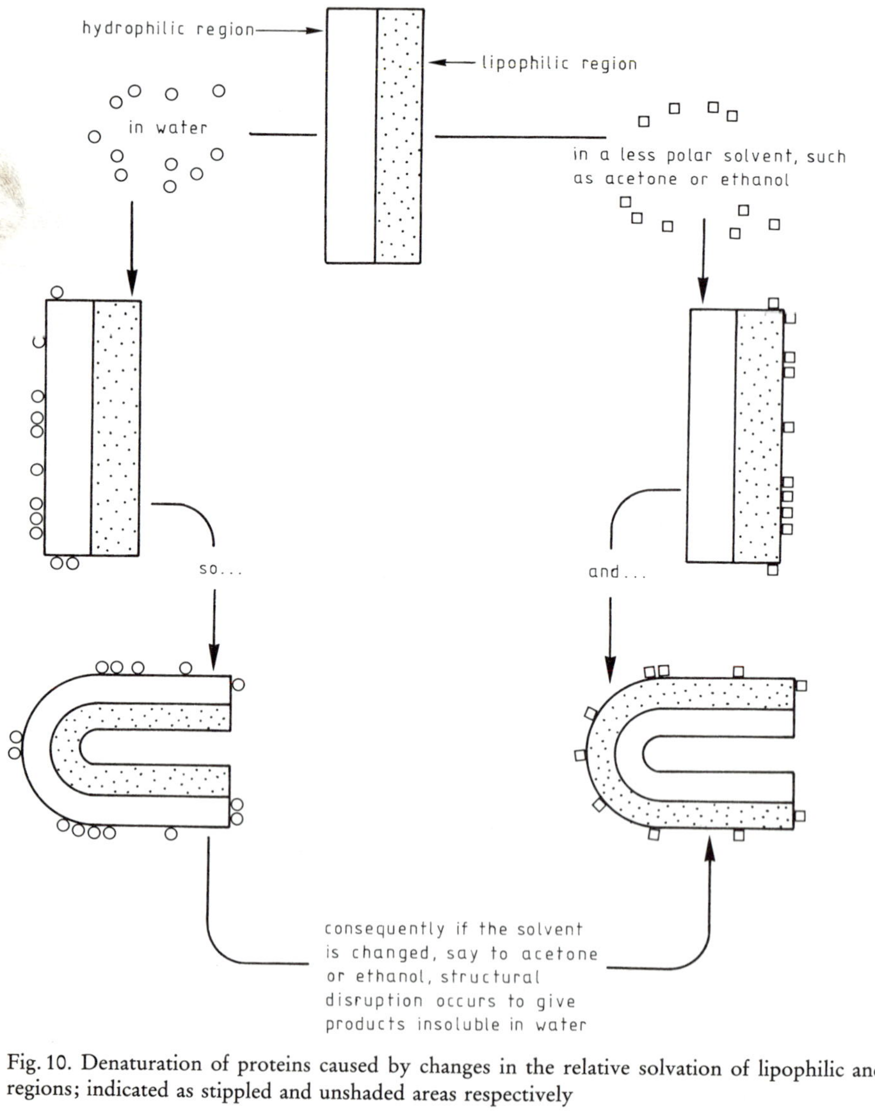

Fig. 10. Denaturation of proteins caused by changes in the relative solvation of lipophilic and hydrophilic regions; indicated as stippled and unshaded areas respectively

Table 1. The action of a variety of common primary fixative agents on proteins

Fixative agent	Chemical action	Structural consequences	For further information . . .
Organic solvents, such as alcohols or acetone	Ionisable groups solvated less well than in water, lipophilic rather better	Lipophilic/hydrophilic balance upset, leading to denaturation	See accounts of protein denaturation in the biochemical literature, e. g. PUTNAM (1953), TANFORD (1968). See also BAKER (1958), TIMASHEFF (1970), EAGLAND (1975)
Heat	Generalised structural disruption		
Strong acids, such as picric and trichloroacetic	Alter charges on ionisable groups, e. g. $-NH_2 \xrightarrow{H^+P^-} NH_3^+P^-$ $-COO^- \xrightarrow[H^+]{} -COOH$	Disrupts electrostatic and hydrogen bonding interactions Inserts lipophilic anions (e. g. P–) into hydrophilic regions	
Aldehydes such as formaldehyde and glutaraldehyde	React with ionisable amino groups giving less basic higher amines. This reaction may proceed to give crosslinks React with lipophilic groups, e. g. tyrosine, to give more hydrophilic products React with thiols, could crosslink	Reduces solvation of hydrophilic regions by water, increases that of lipophilic regions Disrupts hydrogen bonding and electrostatic interactions: proteins become less cationic	General accounts in HAYAT (1970, p. 65–95), PEARSE (1968, p. 70–76). For formaldehyde see also BAKER (1958, p. 104 & 126), ELLEDER and LOJDA (1968), WALKER (1964)
Chromium (VI) compounds such as potassium dichromate and chromic acid	At low pH split disulphide bridges, typically found in the core, into ionic sulphonates Attack lipophilic residues, e. g. tyrosine and methionine Above reactions reduce Cr(VI) to Cr(II) & (III), which react with (anionic) carboxylates giving cationic Cr(III) complexes	Increases solvation of liphophilic regions by water Disrupts van der Waals attractions and electrostatic interactions, proteins become less anionic	Reviewed by HARTLEY (1969) for the case of keratin. See also BAKER (1958, p. 104–110, 126–133), HOPWOOD (1969 c)
Mercury salts, usually mercuric chloride	React with ionisable bases, especially heterocyclic amines React with thiol groups, typically found in the core. Can crosslink React with lipophilic aromatic residues, e. g. tyrosine	Disrupts the lipophilic region, may introduce crosslinks, decreases solvation of hydrophilic regions Disrupts van der Waals attractions, hydrogen bonding and electrostatic interactions, proteins become less cationic	General account by BAKER (1958, p. 99–104). See also GURD and WILCOX (1956, p. 365), KITCHING (1968)
Osmium tetroxide	Reacts with ionisable amino groups, yielding carbonyls Reacts with lipophilic residues, e. g. tryptophan and tyrosine, to give polar products. Could crosslink Reacts with thiol groups giving ionic sulphonates	Lipophilic region becomes better solvated by water and hydrophilic region less so Disrupts van der Waals attractions, hydrogen bonding, and electrostatic interactions, proteins become less cationic	General accounts by HAYAT (1970, p. 36–58), HOPWOOD (1969 a), PEARSE (1968, p. 79–85). See also BARRNETT and ROTH (1958)

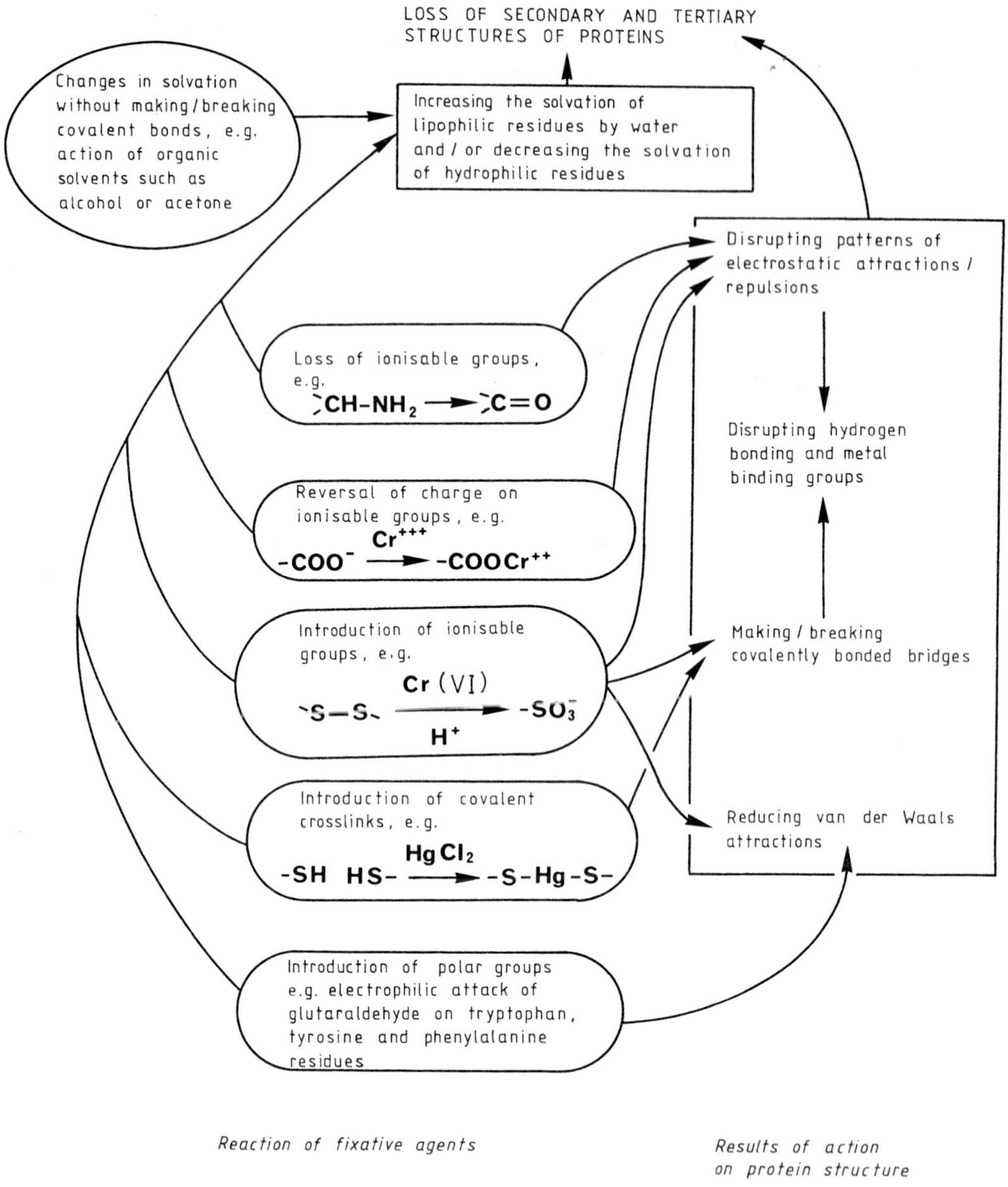

Fig. 11. Destruction of the secondary and tertiary structure of proteins by various fixative agents. Fixative agents always upset the lipophilic/hydrophilic balance in addition to disrupting other structure-stabilising factors

disruption of the lipophilic-inside/hydrophilic-outside protein structure make the molecules insoluble in water? It is because on disruption the proteins become rearranged, as shown diagrammatically by Fig. 9, in such a way as to expose the lipophilic amino acid residues to the aqueous solvent. With the hydrocarbon groups thrust at the water, the solubility of the proteins in the water falls sharply.

From this standpoint consider the *action of various types of fixatives* of proteins. Protein structures can be disrupted without any covalent bonds being made or broken, as in the process of denaturation. Fixatives probably acting in this way include organic solvents such as alcohols and acetone. This topic has been widely studied biochemically, see for example a review of TANFORD (1968) and also PUTNAM's (1953) classic account. Such fixatives act, as illustrated diagrammatically in Fig. 10, by increasing solvation of the lipophilic core and decreasing that of the hydrophilic surface region, so allowing the protein structure to open up. One may ask at this point if, as no covalencies are involved, the denaturation of proteins is reversible? For some proteins, such as insulin and serum albumin, denaturation caused by some organic solvents can indeed be reversed, and so such proteins are not readily preserved in tissues. In most cases however the protein is trapped in the denatured state (or more precisely, in one of many such states) due to the excessively slow rate of the reverse reactions from disordered state or states to the highly organised native state.

Most other fixatives do make and break covalent bonds in proteins, though this process may still be reversible as with the case of strong acids. Such «additive» fixatives will variously involve interference with stabilising electrostatic attractions and repulsions, covalent bridges, hydrogen bonds and van der Waals attractions. This stated, the common mechanistic factor for all fixatives is the disruption of the secondary and tertiary structures of proteins via change in the balance of lipophilic/hydrophilic regions. This is shown diagrammatically in Fig. 11. This figure gives only selected examples and most fixative actually produce more than one chemical effect. Hence it is

Table 2. Examples of the crosslinking and fission of proteins by fixatives

Fixative agent[a]	Was cross-linking detected?	Was fission detected?	Proteins studied (and method of study)	References
Glutaraldehyde	Yes: in quantity	not studied	Bovine serum albumin and casein (acrylamide gel electrophoresis, gel filtration, viscometry); bovine serum albumin, bovine γ-globulin and ovalbumin (gel filtration); collagen (biochemical analysis)	HOPWOOD (1969b, 1970), HABEEB and HIRAMATA (1968), BOWES and CATER (1966)
Formaldehyde	Yes	not studied	Bovine serum albumin and casein (acrylamide gel electrophoresis, gel filtration, viscometry); collagen (chemical assay); gelatin (chemical kinetics); and haemoglobin (electrophoresis)	HOPWOOD (1969b, 1970), CATER (1963), DAVIS and TABOR (1963), TOOZE (1964)
Mercuric chloride	Yes	not studied	Human serum albumin and insulin (biochemical analysis); and keratin (mechanical properties)	GURD and WILCOX (1956, p. 365), MENKART and SPEAKMAN (1947)
Osmium tetroxide	A little	A little	Bovine serum albumin and casein (acrylamide gel electrophoresis, gel filtration, viscometry)	HOPWOOD (1969c, 1970)
Potassium dichromate	A trace	Yes	Bovine serum albumin and bovine γ-globulin (acrylamide gel electrophoresis, gel-filtration)	HOPWOOD (1969c)

[a] Usually from dilute neutral aqueous solution, except for mercuric chloride on keratin where the solvent was acetic acid

convenient to list the chemical effects of common fixatives on proteins, as in Table 1. This table excludes mention of crosslinking and fission of polypeptide chains, for which see Table 2, but includes the probable effects on protein structure of these chemical reactions.

Though the structure-breaking model of protein fixation is powerful, the *protein crosslinking aspect of fixatives* must be given attention. The crosslinking action of several fixatives on pure proteins in solution and also on insoluble proteins has been investigated. Some results are displayed in Table 2. The chemical mechanisms of the crosslinking processes are suggested in Table 1 above. Various compounds considered satisfactory fixative agents vary widely in their crosslinking abilities, glutaraldehyde being exceptionally effective. Also displayed in Table 2 is information on *the fission of proteins by fixatives.* Loss of proteins from tissues need not, it seems, be due to *failure* of the protein to react with the fixative. In addition to substances mentioned in the Table, fixatives such as chromic and picric acids would be expected to cause fission of proteins by acid hydrolysis of peptide links.

3.5 The fixation of substances other than protein

The principles underlying the fixation of, for example, lipids, nucleic acids and polysaccharides are varied. Some fixatives act by denaturation or by crosslinking. Other methods of fixation involve the trapping of non-protein materials inside fixed protein, or the formation of insoluble precipitates with the fixative agent.

The *structure-breaking type of fixation* can occur with some high molecular wight nucleic acids which possess a hydrophilic/hydrophobic structure analogous to that found in proteins and discussed above. The essential factors maintaining the secondary and tertiary structures in these nucleic acids are hydrophobic bonding (see Supplement 7) plus van der Waals attractions and hydrogen bonding (see Supplements 4 and 5 respectively) linking base pairs. As with protein, destruction of the secondary and tertiary structures of native nucleic acids will lower water solubility and result in aggregation and precipitation. This may again be achieved without making or breaking covalent bonds, for example by denaturation with heat or organic solvents. Alternatively covalent bond changes which also give rise to disruption of stabilising forces and change the lipophilic/hydrophilic balance may be significant.

The possible roles of several fixatives of nucleic acids viewed in this light are shown diagrammatically in Fig. 12. Some probable chemical changes produced by a number of common primary fixatives on nucleic acids or their constituents are illustrated in Table 3. This information is more speculative than in the analogous discussion of protein fixation, as less histochemical work has been done. The present account emphasises the structure-breaking role of fixatives for nucleic acids, though additional mechanisms are probably also of importance.

Fixation by crosslinking appears to be involved in the preservation of a variety of substances. Taking lipids as an example, crosslinking could and probably does involve both lipid-lipid cross links and lipid-fixed protein links. For instance it is generally considered (and reviewed by PEARSE, 1968, p. 79, and HAYAT, 1970, p. 36) that osmium tetroxide can crosslink unsaturated lipids by the formation of cyclic osmate diesters. Work confirming this indicates moreover that such diesters could themselves react with amines (GRIFFITH and ROSETTI, 1972; NIELSON and GRIFFITH, 1979). The reaction of

Table 3. The action of a variety of common fixative agents on nucleic acids or their derivatives

Fixative agent	Chemical action	Structural consequences	For further information...
Organic solvents such as alcohols or acetone	Decreased solvation of sugar and ionic phosphate groups, increased solvation of bases	Lipophilic/hydrophilic balance upset, leading to denaturation	Introductory accounts of the denaturation of nucleic acids are available in most undergraduate biochemistry texts, for more details see, for example, JORDAN (1960, chapt. 11), MARMUR et al. (1963), EAGLAND (1975)
Heat	General structural disruption		
Acids, for instance acetic or chromic	Alteration in charges of ionogenic groups of bases, insertion of hydrated counter ions	Lipophilic region becomes better solvated by water	
		Polymer broken up	
	Hydrolysis of sugar residues and bases	Hydrogen bonds between bases disrupted	
Aldehydes such as formaldehyde or glutaraldehyde	Conversion of amino groups to methyl or analogous derivatives	Lipophilic region becomes better solvated by water	Chemical literature is reviewed in standard texts, e. g. BROWN (1962), LISTER (1971). Biochemical literature reviewed by HOPWOOD (1969a, 1972), KOCHETKOV and BUDOWSKY (1969). For a surprising study, see HOPWOOD (1975) on glutaraldehyde
	Possibly substitution in the heterocyclic rings	Hydrogen bonds & van der Waals attractions between bases disrupted	
Chromium (VI) compounds, e. g. chromic acid	Effects of acids described above	Lipophilic region becomes better solvated by water	See accounts in standard heterocyclic chemistry texts, (e. g. BROWN, 1962; LISTER, 1971) and reviews (WATERS, 1958)
	Bases can be oxidised to aldehydes and carboxylic acids	Hydrogen bonds and van der Waals attractions between bases disrupted	
Mercury salts, usually mercuric chloride	Reacts with the nitrogen of the heterocycles giving -NHgCl derivatives, then can repeat to give crosslinks	Disruption of hydrogen bonds and van der Waals forces between the bases	See for instance YAMANE and DAVIDSON (1961), LIPSETT (1964), KOBE and DOUMAN (1941)
	Could give nuclear mercuration		
Osmium tetroxide	Can react with carbon-carbon double bonds of bases to give diol derivatives	Lipophilic region becomes better solvated by water	BURTON (1967)
		Hydrogen bonds and van der Waals attractions between bases disrupted	

osmium tetroxide with aliphatic primary amines and with thiols has also been reported (MILAS and ILIOPULOS, 1959; HANKER et al., 1967). So in the tissues it is possible that with unsaturated lipids osmium tetroxide produces lipid chains crosslinked via osmate diesters, in turn linked to fixed proteins.

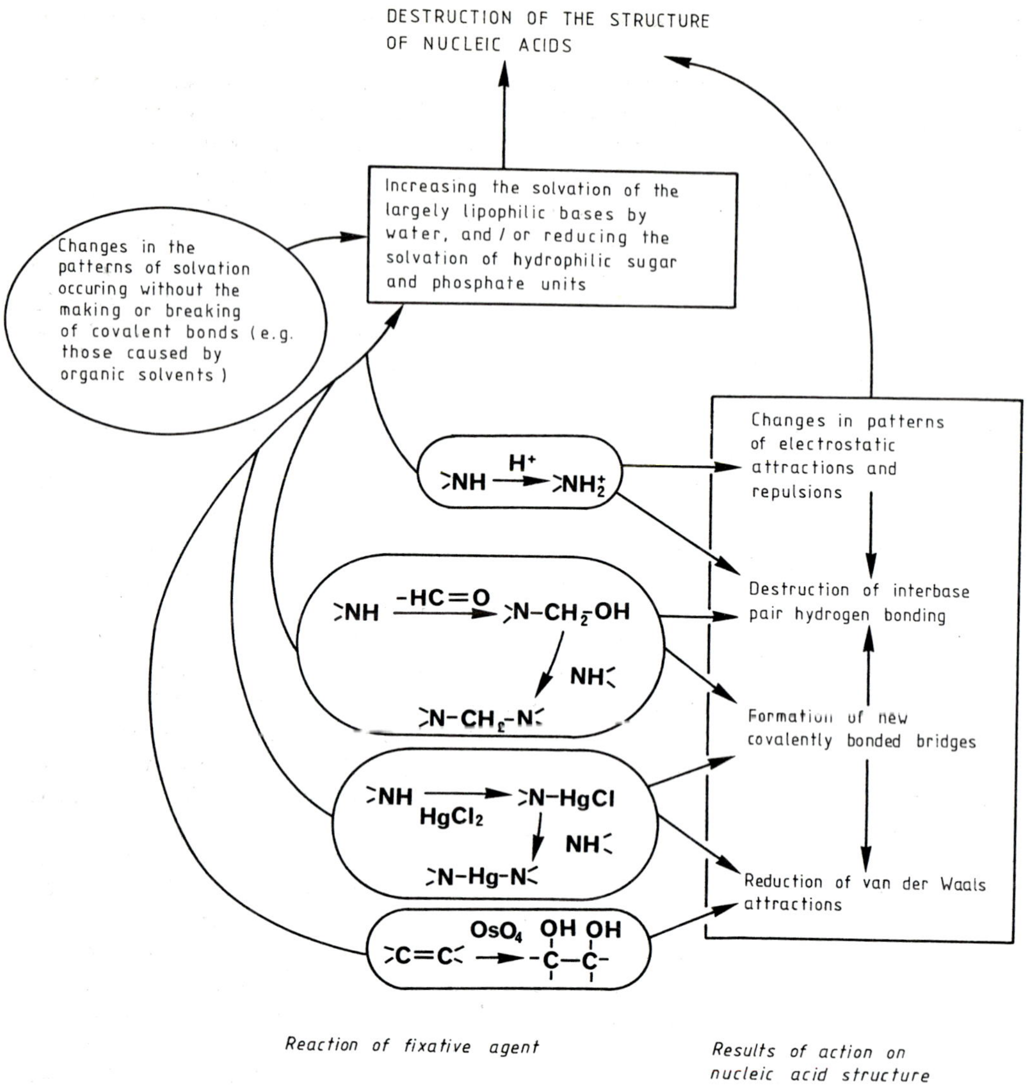

Fig. 12. Probable structure-disrupting actions of some common fixative agents on nucleic acids

It seems that analogous events may also occur during the fixation of unsaturated lipids by dichromates or chromic acid. Lipids fixed by such Cr(VI) salts are still non-polar, and may still be stained by Sudan dyes (CHOU, 1958). It is noteworthy that such fixed lipids contain no more chromium than do adjacent regions not expected to contain carbon-carbon double bonds (CHOU, 1958; ROOZEMOND, 1971). These observations can be explained as follows. One of the common groupings resulting from the oxidation of carbon-carbon double bonds by Cr(VI) salts is the epoxide group (WATERS, 1958). This labile species readily reacts, especially under acid conditions, to give metal-free polymers. In addition the epoxides, together with any aldehyde produced, can react with amino groups. Thus in the tissues it is possible that Cr(VI) salts react with unsaturated lipids to produce polymers containing ether crosslinks, in turn linked to fixed proteins.

There are also some indications that other classes of compounds, saturated phospholipids and polysaccharides for instance, can be crosslinked, either to themselves or to protein or to both. For example phospholipids containing amino groups are better preserved than amino-free lipids when fixed with formaldehyde or glutaraldehyde, and both these fixatives are capable of crosslinking amino groups (ROOZEMOND, 1969). There is also evidence that formaldehyde can crosslink unsaturated lipids via the carbon-carbon double bond (JONES, 1969). Turning to polysaccharides, it has been noted that hexosamine-containing mucosubstances are more effectively preserved by glutaraldehyde fixation than are compounds lacking amino groups (KORHONEN and LINDHOLM, 1969). Direct crosslinking of polysaccharides can be achieved using chloro-s-triazines and there is indirect evidence that this occurs with mucosubstances when trichloro-s-triazine, cyanuric chloride, is used as a fixative (GOLAND et al., 1967; SORVARI and LAUREN, 1973).

Fixation by trapping, that is mechanical retention due to the fixation of the surrounding material, has often been suggested as the mode of fixation of glycogen (as reviewed by PEARSE, 1968, p. 89). SMITHERMAN and colleagues (1972) have substantiated this, and find that ineffective fixatives may precipitate but not permanently insolubilise glycogen. Furthermore though several fixatives are equivalent in their abilities to precipitate glycogen from aqueous solution and are able to retain glycogen in fixed blocks of tissue, fewer fixatives ensure the retention of glycogen in tissue immersed in an aqueous solution. It is quite possible that this type of effect is more widespread than is generally appreciated. For example the DNA and protein of chromosomes are intimately entwined and a trapping mechanism might in part explain the ability of fixatives such as alcohol to cause the retention of DNA. See also Section 3.6 for comments on the fixation of insulin.

The formation of insoluble salts has been used, for instance to aid the fixation of anionic mucosubstances. The common precipitant is the quaternary salt cetyl pyridinium chloride, though some cationic dyes have also been used in this role (WILLIAMS and JACKSON, 1956; CONKLIN, 1963). It has been suggested, though only on the basis of morphological appearance, that fixation of nucleic acids is also improved, perhaps again due to the formation of insoluble DNA-large cation products. A snag with these methods is that though mucosubstances are retained, morphological damage is accentuated by these detergent-like fixative agents (GOLAND et al., 1967). Another example is the retention of copper ions by adding sulphide ions to the fixative (SIMKISS, 1979).

3.6 Effects of fixation on biological staining: limitation of losses of tissue materials

For convenience effects of fixation on staining will be considered under the following headings:

1. Losses of material from the tissues. A fundamental concern, since what is not retained cannot be stained. This is the topic discussed in the present Section 3.6.

2. Chemical modification of the tissues. As was shown in Section 3.4 this is both commonplace and various. Different fixatives give different modifications, as discussed in Section 3.7.
3. Physical modifications of the tissues. Different fixatives leave the tissues in different physical states, influencing some modes of staining as discussed in Section 3.8.
4. Rates of fixative penetration. Tissue can only be fixed if the fixative agent reaches it, as discussed in Section 3.9.

In practice biological staining can be influenced by any or all of these four factors. Unfortunately few studies of the influence of fixation on staining actually take this general view. It has been more common to study losses while disregarding chemical modification, or to study chemical modification whilst disregarding the possibility of selective losses. The physical modification of fixed tissues is usually disregarded entirely, and the effects of fixative penetration are known but are usually glossed over with the suggestion that small pieces of tissue should be used.

Many specific consequences will be dealt with in detail later when discussing different types of staining, e. g. reactive methods in Chapter 5, enzyme histochemistry in Chapter 7, labelled-antibody staining in Chapter 8, and the question of the localisation of staining in Chapter 11. In this section we will consider general points not requiring any detailed understanding of staining mechanisms. One may note before commencing this discussion that in the everyday situation variables such as the pH of the fixative, the particular buffer system used and the period of fixation will be extremely important. Due to paucity of information we will, however, be concerned here primarily with the differences between fixatives.

To what extent is the material under study retained in the tissue?

Many studies on this topic have been published but much of the work is of unknown significance. In quantitative studies the amount of a substance retained in cells or tissue blocks immediately after fixation has often been assayed, rather than the retention in a section which has been exposed to the staining bath or even the embedding medium. Examples include Kugler and Wilkinson (1964) who studied fixation of glycogen, and Saunders and associates (1968) who studied lipid fixation. However quantitative work, whatever its limitations, is not the norm. Assessment of the retention of substances by different fixatives has commonly been made merely by comparison of staining intensities. In the following account the retention of various classes of materials will be considered separately. As noted in Section 3.5 this could mislead, since the fixation of one class of compound may aid the fixation of another.

The retention of proteins will be discussed first since losses of protein so markedly affect the fixation of the other constituents of cells. Studies carried out on various systems have been reviewed by Hopwood (1969, p. 336) and Williams (1973, p. 144). Some of the available data, emphasising differences between fixatives, are displayed in Table 4. With the qualification of the previous paragraph in mind one can nevertheless make certain general comments on these results.

The crosslinking fixative glutaraldehyde is usually outstandingly effective at retaining protein, and other aldehyde fixatives, e. g. formaldehyde or acrolein, are quite effective. One could hope to retain some 80–90% of protein when using aqueous solutions of aldehyde fixatives. With other fixatives, whether reacting covalently with protein as can osmium tetroxide, or acting merely as denaturants as do the organic solvents of Carnoy's solution, much less protein is retained. Any fixative is better than none however.

Table 4. Examples of losses of protein from tissues during various modes of fixation

Tissue or cell type, and mode of fixation and/or processing	Protein lost	Method of investigation and its limitations	Reference
Some comparisons between various fixative agents			
Rat liver fixed in solutions of:			
Osmium tetroxide	24 890 cpm/g	Tissue proteins were labelled by administering tritiated leucine. Tissue was fixed, washed and dehydrated; then labelled protein was assayed by scintillation counting and compared to unprocessed control. This work disregards possible protein losses into processing fluids and on rehydration	Monneron and Moule (1969)
Formaldehyde	30 670 cpm/g		
Glutaraldehyde	32 180 cpm/g (N. B. activity due to *retained* protein)		
Rat liver and kidney treated as described:			
Unfixed	28%	Fixed blocks of tissue were placed in water and the protein extracted was determined by chemical assay. This work thus disregards losses during processing and rehydration and, more significantly, into the fixative	Ostrowski et al. (1961)
Freeze dried	39%		
Freeze substituted	32%		
Fixed in:			
Acetone	13%		
Carnoys fluid	1%		
Ethanol	8%		
Formalin	0%		
A model system consisting of FITC-labelled albumin in gelatin was treated so:			
Unfixed	100%	Pieces of gel were fixed then washed in water. The losses of labelled albumin were estimated fluorimetrically. This work disregards losses into processing fluids and on rehydration	Flitney (1966)
Fixed in solutions of:			
Acrolein	4%		
Formaldehyde	54%		
Glutaraldehyde	20%		
Hydroxyadipaldehyde	80%		
. . . and some comparisons of losses at various stages of processing			
Liver, muscle and cerebral cortex			
When fixed in Carnoys fluid losses were:		Tissues were fixed, processed and embedded, then rehydrated. Dry weights were measured at each step. The losses into the solutions were, when possible, determined colourimetrically	Merriam (1958)
Into fixative plus processing fluids	ca. 10%		
On rehydration	0.5%		
When fixed in formalin losses were:			
Into the fixative	2.4%		
Into the processing fluids	0.5%		
On rehydration	0.2%		
Cell cultures, variously treated but typically:			
When unfixed the losses were:			
Into wash solution	14%	Cells (labelled by growth in media containing tritiated amino acids) were fixed, washed and dehydrated. Losses of tritiated material were assayed at each step using scintillation counting. This work disregards losses occurring after rehydration	Vanha-Perttula and Grimley (1970)
Into processing fluids	0.5%		
When fixed in formalin losses were:			
Into the fixative	8.5%		
Into the wash solution	5.5%		
Into the dehydrating fluids	0.5%		
When fixed in glutaraldehyde solution losses were:			
Into the fixative	0.2%		
Into the dehydrating fluids	0.5%		

The major losses of protein take place into the fixative solution. Rather less protein is lost into the subsequent water washes, and very little is lost into other processing fluids. Certain proteins, though unfixed, may be retained by being trapped in a fixed protein matrix. Insulin for instance may be recovered in a physiologically active form from pancreatic tissue fixed in aldehydes or alcohol even though pure crystalline insulin is not fixed by these reagents (Grillo et al., 1971). It may be noted that retention up to the stage of the embedded block is no guarantee that rapid losses will not occur when tissues are rehydrated. This occurs for example with freeze-dried materials. The possible advantages of combining vapour-phase fixation with freeze-drying have been demonstrated (Gersh et al., 1960) but apparently not followed up.

The consequences of varying the period of fixation differ from one fixative-protein combination to another. For example formalin and Carnoy's solution continue to extract proteins from cells for many hours. However, at least in the case of formalin, the longer the period of fixation the less protein is extracted in subsequent aqueous washes. With glutaraldehye on the other hand such prolonged losses into the fixative do not occur (Merriam, 1958; Vanha-Perttula and Grimley, 1970).

Various practical factors, e. g. the temperature, concentration and pH of the fixative solution, the pH of subsequent washes, the type of buffer, all influence protein retention. As no general comparative study is available this account will be limited to merely illustrating some, perhaps idiosyncratic, variations in retention apparently due to these factors.

Varying the concentration of the fixative agent can have unexpected results. Depending on the fixative agent concerned and on its temperature, increasing the concentration of a fixative can either increase or decrease the losses of protein into the fixative solution (Vanha-Perttula and Grimley, 1970).

The same study showed that increasing the temperature of a formalin fixative solution from 4°C to 25°C increased losses of protein. If the temperature of post-fixation buffer washes was raised through the same range, losses of protein from either formaldehyde or glutaraldehyde fixed tissues into the washes fell. In another study the losses of protein into alcoholic, formaldehyde or glutaraldehyde fixative solutions were much the same whether fixation was carried out at 4°C or 22°C if fixation was carried out over long periods. However short periods of fixation resulted in greater losses at the higher temperature (Grillo et al., 1971).

An investigation of the fixation of red blood cells showed that the precise pH and salt content of a fixative solution can have marked effects on protein retention (Tooze, 1964). Fixation with osmium tetroxide at pH 7.4 resulted in the loss of some 15% of the haemoglobin from the red cells in 45 minutes, whereas at pH 6.2 this loss was cut by half. Losses were reduced to negligible proportions by adding 0.02 M calcium ions to the fixative solution. An even more striking demonstration of the influence of the ionic content of the fixative solution on protein losses was given in a study carried out by Luft and Wood (1963). When tissue was fixed in osmium tetroxide dissolved in various pH 7 buffer solutions, it was found that with most systems, for instance phosphate and sodium bicarbonate-hydrochloric acid, protein losses into the fixative solution amounted to some 8% with a further 4% lost into the subsequent buffer wash. With an s-collidine buffer, however, losses rose to some 15% into the fixative and another 15% into the wash solution.

Table 5. Examples of losses of particular proteins from cells, or of protein from particular organelles

Protein and/or organelle	Mode of fixation and its consequences	Reference
Albumin, taken up into kidney proximal tubules	Using glutaraldehyde fixation, with postfixation in osmium tetroxide, some 3% of the albumin was lost into the osmium tetroxide. Subsequent losses were trivial	MAUNSBACH (1966)
Haemoglobin in isolated red cells	After fixing in a pH 6.2 solution of osmium tetroxide, containing calcium ions, the loss of haemoglobin was negligible. Raising the pH or omitting the calcium caused losses of up to 15%	TOOZE (1964)
Insulin, both pure and in pancreatic islets	Islets fixed in glutaraldehyde, formalin or ethanol for 8 hr at room temperature lost, respectively, 69%, 81% and 84% of their insulin. None of these fixatives insolubilised pure insulin	GRILLO et al. (1971)
Microsomal fraction isolated from liver	Using neutral osmium tetroxide as fixative the microsomal fraction lost some 11% of its protein into the fixative solution. A further 3% was lost into the dehydrating fluids with no subsequent losses	DALLAM (1957)
Mitochondrial fraction isolated from liver	Fixation as above. Losses into the fixative solution amount to some 20% with a further 10% being lost into the washing and dehydrating fluids and rather less than 10% into the embedding media	DALLAM (1957)
Protein of zymogen granules isolated from the parotid	Using glutaraldehyde fixation most of the protein was retained. Using osmium tetroxide more than 75% of the soluble protein was lost	AMSTERDAM and SCHRAMM (1966)

For illustration and convenience Table 5 lists some studies of the retention of particular proteins or of the protein constituents of particular organelles. This is, unfortunately, not a common type of investigation.

The retention of lipids is, due to the almost universal presence of membranes, another factor relevant to the efficacy of the fixation of whole cells. Fortunately this topic has been subjected to numerous effective investigations. For a review, biased towards the needs of practitioners of electron microscopic autoradiography, see WILLIAMS (1969, p. 219).

Various general comments may be made. Saturated neutral lipids are difficult to preserve by any means of fixation, whereas the extraction of unsaturated neutral lipids is largely prevented by osmium tetroxide-containing fixatives. Phospholipids are more easily fixed, especially by osmium tetroxide or by using formaldehyde with post-chromation.

The physical state of the lipid, and the situation and type of the cell containing it, seem to influence the degree of loss. Also, fixation of (associated?) protein by aldehydes can increase retention of lipid. Low temperature processing also reduces losses, at $-20°C$ for instance substantial amounts of lipid are retained even by glutaraldehyde. Studies on the effects of varying the concentration of the fixative or the period of fixation seem lacking.

Major losses of lipids occur into the fixative and also, unlike the situation occurring with proteins, into the dehydrating fluids and the embedding media. Even a water miscible embedding medium such as glycol methacrylate extracts lipids. Such studies are remarkably consistent, with a more recent investigation involving the processing of cardiac muscle giving closely similar results (WARD and GLOSTER, 1976). Typical findings are provided in Table 6.

Table 6. Examples of losses of various lipids during different fixing and processing schedules

Material studied and fixation schedule adopted	Losses	References
Some comparisons of losses of different lipid types, with the mode of fixation kept much the same		
Lipids of amoebae labelled with H^3-palmitic acid. Cells were fixed in osmium tetroxide, dehydrated through alcohols and embedded in Maraglas. Losses assayed chemically	100%	KORN and WEISMAN (1966)
Lipids of intestine labelled with H^3-linoleic acid. Tissue was fixed in osmium tetroxide and embedded in Araldite. Losses assayed by scintillation counting	26%	BUSCHMANN and TAYLOR (1968)
Phospholipids of rat liver labelled with H^3-phosphatidyl choline and ethanolamine. Tissue was fixed in osmium tetroxide, dehydrated through alcohols and embedded in Araldite. Losses were assayed by scintillation counting	3%	COPE and WILLIAMS (1969 a, b)
Some comparisons of lipid losses when using different fixation and embedding schedules		
Phospholipids of rat liver were labelled by H^3-phosphatidyl choline (and *ethanolamine*). Tissue was processed as below, with losses assayed by scintillation counting		COPE and WILLIAMS (1969 a, b)
Tissue was fixed in glutaraldehyde and embedded in water-miscible glycol methacrylate	97 *(30)*%	
Fixation in glutaraldehyde and embedding in glycol methacrylate was carried out at −20° C	36 *(30)*%	
Tissue was fixed in glutaraldehyde, dehydrated through alcohols and embedded in Araldite	100 *(47)*%	
Tissue fixed in glutaraldehyde, post-fixed in osmium tetroxide, dehydrated through alcohols and embedded in Araldite	11 *(1)*%	
Tissue fixed in formalin, post-chromated, dehydrated through alcohols and embedded in Araldite	12 *(1)*%	
Losses of lipid at various stages during fixation and processing		
Rat liver lipids labelled as above, processed as detailed below. Losses assayed chromatographically and by scintillation counting at each step in the procedures		COPE and WILLIAMS (1969 a, b)
Losses into glutaraldehyde fixative solution	6%	
into the buffer wash	2%	
into osmium tetroxide post-fixation solution	1%	
into dehydrating alcohols	1%	
into epoxy propane/Araldite monomer	4%	
into Araldite embedding medium	1%	

The effect of lipid loss, with concomitant destruction of membranes, on the loss of other biopolymers does not appear to have received systematic study. Pointers to its importance include the observation (SMITHERMAN et al., 1972) that neutral Tris buffer solution removes glycogen from fixed, unprocessed tissue more readily than do other neutral buffer solutions, and Tris has been observed to disrupt membranes.

The retention of nucleic acids has received very little quantitative investigation. Some representative information is displayed in Table 7.

Several general comments can be made, but should be regarded cautiously. It seems that, except with acidic fixative agents, DNA is easly retained in the tissues. Most studies, however, have been only semi-quantitative, making use of the Feulgen method applied histochemically (for example HUXLEY and ZUBAY, 1961; PEARSE, 1968). Only high molecular weight RNA species are usually retained, with glutaraldehyde being a more effective fixative than osmium tetroxide. Perhaps due to the synergistic effect of efficient fixation of protein by the former method? Though losses of RNA into fixative or washing solutions were trivial or small with several fixatives losses on rehydration were major. Carnoy's fluid for instance retains RNA excellently until material is rehydrated for staining, at which stage all RNA may be lost. This is of little

Table 7. Examples of losses of nucleic acids during different fixing and processing schedules

Mode of fixation and/or processing	Nucleic acid lost	Method of investigation and its limitations	Reference
With H^3-thymidine (i.e. a DNA precursor) and using formalin *or* glutaraldehyde solution as fixative losses were:		Cell culture labelled by H^3-nucleosides (RNA *and* DNA precursors). Cells were fixed, washed and dehydrated. Loss at each stage assayed by scintillating counting. No allowance made for losses after rehydration	VANHA-PERTTULA and GRIMLEY (1970)
Into fixative solution	ca. 2.5%		
Into subsequent wash	ca. 1.2%		
Subsequently	trivial		
With H^3-uridine (i.e. a RNA precursor) losses were:			
Into formalin	8%		
Into subsequent wash	16%		
Into glutaraldehyde solution	11%		
Into wash solution	11%		
Losses after fixation in Carnoys fluid or formalin	trivial	Tissue was fixed and processed through to paraffin. Nucleic acids were estimated by phosphate assay and spectrophotometrically. No allowance made for losses after rehydration	SANDRITTER and HARTLIEB (1955)
Losses of RNA after fixation in Carnoys fluid	100%	Samples of rat pancreas were fixed in Carnoys fluid and processed through to paraffin. Sections were dewaxed and weighed. Some were then postfixed in formalin. Sections from all regimes were then extracted by aqueous buffer or RNAase solution. The amount of nucleic acid lost was assayed spectrophotometrically	LAGERSTEDT (1956)
Losses of RNA after fixation in Carnoys fluid and post-fixation in formalin	20%		
Losses of DNA after either regime	trivial		
Losses (in cpm) after fixation in:		Ribonucleic acids were labelled by administration of H^3-uridine. Livers were fixed, dehydrated, RNA extracted with perchloric acid and assayed by scintillation counting. No allowance made for losses after rehydration	MONNERON and MOULE (1969)
Glutaraldehyde	37200 cpm		
Osmium tetroxide	25300 cpm		
Both fixatives	19400 cpm (NB: counts reflect retention, not loss)		

consequence for the practitioners of autoradiography, but very disturbing for those using staining procedures.

The retention of polysaccharides and their complexes is perhaps the least studied topic in the field. This is unfortunate, in view of the great variability of these materials. Some of the sparse quantitative results which are available are shown in Table 8. One of the uncomplexed polysaccharides, namely glycogen, has attracted particular attention.

Unfortunately, as with RNA, retention of glycogen in the fixed block is no guarantee of retention on further washing. Comments of authors not appreciating this point (e. g. «it became apparent that some solutions, e. g. absolute alcohol . . . modify the existing glycogen in such a way as to render it unstainable») may thus be viewed with some scepticism.

Mucosubstances are particularly difficult to retain, no doubt due to their high solubility in water. Support for this view is provided by the good retention of mucins after fixation (of the protein component?) with formaldehyde vapour.

Retention of non-polymeric materials. Proteins, many nucleic acids, and polysaccharides are polymeric. In fact, of the classes of compounds considered so far only lipids are not polymeric: and due to their poor water solubility, and to their being often found in membranes, even lipids are not readily lost into aqueous fixative solutions. But consider other constituents of cells such as sugars, metal ions, amino acids. Their small size and high solubility in water means that diffusion, both into processing fluids and through the cell, will occur easily. Their small size also means that fixation by structure breaking or crosslinking is less likely than with proteins or nucleic acids.

Many such compounds are, however, to some extent, retained in the tissues. Work on this topic has rarely been quantitative since the problem has been studied more often by practitioners of the technique of autoradiography. For this procedure *some* labelled material must be retained in a localised form but the degree of retention is less important. Thus various amino acids, e. g. alanine, cystein, leucine and tyrosine (Conger et al., 1978; Fukuyama and Epstein, 1969; Whur et al., 1969; Hodson and Marshall, 1967) are retained after fixation by glutaraldehyde. Formaldehyde seems less effective (Peters and Ashley, 1967), less crosslinking perhaps? The aminosugar glucosamine is also retained after glutaraldehyde fixation and not after treatment with formaldehyde (Whur et al., 1969). Noradrenalin, carrying a primary amino group, is retained well by glutaraldehyde (Coupland and Hopwood, 1966) or by dichromate. As uridine, which does not carry a primary amino group, is fixed by both glutaraldehyde and osmium tetroxide (Monneron and Moule, 1969) perhaps retention may be aided by fixation of surrounding materials. Quite another way of preserving small molecules is to add to the fixative solution a reagent giving an insoluble complex with the substance under study. Cholesterol for instance has been retained in this way. Cholesterol was lost from rat liver almost completely following fixation in either formaldehyde or glutaraldehyde solutions. However, if the alkaloid digitonin was added to the fixative solutions the cholesterol was said to be retained almost completely (Scallen and Dietert, 1969, but see also Vermeer et al., 1978).

Other classes of small molecules have been found to be retained after fixation and processing, for example drugs such as actinomycin D (Simard, 1967) and hormones such as cortisol and aldosterone (Williams and Baba, 1967). Since such materials have an intrinsically high affinity for tissue components it is hard to know whether such preservation should be regarded as the fixation of small molecules or merely the fixation of the macromolecular substrates to which they are bound.

Table 8. Examples of losses of various polysaccharides and polysaccharide-protein complexes during different fixing and processing schedules

Mode of fixation and/or processing	Losses of polysaccharide	Method of investigation and its limitations	References
Amount of *glycogen* lost after various modes of fixation varied, for example:		Rat liver was fixed then assayed chemically for glycogen. Losses into processing fluids or on rehydration are ignored	KUGLER and WILKINSON (1964)
After absolute alcohol:	10%		
After formalin	20%		
After Rossmans fluid	21%		
After Bouins solution	18%		
Amounts of *glycogen* lost after various modes of fixation varied, for example:		Rat liver was fixed then the fixed block washed with aqueous solutions. The amount of glycogen retained was then assayed chemically. This procedure ignores losses into and effects of the dehydrating fluids	SMITHERMAN et al. (1972)
After 95% alcohol	78%		
After formalin	91%		
After alcoholic formalin	54%		
After Rossmans fluid	30%		
After Bouins solution (values approximate, read off chart)	56%		
The losses of *acid glycosaminoglycans* at various stages of two typical schedules varied		Acid glycosaminoglycans (GAGs) of the epiphysial growth zone were labelled by administration of S^{35}. The zone was removed, frozen and sectioned. Sections were treated variously. Some were assayed directly for acid GAGs by scintillation counting. Some were fixed, washed, post-fixed, washed again and dehydrated. This ignores losses on rehydration	ENGFELDT and HJERQVIST (1968)
1. Using straight glutaraldehyde solutions as fixative losses were:			
Into fixative	10%		
Into wash solution	5%		
Into osmium tetroxide post-fixative	50%		
Into wash solution	5%		
Into dehydrating fluids	trivial		
2. Using a glutaraldehyde-cetyl pyridinium chloride solution as fixative losses were:			
Into fixative	4%		
Into wash solution	1%		
Into osmium tetroxide post-fixative and wash	1%		
Into dehydrating fluids	2%		
Losses of *acid glycosaminoglycans* into 5% w/v cetyl trimethylammonium bromide plus 10% w/v formaldehyde in water never exceeded 10%		Acid GAGs lost into the fixative from cow genital tract was assayed by an Alcian Blue method. The insoluble GAG fraction was similarly estimated after digestion of the tissue with papain. This ignores losses into dehydrating fluids, and on rehydration	RADMEHR and BUTLER (1978)
Losses of *mucins* were varied. For example:		Salivary glands of rat were freeze dried. A sample was extracted with water and another sample was fixed in formaldehyde vapour then water extracted. Losses into the water were assayed spectrophotometrically. This does not allow for losses into or effects of processing fluids	TOCK, WOODWARD and PEARSE (1966)
After formaldehyde vapour fixation	ca. 0.5%		
with no fixation	30–60%		

Table 8. (Continued)

Mode of fixation and/or processing	Losses of polysaccharide	Method of investigation and its limitations	References
Of the extensive data on the various *polysaccharides* the following are characteristic: Losses of heparin after fixation were:		Human aorta, skin and cartilage were fixed then assayed chemically for hyaluronic acid, chondroitin sulphate and heparin. The tissues were also assayed directly This ignores losses into processing fluids and on rehydration	JOZSA and SZEDERKENYI (1967)
After 4% formaldehyde (at 20 C)	49%		
After 10% formalin at 20 C	42%		
After 70% alcohol, at 4 C for 12 h	7		
After 10% formalin at 4 C	45%		
After Carnoys fluid	33%		
After Rossmans fluid	21%		
After 4% formalin plus cetyl pyridinium bromide at 20 C	2%		
After 70% alcohol plus cetyl pyridinium bromide at 4 C	2%		

It may be emphasised that many materials are currently regarded as unfixable. For example the adrenaline in chromaffin granules, 90% or more of which was lost after fixation in either formaldehyde or glutaraldehyde solution (COUPLAND and HOPWOOD, 1967; HOPWOOD, 1967). Admittedly this figure was based on chemical assay. When comparing this to information given previously one should consider whether the residual catecholamine, if tritiated, might not be detectable by autoradiography. Metal ions such as Ca^{++} and Rb^{+} are, unless complexed to macromolecules, also mostly lost into aqueous fixatives. Use of alcoholic solutions or of precipitative freeze dissolution can improve preservation (RENAUD, 1957; IREN and BANGE, 1978). Finally note that simple sugars such as mannose and galactose are not preserved by fixatives such as formaldehyde or glutaraldehyde (WHUR et al., 1969), nor are sugar derivatives such as adenosine triphosphate (HOPWOOD, 1967).

3.7 Effects of fixation on biological staining: increases in staining and changes in staining patterns due to chemical modifications of the tissues

Such modifications, changing the structures and reactivities of covalently bonded groupings, occur with most fixative agents as an intrinsic part of the fixation process. The chemistry has already been outlined in the Figures and Tables of Sections 3.4 and 3.5. Here we will consider such chemical changes under the following headings:

- Modifications causing general increases in staining.
- Modifications giving changes in staining patterns, due either to losses of stainable groups or the introduction of new ones.
- Modifications altering the acidophilia/basophilia balance of tissues.

Increases in the staining of tissues after fixation are well known. These are usually regarded as a consequence of the disruption of the secondary and tertiary structures of proteins, resulting in the exposure of previously inaccessible groupings. For instance such typically interior groupings as cystine, tryptophane and tyrosine residues will become more reactive. Also such lipophilic groups will increase dye binding due to the increased possibilities of van der Waals attractions and of hydrophobic bonding, see Section 4.4. The exposure and subsequent reactivity of such buried groups is widely described in the biochemical literature (see reviews by STEINHARDT and REYNOLDS, 1969, p. 313; and KRONMAN and ROBBINS, 1970). The effect has also been described in the histochemical literature, where its significance is perhaps less well appreciated, though see SINGER (1952) and BAKER (1958). This effect is illustrated in Table 9. It should be noted that though the terms increase/decrease/change in staining caused by fixation imply some absolute standard, in fact what is most often done is to compare tissues fixed in different media. So it is quite possible that other factors, such as selective retention, are also involved.

A variation on the theme of increased accessibility of protein groups is the unmasking of the lipid portions of lipoproteins. Native lipoproteins, or at least those occurring in frozen or cryostat sections, are commonly unstained by dyes such as Sudan Black. However, after treatment with such fixative reagents as mercuric chloride or potassium dichromate the lipoproteins are coloured by the traditional fat stains, see Table 9.

Changes in patterns of staining. Any of the fixative agents will inevitably eliminate or reduce in number some chemical groupings aiding some staining processes, whilst simultaneously introducing new groupings which may aid other staining methods.

Table 9. Some instances of increases in staining intensity following the action of fixative agents

Fixative agent and material fixed	Increase in staining seen	Reference
Formalin fixation of fibrin films	Both an acid and a basic dye stained fixed films more intensely than unfixed films	SINGER and MORRISON (1948)
Glutaraldehyde fixation of various human tissues	Compared to formalin-fixed tissues, both acid and basic dyes stained more intensely	CHAMBERS et al. (1968)
Mercuric chloride fixation of most tissues	Cytoplasmic staining by acid and basic dyes greatly enhanced by this fixative	BAKER (1958, p. 104)
Glutaraldehyde fixation of mammalian liver	Cytoplasmic staining by the coupled tetrazonium method (i. e. of tyrosine and tryptophane) only occurred in fixed tissue	HOPWOOD (1967)
Osmium tetroxide and glutaraldehyde vapour fixation of freeze dried salivary gland	Staining of tryptophan, by the dimethyl-aminobenzaldehyde nitrite method, was much enhanced in the fixed tissue compared to unfixed material	TOCK and PEARSE (1965)
Dry heating of formalin fixed tissues	Collagen is intensely stained by the coupled tetrazonium method, even after treatment with benzoyl chloride	Discussed in PEARSE (1968, p. 220)
Insect and mouse tissues treated with various fixatives, e. g. formalin, mercuric chloride and picric acid	Staining of lipid by Sudan Black occurred only after fixation and not on fresh tissue	CLAYTON (1959)

Table 10. Some examples of decreases of staining intensity probably attributable to modification of particular chemical groupings by the fixative agent

Mode of fixation	Effect on staining	Reference
Staining of thiols		
Formaldehyde, glutaraldehyde and osmium tetroxide vapours	Staining by the dinaphthyl disulphide method greatly reduced	Tock and Pearse (1965)
Fixation by mercuric chloride solution	The staining achieved with a maleimide method was reduced; that with the dinaphthyl disulphide procedure greatly reduced; and that with a mercaptide-forming technique eliminated	Barrnett and Roth (1958)
Fixation by osmium tetroxide solution	Staining eliminated with all methods used	Barrnett and Roth (1958)
Staining of tyrosine and tryptophane		
Fixation by formaldehyde and osmium tetroxide vapours	Staining by the diazo-coupling and dimethylaminobenzaldehyde methods reduced	Tock and Pearse (1965)
Fixation by formalin, and by formalin followed by potassium dichromate	Staining reduced by post-chroming	Hopwood (1969)
Fixation with mercuric chloride solution	Staining by the coupled tetrazonium method was reduced	Barrnett and Roth (1958)
Fixation with osmium tetroxide solution	Staining by coupled tetrazonium method eliminated	Barrnett and Roth (1958)
Fixation by potassium dichromate solution	Staining with diazonium salts eliminated	Casselman (1955)

Examples of cases where the chemical modification of the retained biopolymers *reduces* staining of various groupings can be seen in Table 10. Note that ionic or ionogenic groups are not specifically considered here since the account of acidophilia-basophilia below, effectively covers this topic. The underlying chemistry needed to rationalise these effects has already been outlined in the Figures and Tables of Sections 3.4 and 3.5.

Note that work of the type cited in Table 10 is usually qualitative. Consequently phrases such as «reduced staining intensity» commonly mean the staining looked paler.

Bearing these cautions in mind, some useful general comments may be offered. Comparing Tables 9 and 10 it is apparent that a balance is struck between the effects of the fixative agent in opening up proteins on the one hand and the destruction of groups from inside proteins on the other. Different fixatives tip the balance in different ways. It can also be seen that reactions of some fixatives may be reversed in the staining bath when chemically vigorous staining methods are used. For further examples of this see Table 11.

The fixation processes often introduce new chemical groupings capable of being stained and occasionally this is useful. Table 11 illustrates these effects together with some more irritating modifications.

Such chemical modifications should always be borne in mind when interpreting subsequent staining results. For the chemical basis of these effects, see Section 3.4.

Closely related to these artifacts are those coloured derivatives produced directly by the action of the fixative on the tissue, or by the processing fluids on fixative retained in the tissue. Again this staining can be useful but is more often a nuisance. Table 12 lists some common examples. The various artifactual pigments so formed are in fact usually removed prior to staining, mercury granules by iodine, formalin pigment with acidic or alkaline alcohol and so on.

Changes in the acidophilia/basophilia balance of tissues due to fixation is a well known and largely comprehensible phenomenon. Thus a fixative which either destroys the potentially cationic amino groups (e. g. formaldehyde) or increases the number of anionic groups, or both (e. g. osmium tetroxide) will decrease tissue acidophilia and increase basophilia. Conversely fixatives destroying anionic groups or creating cations or both (e. g. chromic acid) will increase tissue acidophilia and decrease basophilia. Similar changes could be accomplished by selective retention. It has been suggested for instance that picric acid selectively retains the more basic proteins, and so fixation in that medium produces tissue acidophilia. In Table 13 various simple fixatives are listed in a qualitative acidophilia/basophilia sequence (after SINGER, 1952; BAKER, 1958; and unpublished work from this laboratory) which also displays the probable chemistry.

The effects of fixation on the acidophilia/basophilia balance of the tissues have also been described in terms of the optimum pH for staining with basic dye-acid dye

Table 11. Some examples of fixative – introduced stainable groups

Method of fixation	The effect on the tissue	For further information
The new staining may be useful . . .		
Acrolein solution	This unsaturated aldehyde reacts with certain protein nucleophilic groups, notably amino, yielding aldehyde derivatives. These are visualised with Schiff reagent, so staining protein rich sites	VAN DUIJN (1961)
Solutions of chromium (VI) salts	Polysaccharides with 1.2-diol groupings are oxidised to dialdehydes which are visualised with Schiff's reagent	For example CASSELMAN (1955)
Hot formaldehyde vapour	Various monoamines are cyclised to fluorescent heterocyclic compounds	Reviewed by PEARSE (1972, in chapter 27)
Mercuric chloride solution	Certain unsaturated lipids yield mercury addition products. Treatment with Schiff's reagent hydrolyses these to aldehydes, which react further giving red compounds	A good account is that of TERNER and HAYES (1967)
. . . or just an annoying artifact		
Formaldehyde solution	The methylols formed with tissue amines hydrolyse easily, especially at extremes of pH, back to aldehyde. Hence after formalin fixation, expecially if prolonged and if the washing is poor, both the Feulgen nucleal procedure and alkaline silver stains can give nonspecific background staining	For instance see CHAMBERS et al. (1968), KASTEN (1964)
Glutaraldehyde solution	This dialdehyde reacts with the amino groups of proteins and can yield many free aldehyde groups. These stain in procedures using Schiff's reagent	HOPWOOD (1967), CHAMBERS et al. (1968), KASTEN and LALA (1975)

Table 12. Some examples of coloured materials produced in tissues by reactions of fixatives

Fixative agent	Effects produced	For further information
Useful staining methods		
Chromium (VI) salts	Catecholamines are oxidised to various poorly defined coloured products (the «chromaffin» reaction)	Discussed for example by LILLIE (1965, p. 233)
Osmium tetroxide	Sites of hydrophobic unsaturated lipids are stained by the deposition of «osmium blacks», probably mixed co-ordination polymers of osmium	HANKER et al. (1967)
Annoying artifacts		
Chromium (VI) salts	If unreacted Cr (VI) salts are not washed out of the tissues before dehydration, alcohols may reduce them to green insoluble chromic oxide	BAKER (1958, p. 108, 132)
Formaldehyde solution	When blood-rich tissues are fixed in formaldehyde, especially if acidic, haemoglobin may react to give brown crystals of «formalin pigment»	LILLIE (1965, p. 391)
Mercuric chloride solution	Black particles occur, some apparently of elemental mercury	BAKER (1958, p. 103)
Osmium tetroxide solution	«Osmium blacks» can be deposited even in saturated fats, due to dissolved fixative reacting with the dehydrating alcohols	BAKER (1958, p. 125)

Table 13. Simple fixatives ranked in a qualitative acidophilia-basophilia sequence

Fixatives listed in order of their increasing ability to induce tissue basophilia	Possible chemical rationale
Picric acid	Selective retention of basic proteins
Chromic acid	Formation of cationic metal complexes from anionic tissue carboxylates and Cr (III) ions
Potassium dichromate	Reaction of tissue amino groups with epoxides derived from tissue lipids by dichromate oxidation
Acetic acid	
Alcohols	
Glutaraldehyde	Reaction of tissue primary amino groups yielding various possible products, all less basic than the amine
Heat	Oxidative deamination
Osmium tetroxide	Oxidative deamination; also oxidation of tissue thiols yielding suphonates
Mercuric chloride	Reaction of tissue amino groups, forming complexes less basic than the original amines
Cyanuric chloride	Reaction of tissue amino groups, to yield weakly basic triazine derivatives
Formaldehyde	Reaction of tissue amino groups, to yield various weakly basic higher amines and methylols

mixtures. LILLIE (1965, p. 163) provides information concerning Azur-Eosin staining of surgical material. The optimum pH required increased from pH 4.0 after fixation in formaldehyde, to pH 4.5 after Zenker, to pH 5.0–5.5 after Bouin, Carnoy or buffered mercuric chloride, to pH 6.0–7.0 after heavy chromation. This largely parallels the sequence in Table 13.

3.8 Effects of fixation on biological staining due to physical modifications of the tissues

Since this account is directed towards staining, morphological changes will be disregarded unless they also change the staining pattern. Such morphological changes, ranging from the gross (e. g. shrinkage, tears, optically observable disruption) to the ultra-structural (e. g. organelles lost or displaced, membranes disrupted, observable by electron microscopy) have been widely discussed elsewhere. For discussion of ultrastructural changes see for instance HAYAT (1970, many Figures throughout the book are of interest), SJÖSTRAND (1967, especially Chapters 6 and 16) and GLAUERT (1974, many Figures in Chapters 2 and 4), and for discussion of more gross effects see BAKER (1958, Chapters 5 and 6).

In later Chapters of the present book the influence of the porosity or permeability of tissues and tissue components on biological staining will be discussed in some detail. For instance it will be suggested that the effects of adding salts to dyebaths vary with the porosity of the material stained, see Section 4.9. Similarly the staining patterns of the rate-controlled acid or basic polychrome methods (such as VAN GIESON's or Methyl Green-Pyronin) are controlled by differential substrate porosity, see Section 4.18. Hence the necessity to discuss the influence of fixation on tissue porosity. Such discussions, however, will be brief, even speculative, as the importance of the topic has not been fully appreciated and so has received little deliberate study.

What is meant by *porosity?* Are we to be concerned with cracks and holes in clumps of cells, visible optically? Or with the pores, natural or induced, between or inside organelles, seen in the electron microscope? Or with the shrinking or swelling of macromolecular networks? No doubt the answer is all three, and possibly with all three at once, even inside a single tissue or cell. For example a coagulant fixative such as picric acid or alcohol could leave the cytoplasm of a cell riddled with microscopic pores and yet the solid mass of biopolymers between the pores might be in a highly condensed, impermeable state. Some of these complications are expressed visually in Fig. 13. When considering staining, porosity of tissue when immersed in a particular staining solution is an important factor. Hence physical effects should be considered in terms of the mechanisms and practicalities of particular staining procedures.

Some tissue characteristics have been investigated which are relevant to the problem of estimating porosity. Firstly there are several studies of the effects of various fixation and processing regimes on the volumes of pieces of tissue or of whole cells. Such work has been reviewed by BAKER (1958, p. 35 and p. 75), HOPWOOD (1969, p. 369), GLAUERT (1974), and PENTTIHA and co-workers (1975). Unfortunately these studies often suffer from two limitations. Firstly that the quantity usually measured is the overall change in volume, disguising possible differential changes in sub-structures, and

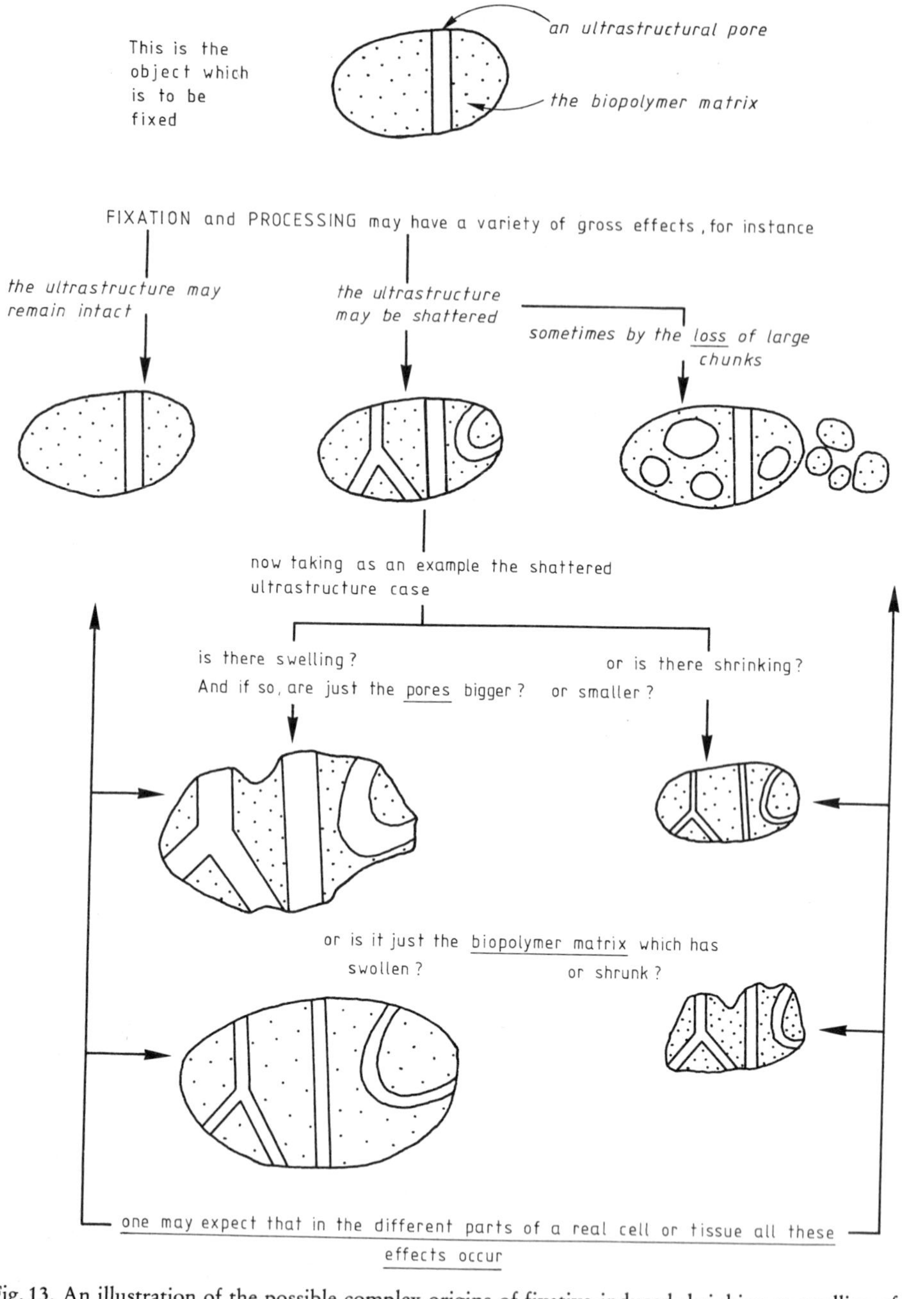

Fig. 13. An illustration of the possible complex origins of fixative-induced shrinking or swelling of cells and tissues

secondly that the volumes measured are not usually the rehydrated (i. e. dyebath) volumes. That considerable changes in volume do occur on rehydration is certain (GOLDSTEIN, 1967) but what is not apparently known is whether such increases occur differentially with different modes of fixation. Similar work has also been carried out on insoluble proteins, e. g. by BAKER (1958, p. 35), and on the proteoglycans and collagen of Whartons jelly (MEYER et al., 1976).

The second approach to the problem of tissue porosity is to observe the ultrastructure directly after various modes of fixation. Careful studies using optical microscopy were made, as described by BAKER (1958), but the effect of fixatives could not be fully appreciated until the era of electron microscopy. Early electron microscopical studies, carried out before the achievement of technical sophistication, will be ignored here since assessment of the micrographs is difficult. However the ultrastructural effects of various fixatives have since been observed in technically unambiguous circumstances. For example the effects of fixing with chromium salts, formaldehyde and mercuric chloride

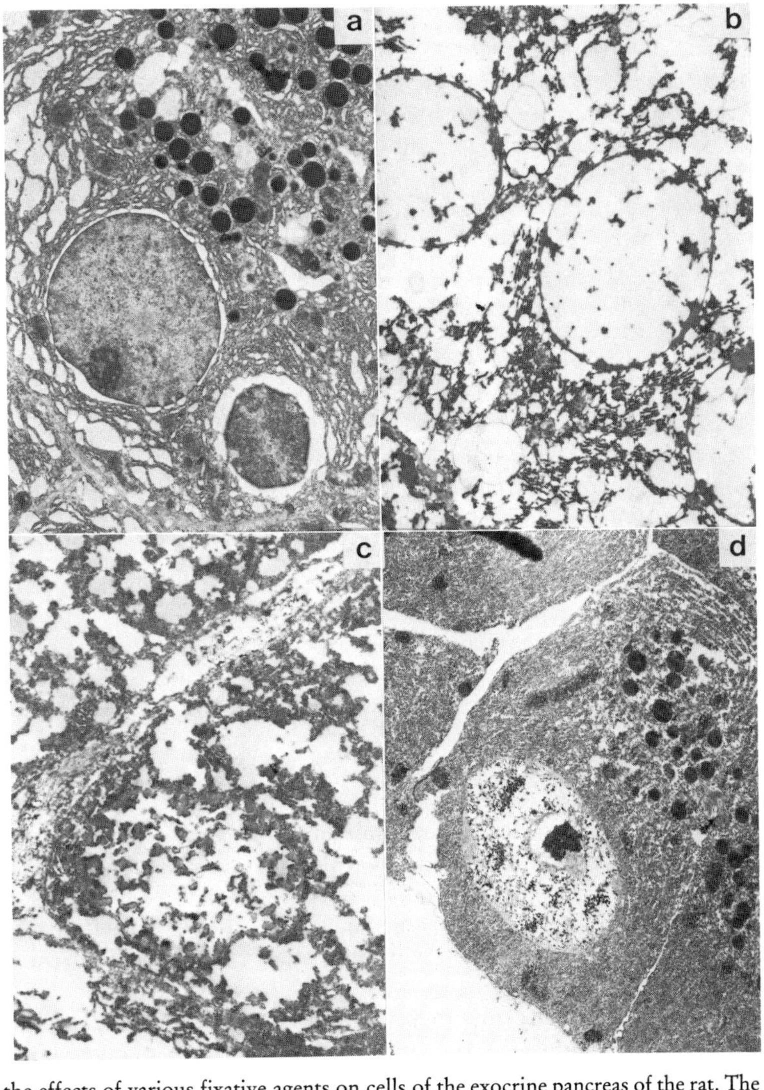

Fig. 14. An illustration of the effects of various fixative agents on cells of the exocrine pancreas of the rat. The fixatives used in Figs. (a) to (d) were respectively neutral formalin, Susa, Carnoy's and Zenker's fluids. ×5000

have been studied by BAKER and his collaborators (BAKER, 1965a, b; BAKER and McCRAE, 1966; BAKER and LUKE, 1963), of permanganate by GRONBLAD and ERÄNKÖ (1978), and of formaldehyde-picric acid by ACCINNI and co-workers (1974). The effects of a variety of aldehyde fixatives, and also of dichromate post-fixation, were classically described by SABATINI and his collaborators (e. g. SABATINI et al., 1964). Osmium tetroxide and its effects needs no specific citation. The effects of alcohol and various other fixatives can be seen in Figs. 2 and 14. Some verbal account of ultrastructural effects occurring with certain fixatives is given in Table 14. In summary one can say that for the fixation of protein-rich structures BAKER's distinction between «coagulant» and «non-coagulant» fixatives, though made on the basis of optical observations, also holds for information derived from electron microscopy. Coagulant fixatives such as alcohol or mercuric chloride produce shattered ultrastructure whereas non-coagulants such as formaldehyde and osmium tetroxide leave the ultrastructure largely intact. All fixatives except dichromates and osmium tetroxide fail to fix lipid rich structures such as membranes and fat droplets, and so can increase porosity even without generalised disruption occurring. Another general point is that if a sequence of fixatives is applied, say glutaraldehyde followed by dichromate, the ultrastructural picture is largely that of the initial fixative agent. For instance in the example cited the picture is largely as that following glutaraldehyde alone, but with improved retention of catecholamine containing granules. Admittedly the information obtained may not directly relate to the staining bath, as the micrographs are of dehydrated tissues, embedded or mounted. In this case, however, one does have some knowledge of the responses of different cell constituents.

Finally the degree of porosity of simple insoluble proteins, and of tissues, has been estimated by measuring staining rates after treatment with the substances present in various fixatives. For example a study carried out on keratin fibres indicated that oxidising agents such as potassium dichromate or permanganate increased the swelling of the protein, apparently by breaking disulphide bridges (WHEWELL et al., 1971). Observations concerning various fixative agents are summarised in Table 14 and illustrated in Fig. 14.

So different fixative agents produce different degrees and patterns of porosity. What are the consequences for biological staining? Starting with the crudest concept of porosity we may note that a simple expedient such as stirring a dyebath can greatly raise the rate of staining (GOLDSTEIN, 1963). So even the gross differences in porosity due to variation in the amount of cellular aggregation probably affect staining rates.

Certainly various degrees of ultrastructural disruption have consequences for staining. For instance, when using Gomori's trichrome stain SWEAT and co-workers (1968) found that after fixation in disruptive, porosity-enhancing fixatives such as acetone or alcohol, tissues were stained too avidly by Aniline Blue, the large dye of the mixture. However, if the tissues were fixed in the less disruptive formaldehyde solution the entry of the large dyes was reduced and a better colour balance ensued.

The degree of swelling of molecular networks also appears to influence staining. For instance treatment of keratin with potassium dichromate to break the disulphide bridges increases the staining rates by and the affinities for several acid dyes (WHEWELL et al., 1971). Again, the effect of mercuric chloride fixation on subsequent dye uptake (« . . . mercuric chloride leaves the tissues more receptive of dyes in general than any other fixative . . .», BAKER, 1958, p. 104) could perhaps be accounted for by increased swelling following disruption of disulphide bridges.

Table 14. Some effects of various fixative agents on the «porosity» of cells and tissues

Fixative agent	Effects[a] on disruption of tissues[b]	Effects[a] on disruption of ultrastructure	Effects on the swelling of proteins
Acetic acid	Clumps of cells split widely apart	Gross disruption of cytoplasmic and, except for chromosomes, nuclear contents Loss of lipid rich structures such as mitochondria	Causes enormous, but largely reversible, swelling of proteins
Solutions of chromium (VI) salts			
1. In strongly acidic solutions below about pH 3.5	Cells remain attached to one another	Gross disruption of cytoplasmic contents, except for lipid rich mitochondria and fat drops; nuclear contents, nucleoli, chromosomes, intact	Causes swelling (due to oxidative fission of disulphide bridges?)
2. In more neutral solutions	Groups of cells pull apart from one another	Appreciable disruption of cytoplasmic contents (except, if long times are used, of lipid rich structures) and gross losses of nuclear contents	As above
Ethanol	Cell aggregates separate from each other	Gross disruption of cytoplasmic and nuclear contents. «Residues» may pile up against cell boundaries Loss of lipid rich mitochondria and fat drops	Causes enormous but largely reversible shrinkage
Formaldehyde solution	Clumps of cells split off from each other	General preservation of cytoplasmic and nuclear contents, poorest for chromosomes and lipids	
Glutaraldehyde solution		General preservation of cytoplasmic and nuclear contents Fat droplets lost	Limits swelling (due to efficient cross linking via amino groups?)
Mercuric chloride solution	Clumps of cells separate	General disruption of cytoplasmic and nuclear contents Mitochondria and fat droplets lost	Causes swelling (due to splitting of disulphide groups?)
Osmium tetroxide solution	Groups of cells separate from each other	General preservation of cytoplasmic and nuclear contents. Perhaps poorest for chromosomes, proteins of mitochondrial matrices and xymogen granules	
Picric acid solution	Cells remain attached to one another	Cytoplasmic and nuclear contents disrupted	

[a] The observations described were made on dehydrated (i. e. embedded or mounted) material
[b] Tissue disruption occurs less in embedding media such as epoxy or acrylic plastic than in paraffins

3.9 Rates of fixative penetration

Quantitative studies of the rates of penetration of fixatives into model systems such as protein gels, and into homogeneous tissues such as liver and kidney have been made. General accounts have been given by BAKER (1958, p. 37 und 67) and HOPWOOD (1969, p. 347). Usually the penetration has been found to follow the pattern of Fickian diffusion. An exception occurs with osmium tetroxide, whose reaction products appear to inhibit entry of further fixative. Excluding this latter case, the depth of fixative penetration (d) is thus linearly related to the square root of the time of fixation ($\sqrt{t}$). Consequently the penetrative performance of different fixatives is often expressed in terms of a k value, the proportionality content.

$$d = k\sqrt{t}$$

The precision implied is rather misleading if applied over-casually to the practical situation. So k values have been found to rise markedly with increases in the concentration and temperature of the fixative solution. The nature of the material being fixed is also an important variable. For instance, whereas chromic acid penetrates albumin gel more rapidly than does picric acid, these rates are reversed when a block of liver is fixed. It has been suggested that this is due to the more ready penetration of the less polar picric acid through the lipid – rich cell membranes. Another example is the relatively easy penetration of fixatives through loose connective tissues and along open tubes such as the larger blood vessels (see for instance BAKER, 1958, Fig. 5 A). And as a final point one may enquire: is penetration of fixative to be equated with fixation? There is no simple answer to this (fixation of what for instance . . .?) but some useful discussion is available (CHAMBERS et al.; 1968, and HOPWOOD, 1967).

Bearing these cautionary points in mind, the following general remarks can be made. Fixatives can be arranged in the following rank order of increasing rates of penetration:

Ethanol
Osmium tetroxide
Chromic acid
Picric acid
Mercuric chloride
Formaldehyde
Acetic acid

It has also been shown that when using mixed fixative solutions (e. g. Zenkers fluid, which contains acetic acid, mercuric chloride and potassium dichromate) the constituents each penetrate at the same rate as when applied as the pure substances. Consequently zones of fixation occur, with an outer zone fixed by all components of the fixative, and an inner by acetic acid alone.

What explanations can be suggested for the differing rates of fixative penetration? The lipid solubility is at least of marginal significance, see the comment above. However the factor of over-riding importance appears to be the tendency of the fixative to swell or shrink the tissue. Compare the information on the swelling and shrinking of various model systems and tissues by fixatives collected by BAKER (1958, p. 36 and 76) to the relative penetration rates listed above. It is seen that a fixative which swells tissues, as do formaldehyde or acetic acid, is a fast penetrating agent. Contrariwise those fixatives which shrink tissues, such as picric acid or alcohol, are slow to penetrate. That fixatives will pass readily through swollen tissues and more slowly through tissues which have been shrunk is a concept easy to grasp (but see Section 3–8), though the reasons for such volume changes are less clear.

4 Staining Methods Involving Dyeing

In Chapter 4 there are two major integrating ideas. Firstly, that staining is commonly due to dye-tissue affinities. Secondly, that though affinities are usually needed for staining to occur, the patterns of staining often reflect differential staining or de-staining *rates*. This viewpoint is far from novel, it was expounded for instance by EHRLICH (1879), MANN (1902), HEIDENHAIN (1903) and more recently by JOHN R. BAKER (1958). Nevertheless such a frame of reference remains unpopular. Why? Perhaps because the commonplace is as difficult to notice as the unusual is hard to find.

4.1 What are dyes and what is dyeing? Definitions, names and descriptive terms

Explanations will be attempted here, not definitions. Dyes are coloured organic molecules with large systems of delocalised electrons (q.v., Section 4.2), consequently most dyes are aromatic compounds. Certain coloured entities such as iodine or some transition metal salts, which are capable of staining tissues, are not dyes. Their actions are discussed elsewhere, see Chapters 9 and 6. Operationally, staining tissues by dyeing involves placing the tissues into a solution of the dye (the dyebath) and allowing the dye to become adsorbed onto and into the tissue. During this process the dye undergoes no major chemical transformation and may be recovered unchanged from the stained tissue. Those staining methods not using dyes, such as procedures using colourless organic reagents, metal ions, enzyme substrates or antibodies, are not dyeing methods. Using dyes merely as a source of coloured particles (for injection say, or for phagocytic uptake by living cells) or for vital staining are not regarded here as dyeing methods either.

Should this sound bewilderingly negative consider that common staining procedures such as the use of Haematoxylin and Eosin, van Gieson's stain, and the Sudan procedures, are all dyeing methods.

Names and descriptive terms used for dyes. The nomenclature used in identifying and describing the dyes used in biological staining is complicated and irrational, almost an encouragement to ambiguity by users and sellers alike. Consider a statement such as «Pyronin G is a metachromatic basic dye of the xanthene class». What is the chemical nature of Pyronin G? What do the terms metachromatic, basic and xanthene signify? This Section seeks to clarify such points.

Firstly consider the naming of dyes. Most biological stains were originally manufactured as textile dyes and consequently may be sold under several trade names. Each corresponds, or at least once corresponded, to that used by a different manufacturer. For example Acid Fuchsin is also termed Acid Magenta or Acid Rubin or Fuchsin or Rubin. The suffix letters attached to dye names also vary, as does the significance of such letters. Thus Pyronin G and Pyronin Y are merely synonyms. Contrarily Pyronin G (or Y) is *not* the same chemical compound as Pyronin B, though admittedly they are quite similar. On the other hand Light Green is a homonym and has been attached to Light Green SF (a sulphonated arylmethane, see Fig. 16b) and to Malachite Green (a monocationic arylmethane) and to Methyl Green (a dicationic arylmethane), dyes with three very different structures. The associated letters and numbers sometimes relate to the amount of coloured dyestuff in the sample. For example Solochrome Azurine BS or Eriochrome Azurol B200% denote respectively a dye sample of standard commercial dye content (100% or S) and a double strength dye (200%). Alternatively, nomenclature such as Diamond Cyanine R 40 : 100 indicates that 40 parts of this dye sample are to be used in place of 100 parts of the standard grade. Confusing, but do not despair, since an exit from this maze of nomenclature exists in the form of *The Colour Index.* The Society of Dyers and Colourists, Bradford, U.K., in collaboration with the American Association of Textile Chemists, published the third edition in 1971. This work lists all the dyes in large scale commercial use and so includes most, though not all, biological stains. Each dye is assigned a five digit Index Number often abbreviated to «C.I. number». The use of Colour Index Numbers to clarify the ambiguities mentioned above is illustrated in Table 15.

Table 15. Illustrations of the advantages accruing from use of Colour Index Numbers

Name(s) of the dye	Colour Index Number
Acid Fuchsin, Acid Magenta, Acid Rubin, Fuchsin, Rubin	42685
Pyronin G, Pyronin Y	45005
Pyronin B	45010
Light Green:	
When sold as Malachite Green	42000
When sold as Light Green SF	42096
When sold as Methyl Green	42585

So if users and sellers quoted Colour Index Numbers life would be less difficult. More than three quarters of the major suppliers do quote the C.I. Numbers of dyes. However more than three quarters of the users do not, as judged by inspection of recent numbers of the major histochemical journals.

Now to consider various descriptive terms applied to dyes, starting with those relating to the origins of the dyestuff. A few dyes directly derived from animal or plant materials are still used. These *natural dyes* include Haematoxylin, still perhaps the most commonly used dye in biological staining. The structure of this dye and also of another natural dye, namely Carminic acid, is given in Fig. 15. Other dyes used by biologists are synthetic in origin and are sometimes called *aniline dyes* or *coal tar dyes,* in memory of the days when certain of these dyes were synthesised from aniline and all were produced using some coal tar derived precursor. Today distillation and processing of oil has largely replaced that of coal tar. Incidentally the distinction between natural and synthetic dyes is not rigid. Thus Indigo when originally used as a biological stain was

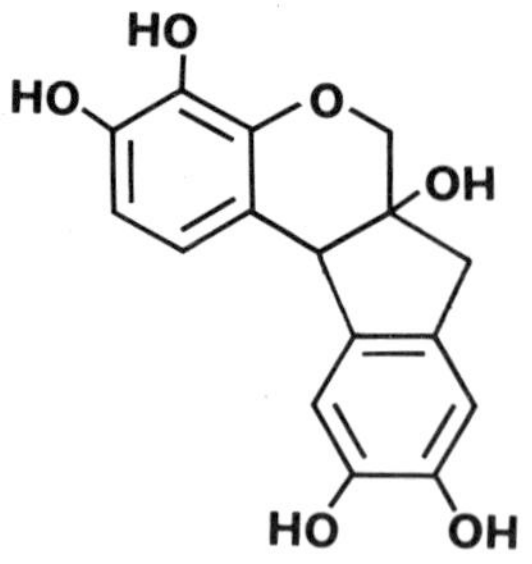

(a)

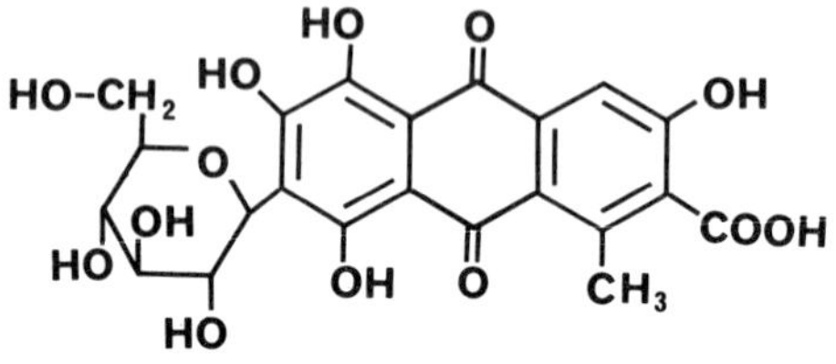

(b)

Fig. 15. The chemical structures of certain natural dyes. Structure (a) is that of Haematoxylin (C. I. 75290), obtained by extraction of logwood. Structure (b) is that of Carminic acid (C. I. 75470), which dye is a C-glycosyl anthraquinone obtained from the cochineal insect

(a)

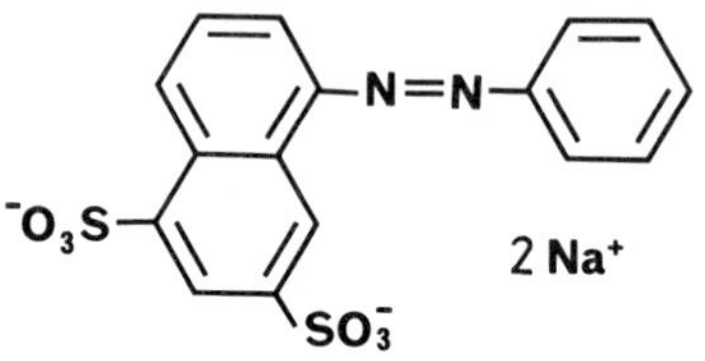

(b)

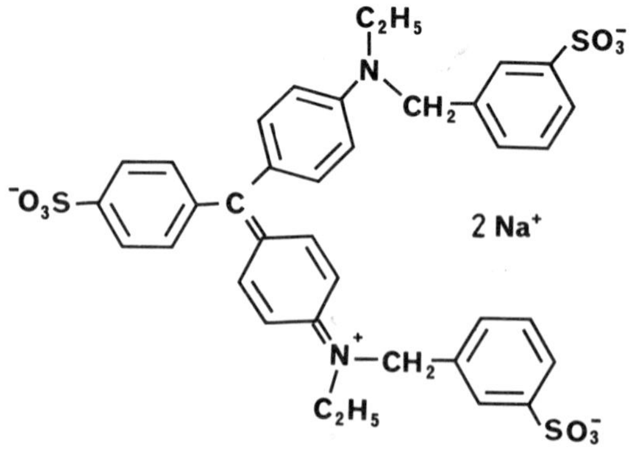

Fig. 16. The chemical structures of some typical acid dyes. Structure (a) is that of Orange G (C. I. 16230), the sodium salt of a disulphonated monoazo dye. Structure (b) is that of Light Green S F (C. I. 42096), the sodium salt of a sulphonated arylmethane dye

obtained from plants of the genus *Indigofera* though nowadays the dye is largely produced synthetically. Another example of ambiguity is Orcein, produced by chemical processing with air and ammonia of phenols contained in certain lichens.

Leaving origins, we consider terms related to usage. The most widely applied descriptions are to say that *a dye is acid or basic*. In fact both classes of dyes are ionic salts. So acid dyes, except at extremes of pH, are not acids and indeed some, such as carboxylates, are chemically speaking weak bases. Few basic dyes are bases, though some chemically speaking are weak acids. The names in fact derive from the original use of such substances for textile dyeing.

So *acid dyes* are capable of dyeing protein, originally wool and silk but also most cell cytoplasms and connective tissue fibres, from acid dyebaths. Such dyes contain *coloured anions* together with colourless cations, as illustrated in Fig. 16. Common substituent

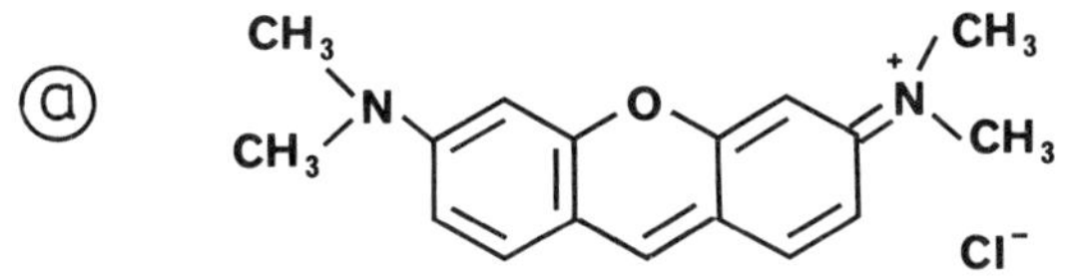

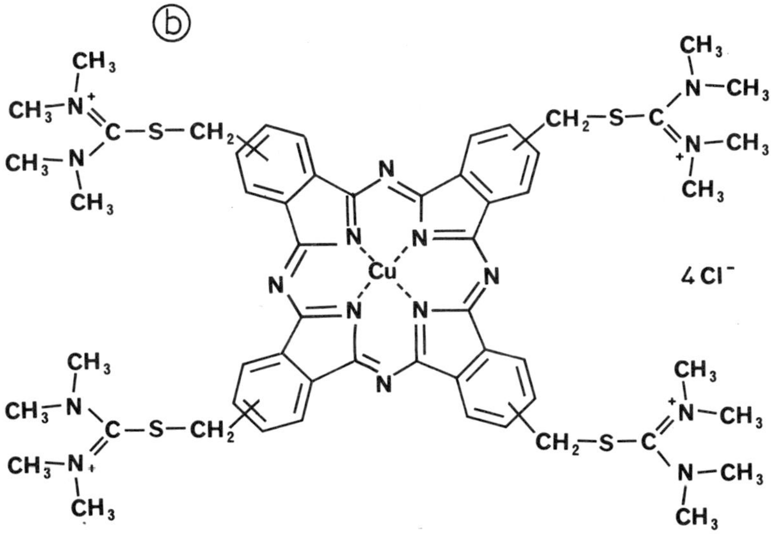

Fig. 17. The chemical structures of examples of basic dyes. Structure (a) is that of Pyronin Y (C. I. 45005). The figure shows a single resonance structure of this xanthene dye with the positive charge localised to a dimethylamine grouping. Structure (b) is of Alcian Blue 8G (C. I. 74240), a copper phthalocyanin derivative whose composition has only recently been published (Scott et al. 1972) and whose precise stereochemistry is still not known. The positive charges of this dye are localised to the thiouronium side chains (pendant cationic groups) and so, unlike the previous example, do not interact strongly with the chromophore

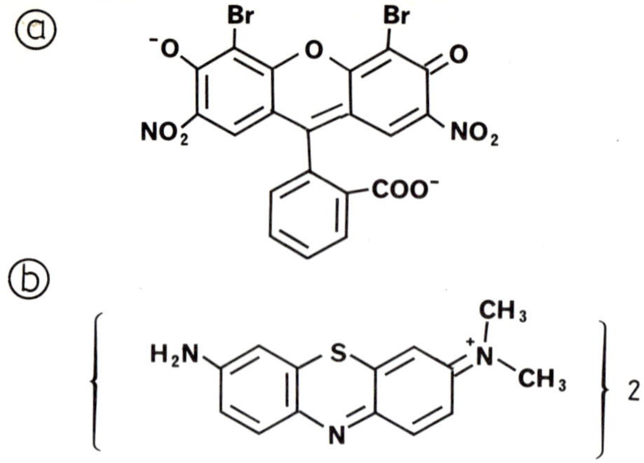

Fig. 18. The chemical structure of a neutral dye, Azur Eosinate. The coloured anion (a) is that of a carboxylated and phenolated xanthene acid dye Eosin B (C. I. 45400). The coloured cations (b) are those derived from the thiazine basic dye, Azur A, which carries variously substituted amino groups

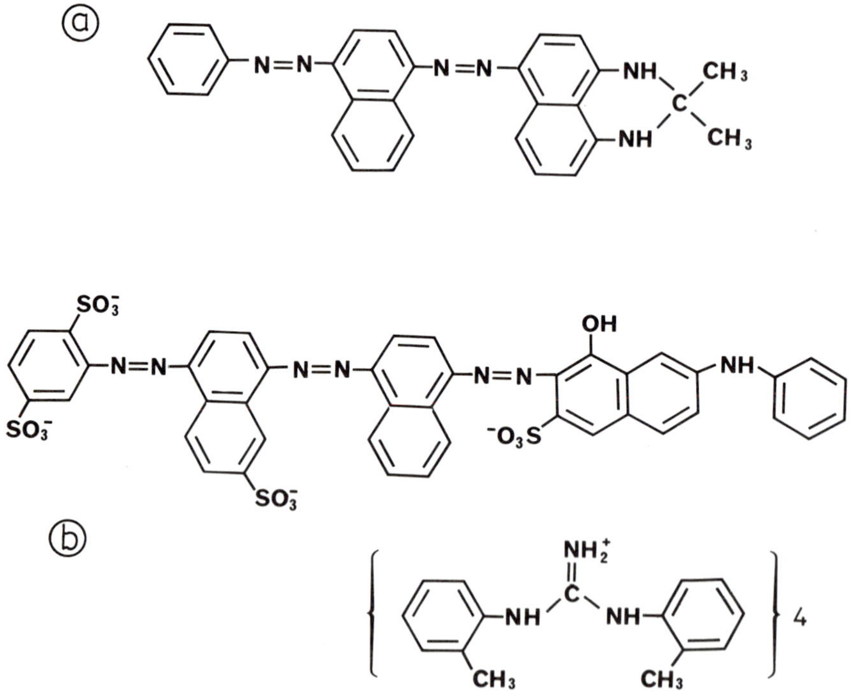

Fig. 19. The chemical structures of various fat stains. Structure (a) is that of Sudan Black B (C. I. 26150), a bisazo dye lacking ionic substituents, which is a non-selective lipid stain. Structure (b) is that of Luxol Fast Blue G (C. I. Solvent Blue 34), a trisazo acid dye with large organic counterions. This dye is a selective stain for myelin sheaths, rich in polar lipids

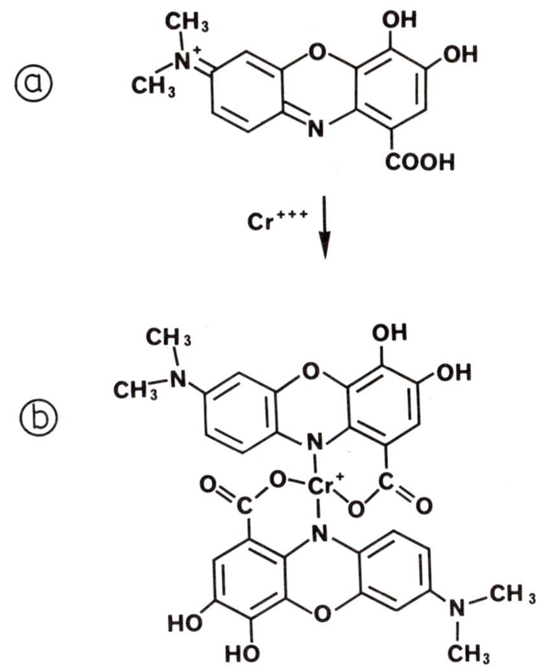

Fig. 20. The amphoteric oxazine dye Gallocyanin (C. I. 51030), structure (a), is substituted by both carboxylate and dimethylamine groupings. After being reacted with a hot aqueous solution of a chromium (III) salt to form Gallocyanin-chrome alum, Gallocyanin is used as a selective stain for nucleic acids. Such a staining solution contains a 2:1 Gallocyanin-chromium complex, structure (b) (Horobin and Murgatroyd, 1968; Marshall and Horobin, 1972)

anions are carboxylate, phenolate and sulphonate. The usual counterions are small inorganic cations such as ammonium, potassium and sodium. The chemical natures of the dye ions vary from simple aromatic nitro compounds to high molecular weight stilbenes of uncertain structure.

Conversely *basic dyes* are capable of staining protein from alkaline, that is basic, dyebaths. Such dyes contain *coloured cations* together with colourless anions, as illustrated in Fig. 17. The origin of the positive charge is commonly a quaternary nitrogen grouping, usually an integral part of the chromophore, as with Pyronin Y, see Fig. 17a. Occasionally the charge and the chromophore are separated, as in Alcian Blue 8G, see Fig. 17b. The term pendant cationic group is used to describe this. The usual colourless anions are acetate, chloride or sulphate. The chemical structures of the dye ions vary widely, from simple heterocycles to high molecular weight phthalocyanine derivatives.

It is convenient to consider *neutral dyes* at this point. Neutral dyes are not, in biological staining nomenclature, dyes staining protein from neutral dyebaths but ionic salts, as are acid and basic dyes. The term neutral arises from their mode of preparation, which involves mixing acid and basic dyes so that both the anion and cation of a neutral dye are coloured species, as illustrated in Fig. 18 for the Azur A Eosinate.

Fat stains (or *lysochromes* in the nomenclature of Baker, 1958 p. 298, though this is not to imply acceptance of his rationale) are of two types: those staining neutral lipids,

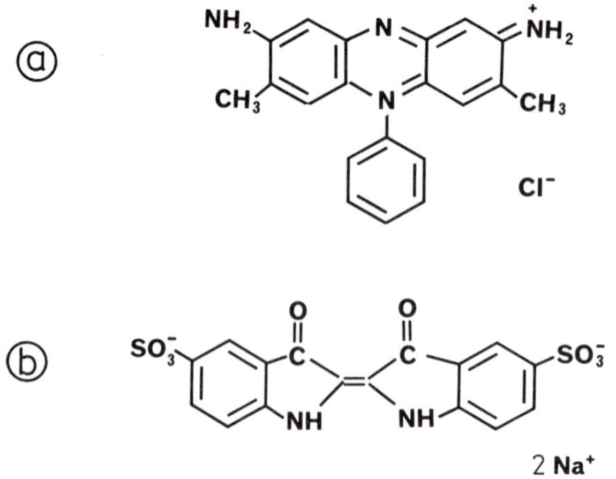

Fig. 21. Chemical structures of metachromatic dyes. Structure (a) is that of a basis azine dye, Safonine 0 (C. I. 50240); and structure (b) that of the metachromatic acid dye Indigocarmine (C. I. 73015), an indigo derivative

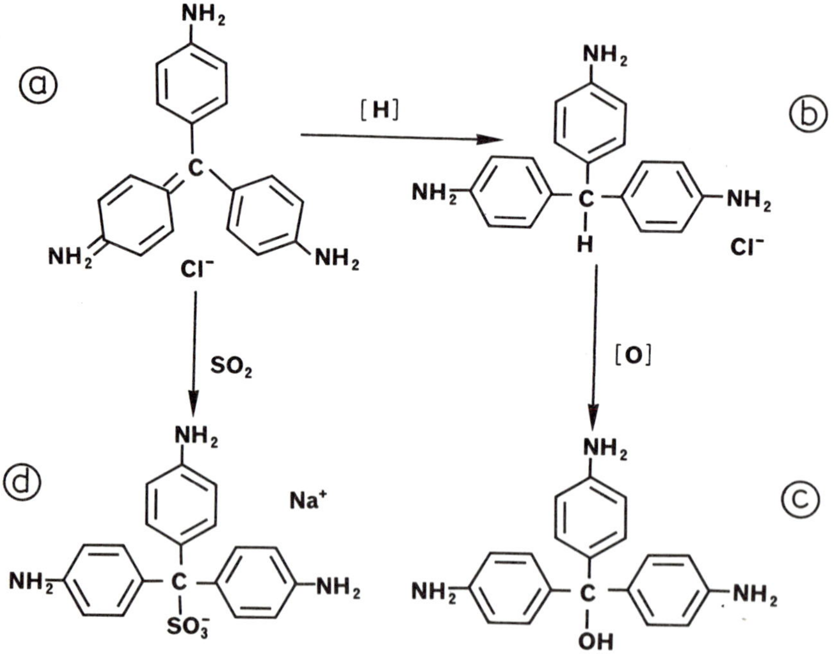

Fig. 22. The chemical structure of various leuco-derivatives of Pararosanilin (C. I. 42500), a common component of Basic Fuchsin. Structure (a) is that of the coloured dye itself, a basic triarylmethane. Structure (b) is the colourless leuco-compound formed from the dye by the action of reducing agents. Structure (c) is that of the dye-carbinol, produced by oxidation of the leuco-compound. Structure (d) is that of the leuco-sulphite derivative contained in Schiff's reagent, formed on treating an aqueous solution of Pararosanilin with such reagents as sulphur dioxide

which dyes are almost always non-ionic and of various structural types such as that shown in Fig. 19a; and those staining polar lipids such as myelin, at least some of which dyes are ionic with colourless organic counterions, as illustrated in Fig. 19b.

A final descriptive term concerning usage considered here is «mordant». *Mordant dyes* may be anionic, cationic, amphoteric or uncharged but in any event only give their characteristic staining patterns when used in conjunction with metal ions. Often stable metal-dye complexes are known to be present in the staining baths, as illustrated for Gallocyanin-chrome alum in Fig. 20.

There remains a mixed bag of terms in fairly common use. *Metachromatic dyes* are those which, even if chemically homogeneous, will stain different substrates different colours. Metachromatic dyes in common use are basic, such as Safronine O shown in Fig. 21 a. The phenomenon is not however limited to basic dyes and Fig. 21 b illustrates a metachromatic acid dye.

The term *leuco-compound* is also found in the biological staining literature. This description is often rather loosely applied to any colourless compound readily obtained from or convertible into a coloured dyestuff. Various leuco-derivatives of Pararosanilin, often the major component of commercial samples of Basic Fuchsin, are illustrated in Fig. 22. For the precise usage of the term leuco and for the explanation of the more specialised terminology of dye chemistry, such as colour base and Homolka base, the reader is directed to such standard monographs as those of VENKATARAMAN (1952 and in progress).

4.2 Why are dyes coloured? The colour, dichroism, fluorescence and metachromasia of dyes

What is colour? In the electromagnetic spectrum the visible region comprises radiation with wavelengths between about 400 nm and 800 nm. Radiations of longer or shorter wavelengths, respectively termed infrared and ultraviolet, are invisible to us. If radiation is fairly evenly distributed over the 400–800 nm range it is perceived as white light which may be split, using a diffraction grating or a prism, into the visible spectral colours of violet, indigo, blue, green, yellow, orange and red.

Black substances absorb light of all wavelengths in the visible range, whilst coloured substances absorb radiation selectively. Thus if white light is used to illuminate a substance which absorbs radiation chiefly in the 500–560 nm region, then the light transmitted or reflected is relatively enriched at both the blue and red ends of the visible spectrum and so the substance is perceived as purple.

The light absorption characteristics of dyes are usually more complex than is implied by the account above. The complexities are well displayed by absorption spectra, resulting from plotting absorption of light by a dye against wavelength. Many examples of such spectra (including those of black and purple dyes) may be found in *Conn's Biological Stains* (LILLIE, 1969a).

Mode of absorption of visible light by dye molecules. Having found the physical origins of colour to be due to selective absorption and seen how this may be described in terms of absorption spectra, we are left with the questions: how is the light absorbed, and why are different dyes different colours? A traditional approach, surviving in the biological staining literature, is sketched in Supplement 1.

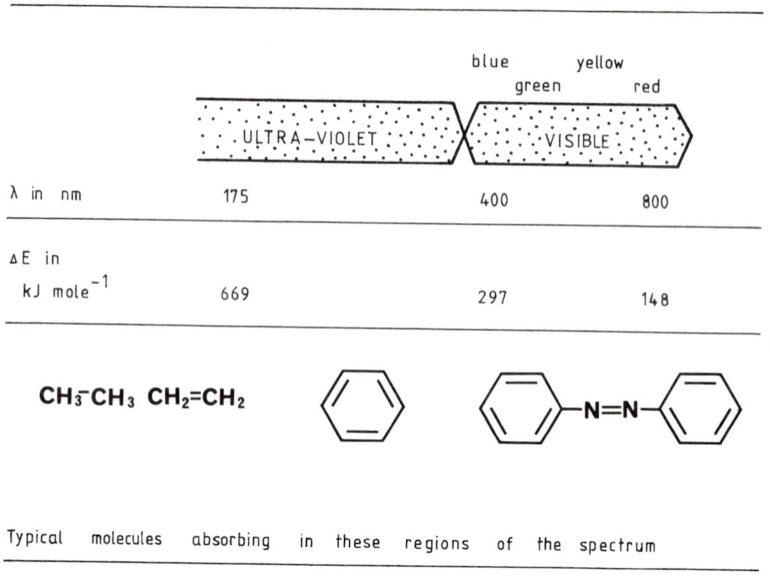

Fig. 23. An illustration of the relationship between absorption of radiation by a molecule and the molecule's degree of electron delocalisation

Here we will focus on more recent views, and firstly the question of the absorption of light. The total energy of a molecule may be defined as:

$$\text{Total energy} = \text{translational energy} + \text{rotational energy} + \text{vibrational energy} + \text{electronic energy}$$

Molecular rotations are responsible for absorption in the far infrared, and a combination of vibrations and rotations cause absorption in the near infrared. However absorption in the visible and ultraviolet, the absorptions giving rise to colour, is electronic in origin.

According to the quantum theory (e.g. see Mason, 1968), when a molecule absorbs visible or ultraviolet radiation with a frequency ν a valency electron is raised from a lower to a higher energy level, each of which levels is associated with a particular molecular orbital (MO). The change in energy, ΔE, is given by the relationship $\Delta E = h\nu = h\,c/\lambda$ (h is Planck's constant, c the speed of light, λ the wavelength corresponding to frequency ν).

In saturated organic compounds, with no double bonds, the electrons concerned with bond formation are all held tightly between the atoms, so ΔE is large, and radiation is absorbed only in the far ultraviolet. In the simple unsaturated molecule ethylene ($CH_2 = CH_2$) the binding energies of the pi electrons of the carbon-carbon double bond are lower, so ΔE is lower, around 670 kJ mole^{-1}, and radiation is absorbed in the ultraviolet at about 175 nm. For absorption to occur in the visible spectrum the ΔE value must be less that 297 kJ mole^{-1} (and of course more than 148 kJ mole^{-1}, see Fig. 23) and this occurs for example as electron delocalisation increases in extended pi-orbital systems. This latter term is explained more fully in Supplement 1.

In dye molecules, with their extensive delocalised electron systems, the energy levels concerned with light absorption are pi-orbitals as shown in Fig. 24. The electronic transitions involve promotion of an electron from a lower-energy π-MO to a higher level π*-MO. This is sometimes described as a π→π* transition.

So light absorption in dyes may be described in terms of the transition of electrons from lower to higher energy levels. Furthermore transitions of the appropriate energy for absorption to occur in the visible region require organic molecules possessing extensively delocalised systems of pi-electrons. We can now turn to the question: how does a particular colour relate to chemical structure? This complex topic will be dealt with rather superfically by considering a few specific examples.

Firstly the colour of a dye is related to the extent of the system of *delocalised electrons.* If a series of structurally similar dyes are considered, differing only in the size of the conjugated system, then as the size of this system increases so will the wavelength of maximum absorption rise. For example certain cyanine dyes whose chromophores are linear polyenes differing from each other only in the number of =CH-units, show this effect clearly (Brooker et al., 1953). Another, more drastic, illustration of this type of effect is seen with arylmethane dyes. For example the Pararosanilin cation (see Fig. 20 a) possesses π-orbitals covering all three phenyl groups and is strongly coloured. However addition of groups such as −H, −OH or $-SO_3^-$ to the central carbon atom (see Fig. 20 b–d) reduces the size of the delocalised units to that of single phenyl groups, and these addition compounds are colourless.

A further important structural factor influencing spectra is *symmetry.* For example the dye Crystal Violet (see Fig. 25 c) has a plane of symmetry such that the molecule's polarisability is the same in all directions in that plane. The dye has a single absorption maximum in the visible region at 591 nm. Malachite Green on the other hand is unsymmetrical with two axes of symmetry, see Fig. 25 b, and in addition to the major absorption at 625 nm there is also strong absorption at 423 nm.

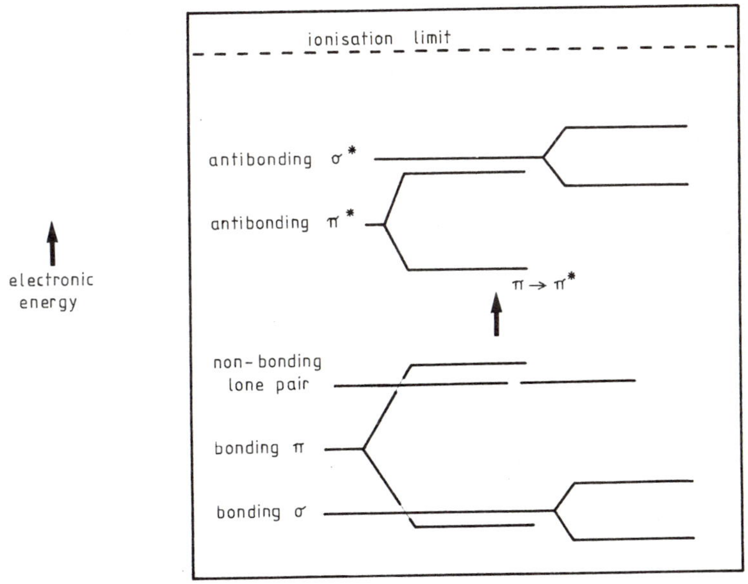

Fig. 24. Relative energies of the sigma (σ), pi (π) and lone pair (η) orbitals of a complex unsaturated molecule such as a dye. A pi→pi* transition is shown

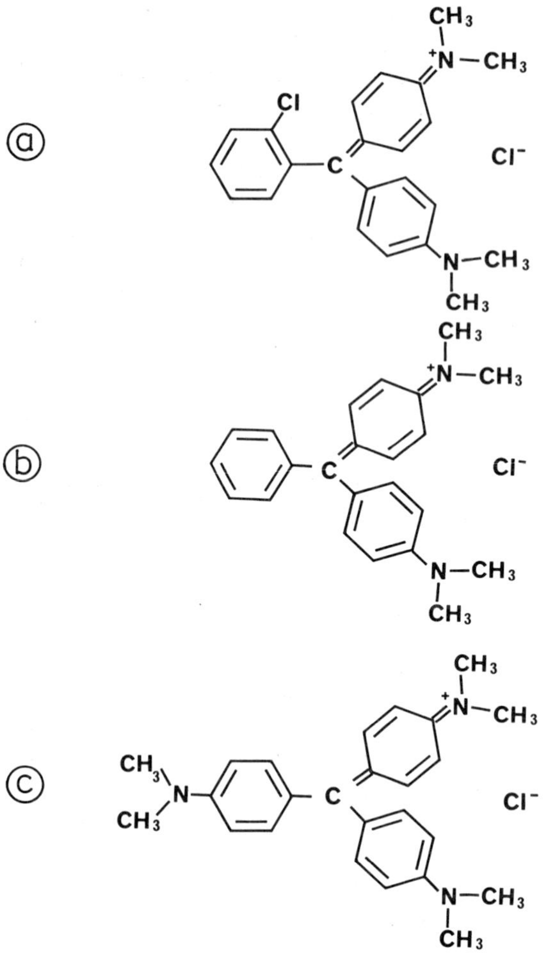

Fig. 25. Chemical structures of three arylmethane basic dyes: (a) is that of New Fast Green 3B (C. I. 42025), (b) that of Malachite Green (C. I. 42000), and (c) that of Crystal Violet (C. I. 42555). In spite of their structural similarity the absorption spectra of these dyes, for reasons discussed in the text, varies appreciably: the absorption maxima are at 633 nm, 619 nm and 591 nm respectively

Yet another structural factor influencing visible spectra is the *presence of substituents* which interact with the delocalised π-electron system. Thus if the N-methyl substituents of Crystal Violet are successively removed the colours of the resulting dyes become steadily less blue and more red. Such electronic substituent effects were classically systematised by Lewis (1945) and are now within the fold of the MO rationale, see for example Venkataraman (1971, volume 3, chapter 4).

Dichroism of stained tissue sections. An object is described as dichroic when the absorption of light passing through it, and so its colour, varies with the plane of polarisation of the light. A dye molecule will only absorb light strongly when its upper occupied π-orbital is suitably oriented to the plane of polarisation of the incident radiation. So dichroism of stained tissues may be strong if dye molecules are orientated, for instance by some oriented substrate in the tissues or due to an oriented mode of dye-tissue binding. In fact dichroism is exhibited by many dyed, or metal impregnated, components of tissue sections. A technically valuable instance is the dichroism of amyloid deposits after staining with Congo Red, iodine or Toluidine Blue (FRANCIS, 1977).

Fluorescence of dyes. After a dye molecule has absorbed radiation and achieved an electronically excited state this energy must be dissipated. Just how this occurs depends on the precise structure and environment of the dye. Possibilities are summarised in Fig. 26. Thus all the energy may be released inside the molecule, either chemically with bonds being formed or broken, or eventually in the form of heat. Or the energy may be transferred to another molecule. Or the energy may be wholly re-emitted as radiation, or a little may be lost internally before the remainder is re-emitted. This latter is fluorescence.

Thus Primulin absorbs in the near ultraviolet but fluoresces in the blue, Fluorescein absorbs in the blue but emits in the green, and Eosin absorbs in the blue-green but fluoresces in the yellow-green. This characteristic of fluorescence occurring at longer wavelengths than absorption is explained by the quantum theory as indicated in Fig. 26. Each electronic energy level actually comprises a group of sub-levels, due to the different vibrational energy states of the molecule. An unexcited molecule is usually in its lowest vibrational state (the ground state) but on excitation to a higher electronic level the molecule will often find itself in an upper vibrational sub-level. Before re-emission the molecule dissipates excess energy by a radiationless process. After emission the molecule will again find itself in an upper vibrational level, this time of the lower electronic state. Return to the ground state is achieved by further radiationless process. So whilst absorption is achieved by a single process, re-emission occurs sandwiched between two radiationless stages. Since less energy is dissipated as radiation than was absorbed, the wavelength of fluorescense is longer than that of absorption.

Fluorescence occurs only when a excited molecule does not lose its excess energy by some other process. Consider the staining of cells and tissues with dyes which are fluorescent in solution. Will the stained tissues be fluorescent? Quite often not. For example the fluorescent acid dye Eosin Y is not regarded as a fluorochrome. However

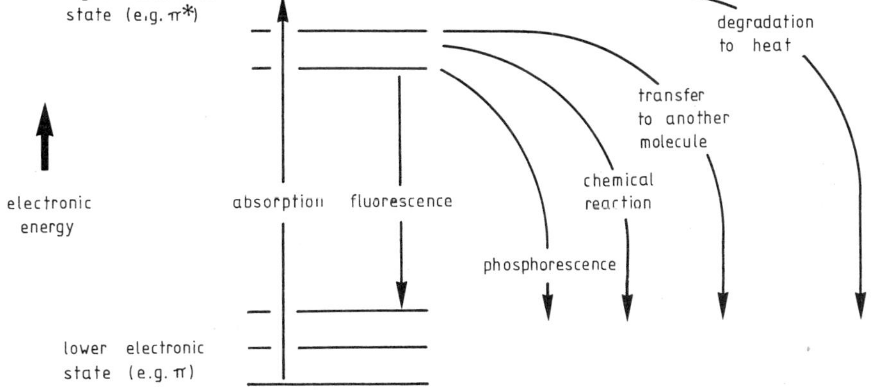

Fig. 26. Diagrammatic representation of possible fates of a dye molecule in an electronically excited state

in standard H & E stained tissue sections, viewed with a dark field condenser plus crossed polarisers and excited by visible light, there *is* strong Eosin fluorescence in the elastic fibres. This has been attributed by GOLDSTEIN (1969a) to the high density of these fibres, relative to other tissue components, selectively limiting aggregation of dye molecules inside the fibres. It is known that aggregated dye commonly fluoresces less strongly than do single molecules. A further possibility is that fluorescence is enhanced by the unusually hydrophobic character of the protein. Similarly if dilute acidic aqueous solutions of the dye Acridine Orange are used to stain nucleic acids then DNA containing structures fluoresce yellowish-green but RNA-containing structures fluoresce red. This fluorescent metachromasia may also reflect differing degrees of dye aggregation at the different sites, see THOMPSON (1966, p. 269).

Metachromasia of dyes. As far as biological staining is concerned this is the colouring of different tissue constituents in different colours by a single dye. It is not accompanied by the making or breaking of any covalent bonds in the dye. For example a solution of Methylene Blue, even when pure, will stain cell nuclei and most cytoplasms blue (termed orthochromasia) but cartilage matrix and mast cell granules red (metachromasia). Similar effects occur in solution, so solutions of such metachromatic dyes do not obey Beer's Law. Metachromatic colour changes are quite distinct from those produced by acid-base or oxidation-reduction indicator effects.

It is now usually considered (e.g. SELWYN and STEINFIELD, 1972) that metachromasia in solution is due to the electronic interactions of aggregated dye molecules, and for a brief account of this latter phenomenon see Supplement 10. When dye molecules aggregate, whether surrounded only by solvent or while bound to adjacent sites on a co-solute macromolecule or chromotrope, the resulting dimer, trimer, or higher aggregate forms a new light absorbing entity with light absorption characteristics different to that of the monomeric dye. Relative to the absorption of the monomeric dye (as determined in very dilute, say 10^{-6}M, aqueous solution or alternatively in non-aqueous solution) the absorption of the dimer commonly shows a drop in intensity plus a shift and/or a split in the peak absorption wavelength. With metachromatic dyes used in biological staining such absorption wavelength shifts are commonly into the blue, as with Azure A. Sometimes however, as with Pseudoisocyanine, the shift is into the red.

Metachromasia is strikingly influenced by variations of co-solutes and solvent. The addition of salt will increase the metachromasia of an aqueous dye solution. The addition of water-miscible organic solvents, such as the lower alcohols, dimethylformamide or dimethyl sulphoxide, decreases metachromasia. These effects are probably caused by such additives respectively increasing or decreasing dye aggregation (see Supplement 10). Metachromasia of dye bound to a chromotrope is enhanced with increasing proximity of the charged sites on the macromolecule, that is as the dye aggregation is facilitated. The nature of the charged sites is also important but this effect is complex in origin. Thus the metachromatic shift often increases as sites change from carboxylate to phosphate to sulphate.

Metachromasia in solution may also be caused by dye-substrate interactions, witness the biochemical studies on the binding of dyes to proteins. In such work spectral shifts due to dye-substrate interactions have often been used to distinguish between bound and unbound dye (see STEINHARDT and REYNOLDS, 1969). Since shifts due to dye-dye interactions are unwanted in this context their presence has first to be excluded, as has been done for example when studying the binding of Biebrich Scarlet to chymotrypsin and the binding of Thionin to trypsin. Dye binding to different sites on a protein can however produce different shifts, e.g. when Trypan Blue binds variously to bovine serum albumin.

This work on solutions provides a good model for considering the metachromasia of wet sections just removed from the dyebath, or of sections mounted in an aqueous medium. However once sections have been dehydrated and mounted in a hydrophobic resin they comprise quite different, solid, structures. Is the behaviour discussed above still a good description? Certainly it is well known that metachromasia is greatly diminished as the wet tissue sections are dehydrated. Unfortunately little physical chemical background is available for solid state metachromasia though work by PHILLIPS and his collaborators (BAUGH et al., 1971, 1972) provide useful pointers. These investigators found that the metachromasia of Methylene Blue seen while absorbed onto solid carboxylated, phosphorylated or sulphated celluloses once again depended on such factors as the nature and density of the anionic substituents, and the character of the dyebath. However it was argued that metachromasia in the solid state, unlike that in solution, derived not only from dye-dye interactions (that is from dye aggregation) but also from dye-substrate and dye-site interactions as shown diagrammatically in Fig. 27.

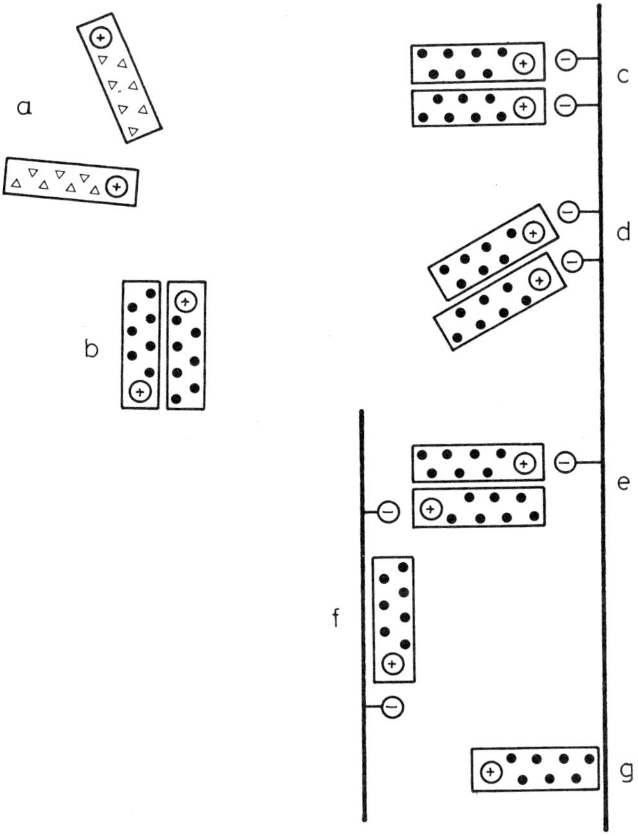

Fig. 27. Various causes of the metachromasia of dyes. Monomeric orthochromatic dye cations (a) are compared to various metachromatic dye molecules (b–g). Dye-dye interaction in solution is shown in (b), with interactions of dyes bound onto a macromolecule in various geometries are shown by (c–e). Dye-site and dye-substrate interactions, as seen in solid state metachromasia, are shown by (f) and (g) respectively

A rather less critical earlier investigation, involving the spectrophotometric study of the uptake of Benzopurin 4B onto cellulose and solid gelatin, assumed that metachromasia is indicative of dye aggregation. Hence it was concluded that the dye is less aggregated on cellulose than on gelatin, that the aggregates build up on the substrate, and that co-solutes such as phenol reduce the solid state metachromasia (CAMPBELL et al., 1959).

So various metachromatic phenomena observed while staining tissue sections, e. g. dependence of metachromasia on substituent charge density and consequently on the pH of the dyebath, on the mode of fixation and on the concentration of stain used (overstaining can cause a reversion to the orthochromatic colour); and the tendency of metachromatic spectral shifts to be into the blue (though bathochromatic shifts are sometimes seen) all have their parallels in the simpler systems discussed above. Whether or not such analogies give a truthful impression is in general untested and unknown. One set of measurements which may be relevant concerns the dye-tissue affinities involved in orthochromatic and metachromatic staining of cartilage matrix and mast cell granules by Azur A (GOLDSTEIN, 1963). For these substrates the metachromatic staining had a higher affinity ($\Delta G°$ values of –22.2 and –18.4 kJ $mole^{-1}$ for metachromatic staining, compared to –15.9 and –16.7 kJ $mole^{-1}$ for orthochromatic), quite in keeping with additional dye-dye or dye-substrate interactions occurring during metachromasia.

A phenomenon for which the simpler systems do not provide a parallel is the widely noted effect of dehydration. If wet sections stained with a metachromatic dye are observed directly, considerable metachromasia is commonly seen. After dehydrating (whether conventionally with alcohols or by placing the slide in a dessicator containing calcium chloride) the metachromasia is no longer seen in most substrates, the exceptions being such hard-to-dry materials as cartilage matrix. In some instances restoration of metachromasia occurs on rehydrating the section, for example after breathing onto it. Though the role of the water remains uncertain it is probably concerned with the dye-dye interactions rather than dye-tissue effects, as may be seen by considering a substrate-free situation. GOLDSTEIN (1969b) for example prepared dried dye smears on glass plates and oriented the dye molecules by polishing the smears. Such a dry, oriented smear was orthochromatic. If breathed on, and so hydrated, the metachromatic colour instantly appeared, only to revert to orthochromasia as the smear once more dried out. The aggregation-disaggregation nature of the effect was supported by measurements of dichroism.

It may be noted here that the favoured metachromatic basic dyes exhibit blue orthochromasia and so, since the metachromatic absorption shift is commonly into the blue, purple metachromasia. BAKER (1958, p. 249) has pointed out that this is no coincidence, as the eye is well adapted to distinguishing blue from purple but is less discriminating for other metachromatic colour pairs, such as red-orange. It is also of interest that these favourite dyes are small and unicationic, so that metachromasia accompanies an increase in dye aggregation. However metachromasia could also result from a generally aggregated dye staining certain tissue components in a monomeric form. Numerous dyes do in fact exist predominantly as aggregates in aqueous solution. An example is the large basic dye Alcian Blue 8G, whose molecular structure is shown in Fig. 17. The orthochromatic colour of the aggregated dye seen in aqueous solution is blue, whereas the metachromatic colour displayed in solvents promoting disaggregation, such as dimethyl sulphoxide or ethanol, is blue-green (SCOTT, 1970). Moreover though Alcian Blue 8G is not regarded as a metachromatic dye by histochemists, since it usually stains tissue sections uniformly blue, sections stained briefly from ethanolic solutions of the dye can show the dense elastic fibres stainding out blue-green against the blue background.

The metachromasia of acid dyes may perhaps also be viewed in this way. The acid dyes reported as staining tissue sections metachromatically, for example Benzopurpurin and Congo Red, are typically large molecules of the type known to be highly aggregated in aqueous solution (e. g. VICKERSTAFF, 1954, p. 68). The tissue elements found to stain metachromatically were usually dense, for example elastic fibres, keratin and red blood cells. So the metachromatic staining could well involve monomeric dye, with the more porous structures containing the orthochromatic dye aggregates. This possibility does not seem to have been studied, probably because the metachromatic shifts of most acid dyes are small compared to those of the better known basic dyes (around 15 nm compared with up to 90 nm). Quite large shifts can sometimes occur however. For example an aqueous-dimethyl sulphoxide solution of the bis-azo dye Direct Blue 152 stains dense elastic fibres red but other substrates blue, this latter being the colour of the dye in aqueous solution (HOROBIN and JAMES, 1970).

Related phenomena are the spectral shifts occurring when a pair of *different* dyes interact electronically. For instance aqueous solutions containing mixtures of acid dyes such as Chrysophenine G and Chlorazol Sky Blue FF contain appreciable amounts of 1:1 and 2:1 heterocomplexes (HOSHI et al., 1965 and cited papers). Such effects could be important in Trichrome stains; and Eosin Y-Azur B complexes formed in the tissues have been suggested as responsible for the Giesma banding of metaphase chromosomes (SUMNER et al., 1973) and for the «Romanowsky effect» in the traditional haematological staining procedures (ZANKER, 1981).

4.3 Why do dyes stain anything? The affinity concept and some thermodynamic formalism

What events occur when dyeing a tissue section or a cell smear? Intially dye is present only in the dyebath but when dyeing is completed some dye has moved from the bath onto components of the tissue. Such components are said to have an *affinity* for the dye, or vice versa. In the present book the term affinity is used to describe this tendency for a dye to move from a dyebath onto a substrate. Factors such as dye-tissue attractive forces are contributions to this process and so contributions to, but not synonyms for, affinity.

So we are concerned with all the factors which drive a dye from the bath onto the tissue substrates. Just what are these factors? In attempting to answer this we will make use of the concepts and formalism of chemical thermodynamics. Below is a general introduction of the terminology, followed by a series of simplified models for different aspects of biological staining.

Just what is meant by affinity? (Some thermodynamic formalism). In the terminology of chemical thermodynamics the affinity, or more precisely the standard affinity, of a dye for a tissue element is the change in the free energy of the staining system as dye moves from the bath to the tissue. This change is expressed symbolically as $-\Delta G°$. For introductions to chemical thermodynamics see texts such as that of EVERETT (1962). Thus a tissue will be stained by a given dye from a given solution when the stained state is more stable, that is of lower free energy, than the unstained state. This concept is shown diagrammatically in Fig. 28 and symbolically below:

When $\Delta G° > 0$ the dye does not have affinity for the tissue and staining does not occur spontaneously.
When $\Delta G° < 0$ the dye has affinity for the tissue so staining can occur spontaneously.

Since the standard affinity is a measure of the tendency of the dye to move onto the section, there will be a relation between the standard affinity and the ratio of the amount of dye in the dyebath to the amount of dye on the tissue. More correctly we should be concerned with the activities of dye in the bath and on the tissue (a_b and a_t respectively). Activities are functions of concentration, which allow for the finite volumes of and intermolecular forces between real solute and solvent molecules. Thus we can write:

$$\text{Affinity (i. e. } -\Delta G°) = a_t/a_b$$

Since the ratio a_t/a_b is merely the equilibrium constant for the staining process, which may be written as K, the relation can be set out explicitly as below, where R and T are respectively the Gas Constant and the temperature.

$$\text{Affinity } (-\Delta G°) = RT \cdot \ln K$$

Following this excursion into physical chemistry there are a number of points calling for particular emphasis. First note that strictly speaking thermodynamic concepts such as standard affinity apply only to systems in equilibrium, and see Section 4.5. Secondly a favourable standard affinity only means that dyeing could take place, not that it does. The rate of occurrance of a possible process may in fact be trivially slow. Thirdly, the approach adopted here (exemplified by the questions «Why do dyes stain anything?» and «Why don't dyes stain everything?») is based on the practicalities of biological staining and not on the logic of physical chemistry. These two practical questions

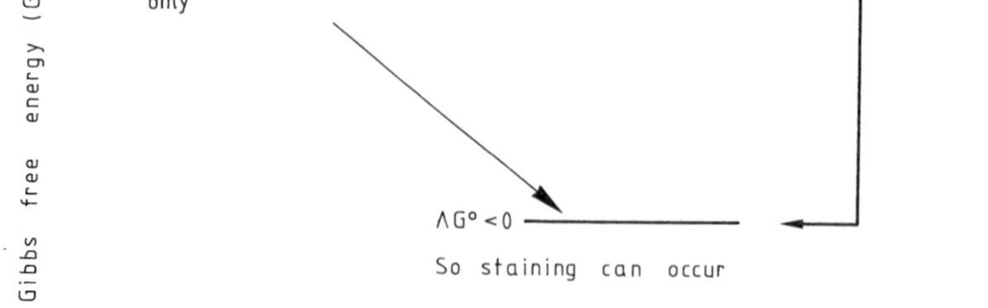

Fig. 28. An illustration of the relationship between the free energy of the staining system and the occurrence of staining. Staining can take place spontaneously only when $\Delta G°<0$, that is when the standard free energy change is negative

provide a useful perspective even though many of the same physico-chemical factors provide both affinity and selectivity.

Finally we return to the essential point that affinity is a measure of the tendency of dye to move from dyebath to tissue. That is, a measure of change in the total staining system (solvent-dye-cosolute-tissue) and not merely a measure of single factors such as dye-tissue attractive forces.

Expanding the focus of our thermodynamic gaze, we can say that the high affinity of a dye for a substrate results from large negative enthalpy changes and/or large positive entropy changes. This is amplified somewhat in Supplement 3. However, what does this mean in practice? The thermodynamic approach, even if it were rigorously applicable to the biological staining situation (and see Section 4.5), yields little information about the physical causes of the phenomena to which it is applied. This is of course merely a reflection of the generality of the thermodynamic approach. If we wish to connect molecular events to the thermodynamic formalism we must introduce models, with particular properties reflecting the action of the various physical factors of interest. It must be remembered though that every commitment to a model means a commitment to a caricature.

In Section 4.4, we will consider a series of models for biological staining. Various parts of the staining system will be considered, e. g. dye, tissue, co-solute, solvent, and their interactions.

4.4 Various affinity models for biological staining

The understanding of a practical staining method will commonly require more than one of these model systems, plus additional information concerning selectivity. The models are introduced in order of conceptual simplicity, not practical importance. Thus although most staining methods use aqueous solutions of reagents, water is a complex solvent and so the intitial models only consider non-aqueous solutions.

Affinity Model 1:

The solvent is non-aqueous, the dye and tissue substrate are non-ionic, no dye-tissue forces are involved. If affinity was due to dye-tissue attractions alone there would be no staining in this case. However, by saying that there is no dye-substrate interaction (and also, since the solution is non-aqueous, no need to assume dye-solvent interactions) we mean merely that $\Delta H° \sim 0$, so that:

$$\text{affinity} = + T\,\Delta S°$$

Hence staining can occur spontaneously if the entropy change is positive, and this is quite plausible. Consider an analogue, the release of a strong odour in a distant corner of a room. In a short time we can perceive the odour throughout the room, yet it is not necessary to say that our noses attracted the odorous molecules. Instead one would say that the final state, with smelly molecules distributed throughout the room, was more random than the initial state, with all such molecules concentrated in one corner. Such increases in randomness will correspond to an increase in standard entropy, so that is to a positive $\Delta S°$ value.

Returning now to the staining situation, consider a dye which forms an ideal solution in the substrate. There will then be no dye-tissue interactions and no specific orientations. The change from dye present only in the dyebath to dye in bath and substrate will then correspond to a positive ΔS° value.

A practical dyeing situation approximating to Affinity Model 1 is the staining of neutral fat droplets with aqueous-alcoholic solutions of Sudan dyes, shown diagrammatically in Fig. 29a. However, even in this apparently simple situation additional factors must be taken into account. For instance MEIER (1959) reported that the proportion of dye in the fat increased as the molecular weight of the dye rose, suggesting the importance of dye-fat interactions, see Affinity Model 2. The same effect occurs after making the dyebath more polar, say by increasing the proportion of water or by using more polar alcohols, suggesting dye-solvent interactions. It is interesting that the dyes were aggregated in the concentrated dyebaths used. From the viewpoint of Affinity Model 1 this would be expected to increase the affinity of the dye for the fat, see Fig. 29b for a diagrammatic display of this, since dye aggregates are less random structures than single dye molecules.

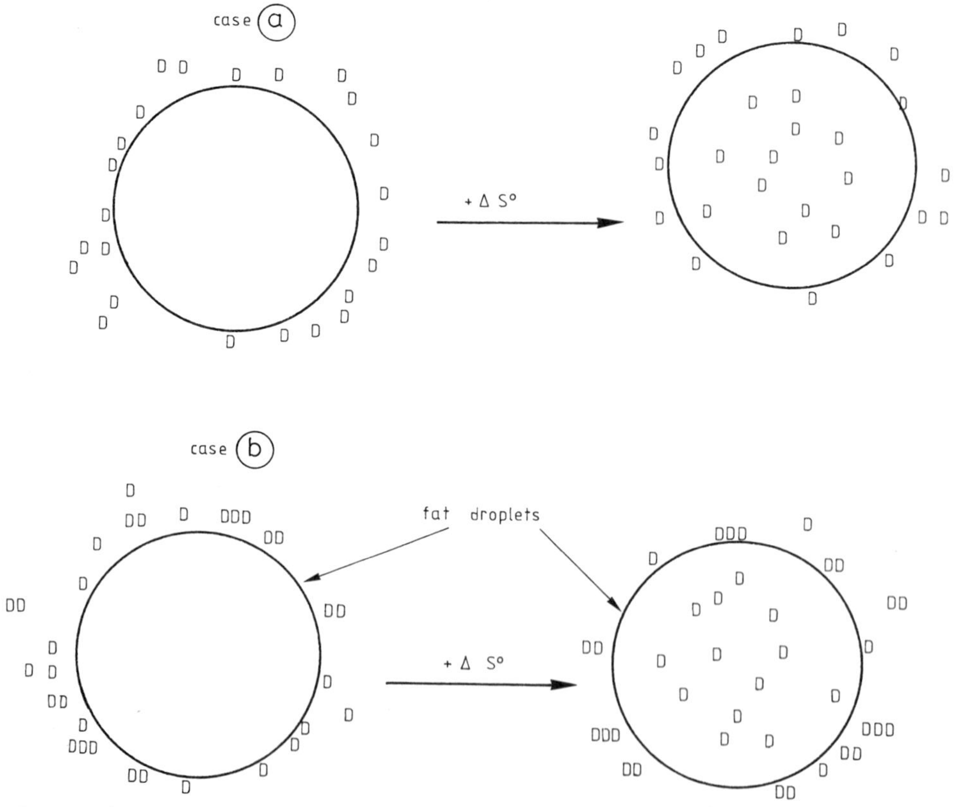

Fig. 29. A diagrammatic representation of Affinity model 1, with the staining process driven by a large $+\Delta S^\circ$ term. The dye in the stained state is distributed more randomly than in the unstained state. This is more marked when the dye in the bath is aggregated (case b). Dye molecules are denoted by D and dye aggregates by DD, DDD etc.

Affinity Model 2:

The solvent is non-aqueous, the dye and substrate is non-ionic, dye-tissue attractive forces occur.

Here we will assume that there is no marked change in the standard entropy. That is $\Delta S^\circ \sim O$, with increases in standard entropy due to effects of the type discussed in Affinity Model 1 balanced by decreases in entropy due to the dye-tissue forces to give rise to specific dye-tissue orientations. So affinity will derive from large negative enthalpy changes, in this Affinity Model taken as:

$$\text{Affinity} = -\Delta H^\circ$$

due to strong dye-tissue interactions. Hence Affinity Model 2 corresponds to one of the traditional views of affinity. Note that coulombic effects are considered separately, in Affinity Model 4 below.

Due to their universal occurrence we will first discuss dye-tissue *van der Waals attractions.* Some characteristics of these forces are set out in Supplement 4. When will such attractions be of major significance for biological staining?

First consider characteristics of the dye. Increases in the extent of the conjugated pi-electron systems (see Supplement 2) will often result in large permanent dipoles. Compare the dipole of aniline, 1.02, to that of a molecule some three times as long, aminostilbene, 1.73. Even for dyes without a significant permanent dipole, increases in the size of the pi-electron systems usually result in polarisable molecules. Hence van der Waals forces would be expected to be most important for dyes with extensive conjugated electronic systems. Examples of such dyes include polyazo compounds such as Congo Red and Chlorazol Black E, or heavily halogenated fluorone derivatives such as Phloxine or Rose Bengal.

Now to consider the role of the tissue substrates in van der Waals interactions. Due to the need for close contact between dye and substrate one factor of importance is the degree of hydration. The more heavily hydrated a substrate, the less effective will be dye-tissue van der Waals attractions. In aqueous dyebaths it will be hydrophilic, ionised biopolymers such as acid glycosaminoglycans which will be most solvated, with non-polar proteins and lipoproteins being least solvated. The dipole moments of the substrate biopolymers are also important. In polar solvents those groupings with large permanent dipoles will tend to be well solvated, unlike hydrocarbon groupings. However, only unsaturated groupings such as the base residues in nucleic acids are readily polarizable. These various structural factors are illustrated diagrammatically in Fig. 30.

A practical dyeing situation fairly well described by Affinity Model 2 is the staining of elastic fibres by the standard methods such as those using Orcein, iron-haematoxylin or Chlorazol Black E. Such dyes can be anionic or cationic, and elastin itself is only slightly ionic: showing that electric charge is not a critical factor. The usual dyebaths contain appreciable quantities of organic solvent so solvent-entropy effects, of the sort discussed in Affinity Model 3, are probably minor. However, elastic fibre stains have extensively conjugated electronic systems, so will be polariable. The major constituent of mature elastic fibres is the protein elastin which is unusually non-polar. A high proportion of elastin's amino acid residues are polarizable or at least lipophilic, for example phenylalanine, alanine and valine, and few of its residues are ionogenic. So several of the factors favouring strong van der Waals attractions occur, as has been argued by several authors (Horobin and James, 1970; Puchtler and Sweat

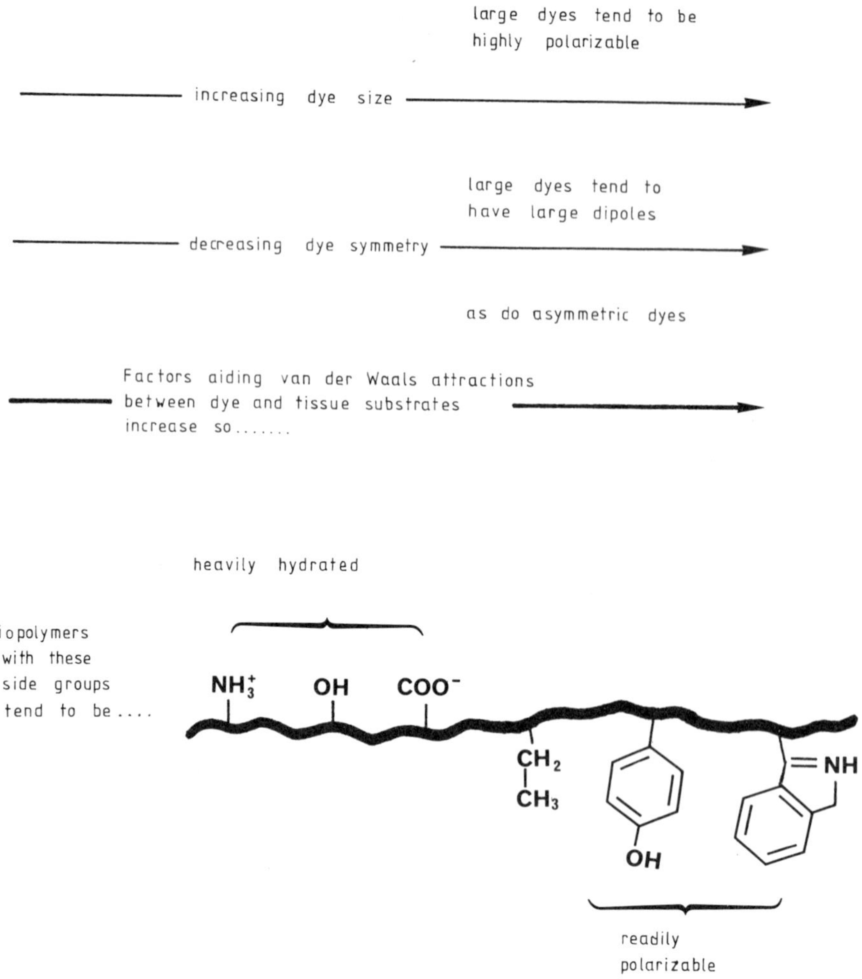

Fig. 30. Some structural factors of dye and substrate which influence dye-tissue van der Waals attractions

WALDROP, 1979; HOROBIN, and FLEMMING, 1980). A discussion of the selectivity of such methods will be given later, under the heading «Why don't dyes stain everything?» (Section 4.7).

Another dye-tissue attractive force evoked to explain dye-binding, especially when everything else fails, is *hydrogen bonding*. A sketch of this phenomenon is given in Supplement 5. Hydrogen bonding certainly occurs between dyes and biopolymers such as proteins and carbohydrates which carry suitable substituents, as reviewed by GILES (1959) for example. However, when aqueous dyebaths are used, hydrogen bonding is unlikely to be an important contribution to dye-tissue affinity, as pointed out by RATTEE (1972). This is because the water, present in great excess, will act as a competing species for hydrogen bonding sites on both the dye and substrate.

Viewed in this light, suggested involvement of hydrogen bonding in biological staining (for example Alcian Blue as a mucin stain and Congo Red as a stain for amyloid, both from aqueous dyebaths) may be regarded as unlikely. However one example which remains plausible is the Best's Carmine stain for glycogen. It has been suggested that Carminic Acid, and other dyes capable of forming basic phenolate anions, form hydrogen bonds with the hydroxyl groups of glycogen and other polysaccharides (HOROBIN and MURGATROYD, 1971; OPTIZ and DORSCHE, 1973). The importance of this factor in staining was indicated by the decreased staining intensity of the glycogen which occurred as the proportion of water in the aqueous alcoholic dyebath was increased.

As a final example of the sources of dye-tissue attractive forces consider *mordanting*. This is the binding of dye to tissue with the aid of metal ions. The role of the metal is traditionally seen as bridge-forming between dye and tissue, so:

Tissue-(metal ion)-dye

That is, the metal is bound co-ordinatively both to groups on the tissue and to groups on the dye. Supplement 6 contains a brief outline of the nature and characteristics of co-ordination complexes, especially those involving dyes and biopolymers.

At various times mordanting has been suggested as the source of the stain-tissue affinities of many of the pre-formed metal complex dyes, such as Gallocyanin-chrome alum for nucleic acids and iron-haematoxylin for muscle protein. In some cases specific structural formulae have been suggested (for example BAKER, 1960, SANDRITTER et al., 1966). However, though there is evidence that mordanting does occur in the staining of keratin by azo dye-chromium complexes (RATTEE, 1953) there have been few direct studies in the biological staining field. Most evidence in this later situation is essentially circumstantial. Mordanting is evoked, with little direct evidence, to explain the great differences in substantivity observed between cationic metal complex dyes and the usual, metal free, basic dyes, for instance as discussed by BAKER (1958, chapter 11; 1962).

Some studies on metal complex dyes used for biological staining have attempted to directly investigate the question of mordanting (MARSHALL and HOROBIN, 1973 a, b, c,). Various metal complex dyes were compared to a number of metal-free basic dyes of varying molecular weights. Quantitative measurements of affinity, histochemical blocking and extraction, and salt-effect experiments were carried out. Most of the metal complexes studied (for example various aluminium complexes of Carminic acid, Gallocyanin and Heamatoxylin) and the iron complexes of Celestin Blue and Gallocyanin showed little evidence of the involvement of mordanting mechanisms. Only with nuclear staining by the chromium-aluminon complex did mordanting of the traditional type seem important. It was suggested that the high substantivity of the metal complex dyes was in fact usually due to a combination of extensive electronic conjugated systems compared to the usual basic dyes (aiding van der Waals attractions, see above) with their low solubility in alcohols, which prevents extraction by the processing fluids. However, the traditional view of mordanting remains the favourite (e.g. LLEWELLYN, 1978).

Mordanting dye-binding mechanisms have also been supposed to act in two-step procedures in which the tissue is treated separately with a solution of a metal salt and the ligand. Proposed examples include the iron- and chromium-haematoxylin stains for, respectively, muscle striations and myelin. As with the pre-formed metal complex dyes specific structural formulae have been suggested (e.g. LILLIE, 1969 b). Again the evidence is usually circumstantial. When treatment with the mordant precedes that of the ligand, the metal ion is perhaps bound to the tissue at least in part via coordinative bonds. This, however, is of no consequence when one is considering the form of possible dye-metal-tissue complexes. Once formed a metal complex dye may bind as other metal-free dyes do.

Affinity Model 3:

An aqueous dyebath, dye and/or tissue substrate uncharged, no appreciable dye-tissue attractions. In a way Model 3 is a varient on Affinity Model 1. As dye-tissue attractions are ruled out (i.e. $\Delta H^\circ \sim O$), the source of the affinity of staining must again be sought in the occurrence of a large $+\Delta S^\circ$ term (affinity = $+T\Delta S^\circ$). However, the great difference is that in Affinity Model 3 the solvent is aqueous. Here the source of the positive entropy change is loss of order in the *solvent* on transferring dye from dyebath to tissue. A brief outline of this effect, often termed hydrophobic bonding, is given in Supplement 7.

The importance of such hydrophobic effects has been convincingly shown for the dyeing of keratin with acid dyes (ZOLLINGER, 1965; MEYBECK and GALAFASSI, 1971). Is this however a general phenomenon in biological staining? Though Affinity Model 3 is unlikely to be a total explanation for the commonly used dyes, there is evidence that hydrophobic bonding contributes to the affinity of a variety of biological stains. For instance with large acid and basic dyes such as Chlorazol Black E or Rhodamine 6G, used from aqueous dyebaths, it has been demonstrated by a pH-signature criterion that appreciable non-coulombic sources of affinity exist. That hydrophobic bonding contributed significantly to this affinity was indicated by the loss of staining after addition of solvents which disrupt water-structure to the dye-bath (see Figs. 2 and 5, HOROBIN and BENNION, 1973). Evidence for hydrophobic bonding has also been found with small, uncharged dyes such as Alizarin Blue.

Such dyes stain most tissue elements, excepting the wetter ones rich in hydrated polysaccharides. Such an effect is expected if hydrophobic effects are a major source of affinity, see Supplement 7.

Affinity Model 4:

Dye and substrate are ionic; non-coulombic dye-tissue forces and the nature of the solvent are taken to be unimportant. This is essentially acid and basic dyeing, as usually conceived. In most theoretical accounts in the biological staining literature the concept used, explicitly or implicitly, to rationalise acid and basic dyeing is that dye ions are attracted coulombically to tissue ions of the opposite charge. Consequently terms such as salt linkage, electrostatic binding and ionic bonding are used. This picture supposes affinity to depend on the negative enthalpy term arising from the coulombic attractions, so:

$$\text{Affinity} = -\Delta H^\circ$$

This unlike charges attract explanation does indeed predict which tissue substrates will be stained with acid and which by basic dyes; and it does avoid the chemically nonsensical «basic dyes stain acid substances whilst acid dyes stain bases». In fact the Donnan and Gouy theories are based on this picture. These sophisticated descriptions have been successfully applied to the colouration of a variety of textile polymers by ionic dyes; for an account of which see RATTEE and BREUER (1974, p.3). However, charged-surface models are rather awkward to use as simple qualitative pictures. Perhaps the most versatile, though simple-minded, qualitative approach is to treat the acid and basic dyes as simple salts, analogous say to sodium chloride or ammonium sulphate. Staining can then be regarded as a process of *ion-exchange* driven by the increase in entropy resulting from the exchange-generated randomisation of ions. Such an ion-exchange picture is illustrated in Fig. 31. This model is of course very similar to

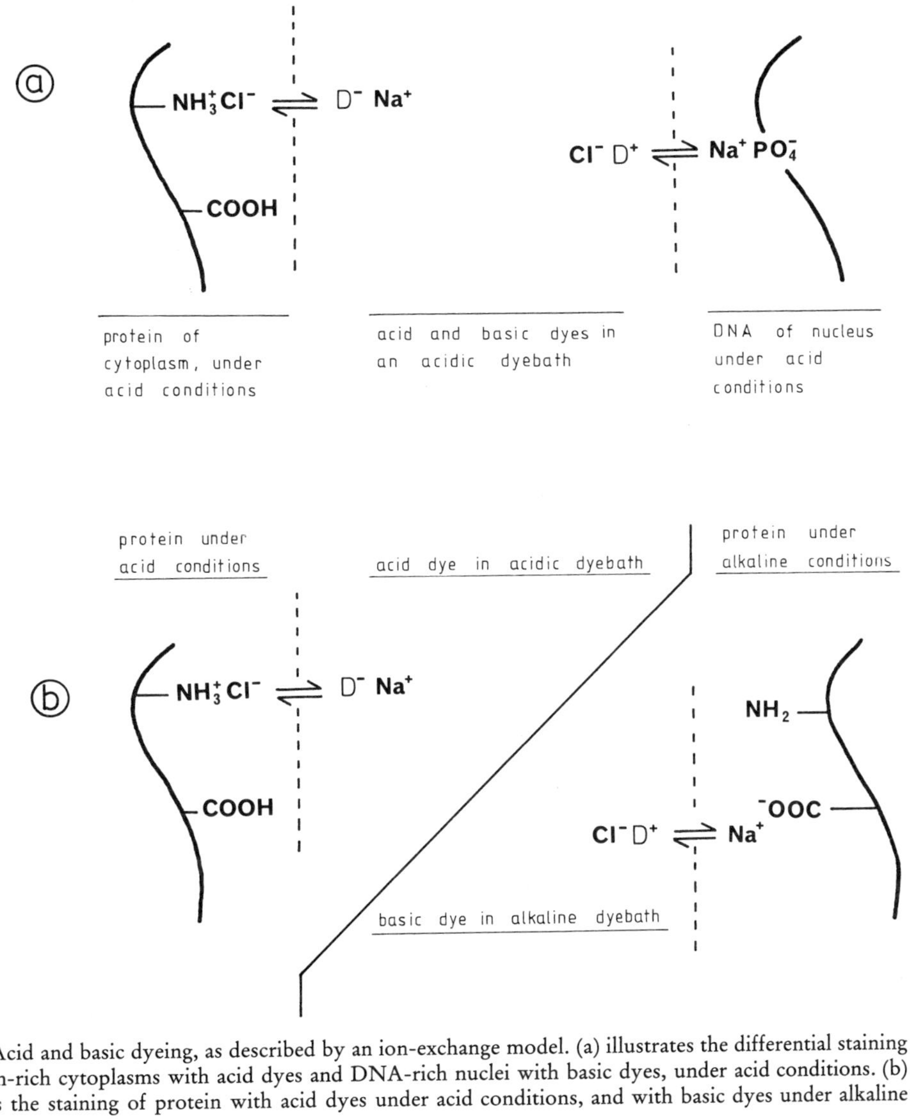

Fig. 31. Acid and basic dyeing, as described by an ion-exchange model. (a) illustrates the differential staining of protein-rich cytoplasms with acid dyes and DNA-rich nuclei with basic dyes, under acid conditions. (b) illustrates the staining of protein with acid dyes under acid conditions, and with basic dyes under alkaline conditions.

Note: The ions of acid dyes are represented as D^-Na^+; the ions of basic dyes as D^+Cl^-. The immobile, fixed macromolecular ions are indicated by heavy wavy lines. Throughout this and related figures Na^+ and Cl^- are representative of the mobile inorganic cations and anions involved in biological staining. In practice the nature of such ions depends on which buffer system has been chosen, and similar everyday matters that described in Affinity Model 1, but now the dyes and substrates are ionic. Hence we need to specify that both before and after staining the solution and the tissue will, overall, be electrically neutral. Both dyes and tissue substrates are salts, not naked ions.

This simple ion-exchange model gives a satisfactory qualitative explanation of the acidophilia-basophilia phenomenon. Thus protein-rich cytoplasms and connective tissue fibres are acidophilic under acid conditions due to the presence of excess immobile NH+ groupings plus mobile hence exchangable anions. Contrariwise DNA-rich nuclei and RNA-rich cytoplasms contain immobile nucleic acid phosphate groups plus mobile hence exchangable cations. Similarly, carboxylated and sulphated mucosubstance-containing cartilage matrix and mucins are basophilic. See Fig. 31 for illustration of this.

The pH control of acidophilia-basophilia can also be understood using the ion-exchange model. Changing the pH may be regarded as titration of the various ionic or ionogenic groups of the tissue biopolymers. It can be seen from the pK values of the groups listed in Table 16 that, as most proteins contain amino and carboxylic acid groupings, protein-rich structures will be acidophilic (i.e. contain mobile anions) at low pHs, but basophilic (i.e. contain mobile, exchangable cations) at high pHs.

Table 16. The pK values of some common tissue groupings

Nature of the group	Approximate pK
Protonated Amines	7–10
Carboxylates	4–5
Phosphate esters	ca. 2
Sulphate esters	ca. 1

It can be seen that, as most proteins contain amino and carboxylic acid groupings, protein-rich structures will be acidophilic (i.e. contain mobile anions) at low pHs but basophilic (i.e. contain mobile, exchangable cations) at high pHs. On the other hand structures containing high proportions of phosphate ester-rich nucleic acids or sulphated mucosubstances will be basophilic (due to the mobile, exchangable ions being cationic) over a wide range of pH, only becoming protonated and so non-ionic under very acid conditions.

A topic missing from the discussions of acid and basic dyeing provided by biological staining authorities such as Baker (1958, chapter 10) and Singer (1952) is the concept of internal pH. The ion-exchange model predicts that the pH inside different components of a tissue may well vary, both from one component to another and also from that of the dyebath. Thus when using an acid dyebath the internal pH of a strongly basic, proteinaceous structure would tend to be higher than that of the dyebath. Conversely the internal pH of a structure rich in sulphated mucosubstances would tend to be lower. An ion-exchange description of this, and also of the further phenomenon of the reduction of these internal pH differences by addition of salt to the dyebath, is given in Fig. 32. Relevant background information on this topic, plus more rigorous theoretical descriptions, may be found in discussions of insoluble proteins (e.g. Rattee and Breuer, 1974, p. 11; Steinhardt and Reynolds, 1969, p. 72) and of synthetic ion-exchange systems (e.g. Helfferich, 1962, p. 83). The implications of variations in internal pH for biological staining could be substantial but do not appear to have been explored, though the concept has received mention in the histochemical literature (Visser et al., 1979).

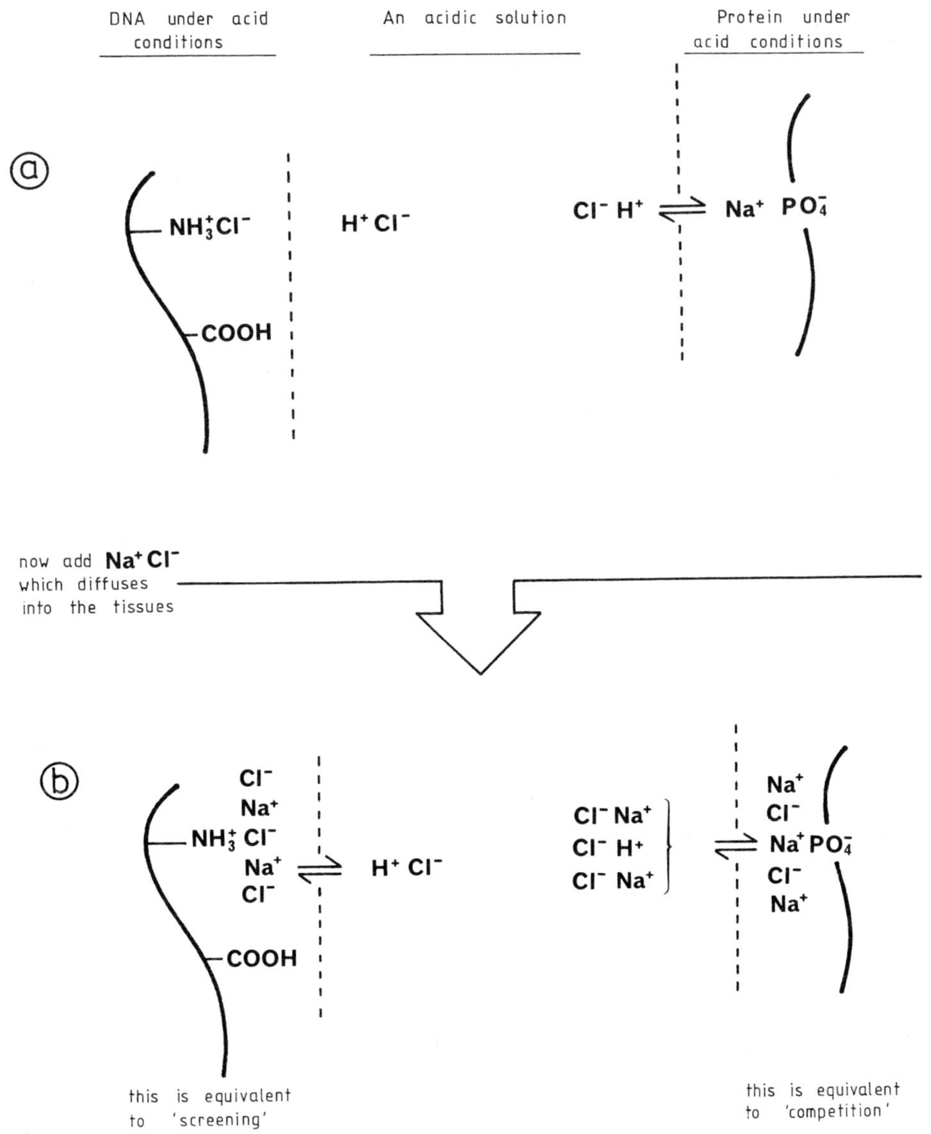

Fig. 32. The internal pH phenomenon, described using an ion-exchange model. (a) illustrates the lowered internal pH of DNA-rich (i.e. basophilic) structures and the raised internal pH of protein-rich (i.e. acidophilic) structures when the external pH is in the acid range. (b) illustrates the equalisation of internal and external pHs after addition of excess salt to the solution phase. *Note* the symbols used are as in Fig. 31

Other dramatic and important phenomena which may be seen when staining with acid and basic dyes are the increases and decreases in staining intensities following the addition of neutral salts to the dyebaths. Such complex effects have been well illustrated (for example BENNION and HOROBIN, 1974a; HOROBIN and GOLDSTEIN, 1974; MARSHALL and HOROBIN, 1973c). Many similar observations can be found in the literature. Typically, addition of salt to salt-free solutions of acid and basic dyes decreases the staining already occurring, such as that of muscle cytoplasms and collagen fibres by acid dyes under acid conditions, and of nuclei and cartilage matrix by basic dyes. Conversely structures previously only poorly stained will stain more intensely after addition of salts, e.g. acid dyes will stain cartilage matrix more strongly, and even under acid conditions basic dyes will tend to stain collagen fibres and muscle cytoplasms. Once again the ion-exchange model can be used to rationalise these effects, as illustrated in Fig. 33.

Such salt-effects have been long discussed, with the explanations offered for salt-induced increases or decreases in staining intensities using the «unlike charges attract, like charges repel» theoretical framework being respectively screening and competition by the added electrolyte (SINGER, 1952).

A more elaborate viewpoint, and one which has attracted widespread interest, application and hope, is that of the Critical Electrolyte Concentration (CEC) theory. As typically described (e.g. SCOTT, 1968; MOWRY and SCOTT, 1967) the CEC theory is fundamentally a mass action model of the interaction between coloured dye cations and anionic tissue groups such as carboxylate, phosphate and sulphate. Inhibition of

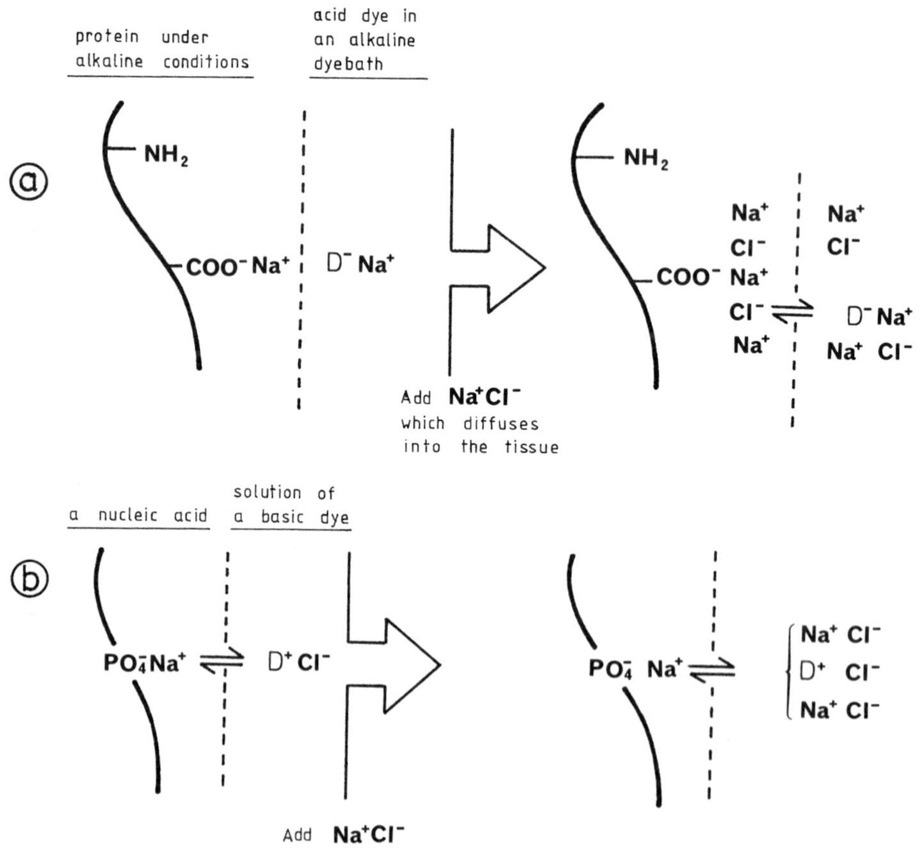

Fig. 33. Salt-induced increases and decreases in acid and basic dyeing, described with an ion-exchange model. (a) illustrates a salt-induced increase in acid dyeing of protein under alkaline conditions. (b) illustrates a salt-induced decrease in basic dyeing of nucleic acid. *Note* symbols are as in Fig. 31

staining by salts is seen as competition between coloured dye cations and the colourless inorganic cations, such as Mg^{++} and Na^{+}, for the anionic sites. The CEC model differs in several respects from the ion-exchange picture adopted in this Section. Thus CEC theory is a single-phase model which regards the tissue as merely a very viscous extension of the dyebath. The ion-exchange model on the other hand regards dyebath and tissue as two distinct phases. One consequence of the single-phase character of the CEC theory is a prediction that, for solutions of polyions, if the concentration of salt is steadily increased the change in solubility or staining characteristics will occur abruptly, the origin of the term «critical» electrolyte concentration. This has indeed been observed with solutions of polyanions, i.e. with single-phase systems (SCOTT, 1962). Such behaviour is, however, not easily perceived during biological staining, and no quantitative studies demonstrating a critical electrolyte effect seem to have been published for tissue sections. Even more dramatically the original CEC mass action model only allowed for decreases in staining intensity due to added salt. The ion-exchange model on the other hand, being a two-phase theory, correctly predicts both increases and decreases in staining intensity, under various circumstances, as salt is added to the dyebath. Since salting-on was soon observed with Alcian Blue in CEC studies, a modification was made to the CEC theory to allow for this effect (QUINTARELLI et al., 1964). This modification, however, requires the dye to be multicationic: whilst salting-on is in fact often seen with monocationic dyes also (BENNION and HOROBIN, 1974a; HOROBIN and GOLDSTEIN, 1974). Some further comments on the selectivity obtained by the CEC methods will be made in Section 4.9, and the present discussion may be closed by noting an irony. Though both CEC theory and the ion-exchange model require staining to have reached equilibrium, the dye most commonly used in CEC work, namely Alcian Blue, stains so slowly and is so chemically unstable that staining equilibrium is quite unlikely in many tissue sites (GOLDSTEIN and HOROBIN, 1974a; TAS, 1977b).

However, the qualitative ion-exchange model of salt-effects used in this book may itself be developed quite rigorously, using the formalism of the Gibbs-Donnan membrane equilibrium for instance. Various such quantitative models have been devised and tested, with some success, by textile dyers. Those interested are directed to the standard texts of VICKERSTAFF (1954), RYS and ZOLLINGER (1970), RATTEE and BREUER (1974) or BIRD and BOSTON (1975).

Before leaving questions of theory we may note that any model which assumes, as does the simple ion-exchange picture, that a dye ion is much the same as a simple inorganic ion involves massive oversimplification. Consider the contributions to dye-tissue affinities emphasised by the other Affinity Models of this Section. In particular think of Affinity Model 5, which emphasises that dyes can form aggregates. Such aggregation is facilitated by the addition of salt to the dyebath, immediately complicating any quantitative, or sometimes even qualitative, predictions.

Affinity Model 5:

The solvent is aqueous, the dye and substrate are ionic, and dye-dye attractions occur. First consider the effect of dye-dye interactions in the dyebath. Formation of dye aggregates is due to hydrophobic bonding and van der Waals forces. So aggregation of dyes occurs most readily in aqueous solutions, at room temperature. If the water structure is destroyed, by adding structure-breaking organic solvents or solutes, or by raising the temperature, then dye aggregation is reduced. Aggregation occurs with both acid and basic dyes, and with large multiply charged dyes as well as with small dyes of unit charge. The rates of interconversion of monomeric dye with aggregated dye, compared to the rates of staining, vary from virtually instantaneous to very slow. The size of the aggregates also varies from dimers to dye particles containing dozens of dye ions. This topic has been reviewed by COATES (1968). The effect of dye aggregation in the dyebath is to reduce the affinity of dye for the tissue substrates since dye-dye interactions provide an alternative to those between dye and tissue.

Dye-dye interactions also occur in the tissue section. Here it is usually supposed that in suitable sites dye molecules interact both with the substrate and with each other. Such dye-dye interactions would be expected to give rise to large spectral changes, as briefly discussed in Section 4.2 under the heading of metachromasia. The factors influencing dye-dye interaction in the tissue will include those acting in solution. For instance metachromatic effects are often diminished when a wet tissue section is taken through the alcohols, perhaps due to the reduction of hydrophobic effects. Since most dyes stain metachromatically only in certain sites the influence of substrate has been extensively studied. Factors such as high charge density and high flexibility of the substrate biopolymers have been found to favour dye-dye interactions in dissolved systems.

In such cases the dye-dye affinity will be expected to increase the affinity of dye for tissue since it is added to and not in competition with the dye-tissue attractions. In keeping with this GOLDSTEIN (1963) found that when staining cartilage matrix with Azur A the affinity of the metachromatic staining was higher than that of the orthochromatic staining.

4.5 Dye-tissue affinities in numbers: various quantitative measures

The dye-tissue affinity ($-\Delta G^{\circ}$) has been defined in Section 4.3 as $-\Delta G^{\circ} = RT \cdot \ln K$, where $K = a_t/a_b$ (a_t and a_b being the dye activities in tissue and dyebath respectively). How may we apply this to the practical biological staining situation, in order to obtain numerical measures of affinity? And given such quantities how can we assess their significance and make use of them?

To date, attempts at quantitation of affinities in the biological staining field use one of two conceptual models. Dye is either considered as dissolving in the tissue substrates and the Freundlich isotherm is applied, or alternatively dye is considered as being adsorbed onto a finite number of tissue sites and the Langmuir isotherm is used. The formalism of these two models is summarised in Supplement 8. A more extensive discussion is provided by GOLDSTEIN (1968).

Both these models require knowledge of the concentration of dye in the tissue (D_t) and in the dyebath (D_b). The latter can usually (but see below) be made a formality since the dyebath comprises a vast excess of a solution of known strength, and the amount of dye removed onto a section is tiny. The D_t values are normally estimated either by comparison of the stained sections to various stained standards (for an account of which ingenious methods, requiring no specialised apparatus other than a comparison eye piece, see GOLDSTEIN, 1963, 1965) or by the direct spectrophotometry of the tissues.

Table 17. Examples of affinity values obtained for biological staining systems

Substrates	*Values of affinities for various dyes (kJ mole^{-1})*		
	Azur A[a]	*chromium-aluminon*[b]	*Orcein*[c]
Cartilage matrix	16.4–22.3	26.5	15.1 (25.2)
Chromatin	13.0	24.8–26.5	12.2–13.9 (25.2–26.9)
Elastic fibres	–	–	15.1–16.8 (29.0–30.2)
Mast cell granules	16.8–18.5	17.6	–
RNA-rich cytoplasm	22.7	23.1–24.8	–
Small gut mucin	14.7	19.3	–

[a] Data from GOLDSTEIN (1963)
[b] Data from MARSHALL and HOROBIN (1973b)
[c] Data from FRIEDBERG and GOLDSTEIN (1969)
Data which was corrected for dye impurity is in parentheses

Some examples of measured affinities of various dyes for various substrates are displayed in Table 17. Such a table prompts us to enquire what is the significance and reliability of this type of data?

The significance is of course limited by the postulates of the conceptual model used, if the model is invalid then so are the conclusions. It is encouraging that when some independent measure of affinity is available the correlation with the measured affinity values is significant, see HOROBIN (1980a). However, even when the fit between model and observation is good (for example see GOLDSTEIN, 1968, Figs. 6.3 and 6.4) it is rarely possible to draw mechanistic conclusions merely from the thermodynamic data, though see below.

The reliability is limited by a variety of factors. For the affinity to be determined the staining system must be at equilibrium. So we must check that sorption and desorption are in balance. For instance a section may be overstained in one dyebath and then destained in the experimental dyebath, though this is not always possible (FRIEDBERG and GOLDSTEIN, 1969). It is also important that no gross chemical or physical changes in dye or substrate occur during dyeing. Such substrate alterations can in fact be seen; and even the dyes may be unstable, e. g. Alcian Blue (GOLDSTEIN and HOROBIN, 1974a). Another source of uncertainty is that since neither a_b nor a_t can be easily measured various assumptions have to be made concerning the relation of dye concentration and dye activity. Consequently affinity values of different dyes for the same substrates may not be comparable since the ratios of a_b values to concentrations will be expected to vary from one dye to another. Hopefully the relative a_t values for a given dye and various substrates will be at least of relative significance. This latter point might be expected to be valid even if the concentration of a dye was uncertain, as illustrated in Table 17. Compare the original and corrected values for the affinity of Orcein for various substrates.

Bearing in mind the cautions regarding significance and reliability, what applications of affinity data have been made in the biological staining field? In combination with extra-thermodynamic information, affinity values have been used to comment on the mode of action of stains for neutral lipids (GOLDSTEIN, 1968, using data of MEIER) and of the elastic fibre stain Orcein (FRIEDBERG and GOLDSTEIN, 1969). Other workers (MARSHALL and HOROBIN, 1973b) have used an extensive collection of affinity measurements to discuss proposed mechanisms for Methyl Green-Pyronin type stains for DNA-RNA, the critical electrolyte concentration methodology, and mordant dyeing.

As a tailpiece, note what a salutary experience it is to dip into the textile dyeing literature on this subject. Though the systems studied are usually simpler than ours, the quantitative models resorted to are usually far more complex.

4.6 Sources of affinity, a practical summary

Now we may profitably consider how the practical staining variables under our control can influence affinities, and so staining. Since variables can have more than a single consequence for staining, each variable will be listed, followed by mention of its effect on affinity, plus some comment on practicalities.

Chemical structure of dyestuffs. The major structural features considered here will be: hydrophobic character, polar character and polarisability, electric charge, and the presence of various covalently reactive groupings.

- *Hydrophobic character.* The more hydrophobic a dye, the greater the significance of hydrophobic bonding in aqueous solutions, see Affinity Model 3. What is important structurally is the hydrophobic-hydrophilic *balance.* For instance a disulphonated monoazo acid dye such as Orange G (C. I. 16230) will be markedly less hydrophobic than a disulphonated bisazo dye like Congo Red (C. I. 22120). However, Orange G will have much the same hydrophobic character as a pentasulphonated bisazo dye such as Trypan Red (C. I. 22850). For a discussion of using Hansch π values to predict hydrophobic character see Section 11.14.
- *Polar character.* Readily polarisable dyes with strong dipoles will be most effective in dye-tissue van der Waals attractions, see Affinity Model 2. Increases of both polarisability and of dipole strength accrue from coupling aromatic sub-units, as occurs in the sequence mono-, bis- and tris-azo dyes. The polyazo dyes may or may not be hydrophobic in character, see above. For a discussion of the use of conjugated bond numbers to estimate polar character see Section 11.14. Another structural feature increasing dipole strength is molecular assymetry.
- *Electrical charge.* With small dyes the factors discussed in the above paragraphs will tend to be small. For such dyes the positive or negative *sign* of the charge can be controlling factor when staining ionic tissue substrates, that is in acid and basic dyeing, as discussed in Affinity Model 4. In addition the *magnitude* of the charge is significant when salting-on and -off occurs, the greater the charge the greater the sensitivity to salt. Furthermore the total number of charged groups, regardless of charge sign, is important since the more ionic substituents the lower the hydrophobic character, see above.
- *Hydrogen bonding groups.* In non-aqueous solvents the presence on the dye of such groupings as phenolic hydroxyl and carbonyl offers the potential of dye-tissue hydrogen bonding, see Affinity Model 2. Naturally such groups are hydrophilic and so will reduce hydrophobic bonding in aqueous solutions, where hydrogen bonding is at any rate unlikely to contribute strongly to affinity.
- *Metal chelating groups.* The presence of such groups as carboxy-, hydroxy- and nitroso- in a dye may facilitate chelation of tissue ions such as Al^{+++}, Be^{++} and Ca^{++}. Such dye groupings may be ionised and are polar, see above for possible consequences.

- *Chelatable metal ions* bound to the dye. Chelation of such ions by tissue ligands can aid dye binding (Mordanting, see Affinity Model 2). Probable examples include the Cr^{+++} and Cu^{++} bound to the o,o′-dihydroxyphenylazophenyl groupings of certain acid azo dyes, which ions interact with carboxyl and amino groups in the tissues. Note that the mere presence of a metal ion does not ensure mordanting. Instances where there is no interaction with tissue ligands include the Al^{+++} of aluminium- «haematoxylins», the Cr^{+++} of Gallocyanin-chrome alum, and the Cu^{++} of Alcian Blue.
- *Covalently reactive groups* present on dyes, in addition to metal ions or ligands, can provide contributions to dye-tissue affinity. Such groupings are discussed at length in Chapter 5. Mention may be made in passing of the chlorotriazine groups (reactive with nucleophilic tissue groupings such as amines) of the Procion dyes, and of the chloromercuri group (reactive with tissue thiols) of Mercury Orange.

Such structural features may all be numerically approximated to, if not defined. Consequently we may, in times to come, hope to inspect the structural formulae of a group of dyes and then say «If used in Blogg's stain for substrate X, *these* dyes will fail to stain anything at all, *those* will stain selectively, whilst *those* will stain excessively». For progress to date in this field see Section 11.14 «The need of structure-staining correlations, practical and theoretical».

Nature of the solvent. An important distinction is between water and organic solvents such as alcohol. An aqueous solvent allows hydrophobic bonding (see Affinity Model 3) but competitively reduces the significance of dye-tissue hydrogen bonding (see Affinity Model 2). The reverse holds true with many organic solvents.

Nature of co-solutes. It is convenient to subdivide these as follows.
- *Acids and bases,* which may change the charge sign on tissue groups such as amino, carboxylate, phosphate. The influence of this on dye-tissue affinity is discussed in Affinity Model 4.
- *Inorganic salts,* of small size and hydrophilic character. These can give rise to increases or decreases in dye-tissue affinities involving acid and basic dyes, again see Affinity Model 4. Such ions may be of low affinity for the tissue, as are the ions of sodium chloride for instance. Others, such as Al^{+++}, Cr^{+++} and Fe^{+++}, may in addition act by interaction with either dye or tissue ligands, or indeed with both.
- *Large hydrophobic salts* including surfactants, and perhaps certain buffer salts such as phthalate, may possess a marked affinity for the tissue. This may competitively reduce dye-tissue affinities.
- *Non-ionic co-solutes* are very varied. For instance dextran, which due to its strongly hydrated character may increase dye activity and so affinity; surfactants, which conversely may enclose dye ions in micelles and so reduce dye activity; and urea, which may compete with dye-tissue hydrogen bonding and so potentially reduce affinities.

4.7 Why don't dyes stain everything?

Inexplicably this question lacks emphasis in most accounts of the modes of action of dyes in biological staining. Given that staining occurs at all, selectivity of staining is vital. Even the oversight stains display cell nuclei and cytoplasms in contrasting colours. The following Sections discuss a variety of ways in which such selective staining is achieved: differential affinities, differential staining rates, and metachromasia.

4.8 Preliminaries concerning substrate density, and the colours of staining

Before dealing with the topics listed in the previous paragraphs, several assorted points relevant to selectivity or apparent selectivity of staining should be discussed.

First consider the *effects of substrate density*. Imagine a given biopolymer present in two structures. In one structure a greater length of the polymer occurs per unit volume than in the other. On staining to equilibrium it will often be the case that the structure containing the close packed substrate will take up the most dye (whether acid or basic) and so will appear darker in the final preparation. Compare for instance the intense staining of red blood cells or zymogen granules with paler staining of collagen fibres or muscle cytoplasm in the everyday H & E preparations. This is a concentration effect. The dye-tissue affinities may well not reflect the staining intensities. Note that there will be no obvious perceptual indication of the cause of such differential dye uptake. A variant on such density effects directly involves *our perceptions*. Imagine staining certain objects too small to resolve optically. Such structures evenly distributed throughout a cell may give rise to staining dismissed as «a pale background». If the structures were clumped together though one might say «look, strongly staining granules». An example of this may be seen in the staining of granules of mucosubstance in the rat salivary gland after various fixative regimes (SIMPSON, 1977).

A different perceptual effect can occur with polychrome staining. We perceive the colours in the middle of the visible spectrum (yellow and greenish-yellow) as being highly saturated, that is diluted with white and so pale looking. The colours at the extremes of the spectrum are however perceived as unsaturated. So a structure stained yellow or greenish-yellow and surrounded by red or blue staining materials will be easily overlooked. Think for example of looking for smooth muscle fibres in amongst collagenous connective tissue after staining by the Van Gieson procedure.

4.9 A dye may prefer one substrate to another: Some affinity models

Such variation in affinity from one dye-substrate combination to another is an old, indeed almost *the* old, favourite in explaining selective staining. For such a mechanism to be significant the staining system must be allowed to reach or at least approach equilibrium. All the sources of dye-tissue affinity discussed in Sections 4.3 and 4.4 can occur differentially. Consequently we will again consider the various staining Models used in Section 4.4, this time looking for their possible differential application.

Affinity Model 1 involves a non-aqueous dyebath, with the staining process driven by a positive entropy term due to the increasing dispersion of the dye occurring as staining takes place. Dyeing of lipids by the sudan dyes was taken as a staining situation approximated to by this Affinity Model. We may now ask why does dye only pass into hydrophobic substrates? And not also into tissue components rich in nucleic acids, polysaccharides or proteins? The answer to this question is probably that those portions of the tissue rich in hydrophilic polymers will contain large amounts of water. Solid proteins for instance can contain up to 90% of water and protein gels may be 97% water (see HAUROWITZ, 1963, Chapter VI). Non-polar solutes such as the sudan dyes have low water solubilities due to these solutes stabilising the highly structured iceberg form of water and so giving rise to a low entropy of solution. So the sudan dyes will dissolve in the hydrophobic lipids but will be much less soluble in the hydrated substances.

Since selective staining explicable by *Affinity Model 3* involves much the same argument its discussion will be placed here rather than in numerical order. Affinity Model 3 described aqueous dyebaths, with no dye-tissue attractive forces. Staining was due to hydrophobic bonding. In this case also the wet substrates will be least stained, since the wetter the substrate the more water structure can be induced by the dye, and so the lower the positive entropy change occurring when dye binds to the tissues. As a probable example of this, it has been noted that certain uncharged dyes stain most tissue substrates, except those which are rich in hydrated polysaccharides (Bennion and Horobin, 1973).

Affinity Model 2 considered a non-aqueous dyebath but this time also allowed for dye-tissue attractive forces, van der Waals attractions, hydrogen bonding and so on. In situations described by this Affinity Model a factor aiding selective staining will thus be the difference in the number of suitable binding sites in the different substrates.

One such is the staining of elastic fibres by dyes such as Orcein or iron-Haematoxylin where strong van der Waals attractions aid dyeing. Since van der Waals attractions are short range forces and are strongest between polarisable groupings, substrates rich in non-polar and often polarisable residues will tend to bind dye more effectively than polar, and so hydrated, substrates. The figures cited in Table 18 show that the proportion of non-polar amino acid residues is much higher in elastin, the major protein of elastic fibres, than in other common but non-staining proteins. So the elastic fibres could indeed contain more binding sites. However, this is not the whole story, see below under the Affinity Model 4 heading. It may be noted that the introduction of additional hydrophobic and polarisable groupings into the tissue, for example by benzoylation prior to staining, gives rise to the staining of additional structures, namely collagen fibres and chromatin, by the elastic fibre stains (Fullmer and Lillie, 1957; Horobin and James, 1970; Puchtler et al., 1978).

A second example where staining selectivity is controlled by relative substrate abundance may be found in the case of Best's carmine, and with the other dyes acting in a similar manner. These dyes form hydrogen bonds with and stain glycogen and other polysaccharides such as cellulose and starch but do not stain other biopolymers carrying hydrogen bonding groups, proteins for instance. Part of the explanation (though not all, see under Affinity Model 4 below) for this selectivity lies in the large number of hydrogen bonding groups per unit length of polysaccharide, as shown by the figures displayed in Table 19. If the number of such groupings is reduced, say by benzoylation or periodate oxidation, prior to staining then the staining intensity is reduced (e. g. Horobin and Murgatroyd, 1971).

Affinity Model 4 describes consequences of Coulombic dye-substrate interactions. Added electrolyte could increase or decrease the staining intensity, and this behaviour

Table 18. The non-polar amino acid content of some common proteins

Protein	Percentage non-polar amino acids
Collagen	53
Elastin	92
Haemoglobin	52

Table 19. Number of hydrogen bonding groups in various tissue macromolecules

Biopolymer	Approximate number of groups able to form strong hydrogen bonds
Glycogen	3 for each monosacharide residue
Collagen	2 per 10 amino acid residues
Haemoglobin	1 per 10 amino acid residues

can in part be rationalised using a qualitative ion-exchange model, or by quantitative theories such as the Donnan equilibrium. Selective stainings explicable using Affinity Model 4 are numerous and mostly due to the simultaneous occurrence of Coulombic replusions and attractions in different parts of the tissue. A number of examples of this phenomenon will be discussed.

As an introduction consider using anionic dyes such as Biebrich Scarlet for staining basic proteins, i. e. proteins with a high amino/carboxyl ratio. At high pH such proteins will still possess cationic backbones. Consequently these proteins will attract anionic dye coulombically, whereas dye-binding to other proteins which now have anionic backbones and to other anionic biopolymer chains such as nucleic acids and acid mucosubstances, will be hindered by Coulombic repulsions. The traditional acid dye-basic dye pairs (such as H & E, Azur-Eosin, Phloxine-Methylene Blue) used as general oversight stains involve a slight complication of this approach. Thus at some optimum pH, depending on such variables as the fixative and dye concentrations adopted, the acid dye binds to those substrates with predominantly cationic, immobile (fixed) ions, aided by coulombic attractions. However, the tissue components with anionic immobile ions are not stained, since Coulombic repulsions occur. Exactly the reverse holds for the basic dye, hopefully resulting in the crisp two-tone effect desired by histopathologists and teachers of histology.

A further variation on this theme is the selective staining achieved by a combination of Coulombic repulsions plus some non-Coulombic attractive process. Consider for example the staining of elastic fibres. Suitable dyes bind to the poorly ionic elastic fibres due to van der Waals attractions. Staining of the markedly ionic non-elastin proteins is minimised by Coulombic repulsions, due to the pHs adopted for the staining solutions, as is illustrated by Table 20. This has been discussed at greater length, with particular attention being paid to the effects of varying the structure of the dye, by HOROBIN and FLEMMING (1980).

For the final examples we will consider some effects of salt on affinity and so on staining selectivity.

Table 20. Ionic parameters of some elastic fibre staining systems

Dyes used to stain the elastic fibres	Nature of the dyebaths used	Ionic character of the backbones of common proteins in such dyebaths
Anionic dyes: e.g. Chlorazol Black E, Congo Red, and Direct Blue 152	alkaline	anionic
Cationic dyes: e. g. Aldehyde-Fuchsin, Orcein, and Resorcin-Fuchsin	acidic	cationic

As discussed in Section 4.4, Affinity Model 4, an ion-exchange model can be used to rationalise many salting on and salting off effects. This model predicts that addition of salt to a dyebath containing a basic dye will increase the affinity for acidophilic substrates; with analogous consequences for acid dyes. A case in point is Rosa's Aldehyde-Fuchsin procedure, the staining component for which may be preciptated from solution and stored in dry form. A staining solution is prepared by dissolving this, basic, dye in acidic aqueous alcohol. That the presence of electrolyte is required, rather than specifically of an acid, is shown by the observation that a solution made up in aqueous alcohol containing an equivalent amount of a neutral electrolyte gives almost identical results. However, omitting electrolyte altogether produces a useless staining solution which overstains various basophils but fails to stain the, weakly acidophilic, elastic fibres (HOROBIN, unpublished results).

A further type of salt effect currently attracts interest. Consider the effect of salt on the staining of groups of substrates of the same overall charge, that is acidophils or basophils. As the salt concentration is steadily raised, first one substrate will start (or stop) staining, then another, then another, and so on. As an example consider staining with basic dyes from acid dyebaths. Addition of salt characteristically greatly increases the staining of elastic fibres and red blood cells, has less effect for muscle, and rarely increases the staining of collagen. Effects in diminishing staining are most marked with substrates such as goblet-cell mucin, and much less with basophilic RNA-rich cytoplasms (HOROBIN and GOLDSTEIN, 1974).

One suggestion offered to explain such substrate-specificities is that they reflect the dye-tissue affinities in the salt-free situation. The Critical Electrolyte Concentration (CEC) theory is an example of this viewpoint, as discussed by SCOTT (1968). SCOTT regards such substrate-specificities as reflecting underlying specific ion effects. A basic dye cation is considered to interact most strongly with sulphate ions and least strongly with carboxylates, with phosphate ions occupying an intermediate position. For metal complex dyes, SCOTT regards the sequence to be more or less reversed. Tissue staining specificities are thus seen as dictated by the ionic character of the predominant tissue polyions, e.g. nuclear chromatin contains much phosphate in the form of DNA, cartilage matrix much sulphate in the form of chondroitin sulphate. Single-phase solutions of ionic solutes do indeed display such effects (PRUE et al., 1974). The published data from studies of two-phase model systems, e.g. of DNA smeared onto paper and immersed in solutions of basic dyes, are also considered by SCOTT to support the specific-ion theory (SCOTT, 1967; SCOTT and DORLING, 1965; SCOTT and WILLETT, 1966). The present author has, however, failed to see any convincing *overall* correlations between the published CEC values derived from the 2-phase models and the staining of tissue sections. Noting that the CEC theory assumes that relative salt inhibition patterns should correlate with relative affinity patterns for the salt-free case, MARSHALL and HOROBIN (1973b) made quantitative measurements of dye-substrate affinity using stained sections. They studied a wide range of basic and metal complex dyes, but overall failed to find any convincing correlation between published CEC values and quantitative affinity values. It is interesting that the model system derived CEC values for Alcian Blue, undoubtedly the dye most widely used in CEC work, do (in my eyes anomalously) correlate with the staining of tissue sections.

How then to explain the observed substrate differences? One possibility lies in the fact that the salt sensitivities appear to parallel the porosities of the substrates (for a discussion of the nature and assessment of tissue porosity, see Sections 4.11–4.13). Thus the porous substrates, for example goblet-cell mucin and collagen fibres, are usually the first to salt-off and the last to salt-on. The impermeable substrates such as elastic fibres and basophilic RNA-rich cytoplasms are on the other hand the last to salt-off and the first to salt-on. This is very reminiscent of the situation which arises with ion-exchange chromatography carried out on sulphonated polystyrene resins, i. e. on cation exchange resins. In this chromatographic situation the more cross-linked the polystyrene (i. e. the less able to swell) the greater the selectivity of the resin. Thus if a mixture of lithium, sodium and potassium chlorides is run down a column of weakly cross-linked, highly swollen resin then all the cations emerge together without any separation having occurred. On the other hand if the mixture is run down a column of a highly crosslinked, poorly swollen resin (of otherwise similar character) then the different cations of the mixture will be separated and will emerge from the column sequentially (HELFFERICH, 1962). If we make an analogy between the cross-linking of the resins with the porosity of the tissue substrates, then we can anticipate the following: when staining tissue sections the selectivity for the preferred species (i. e. the dye ion) rises as the tissue porosity decreases. As stated previously this is what is seen to occur. Though a detailed explanation of such effects is no doubt extremely complex, a plausible qualitative sketch can be offered. Namely, as the porosity decreases, dye and inorganic electrolyte are forced closer to the substrate. Then the short range non-coulombic forces, characteristically strong for the dye ions, will become relatively more important.

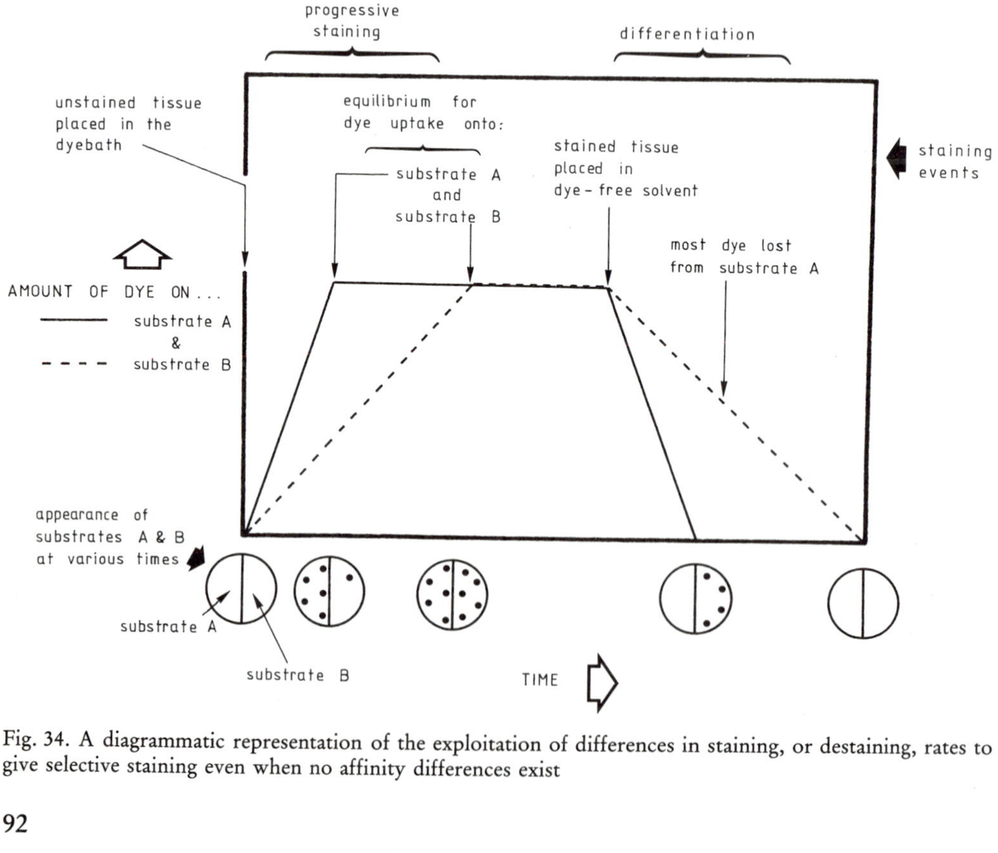

Fig. 34. A diagrammatic representation of the exploitation of differences in staining, or destaining, rates to give selective staining even when no affinity differences exist

4.10 Some substrates stain or destain faster than others

Up to this point it has been assumed that the staining systems are at, or at least approach, equilibrium. So our attention has been focussed on affinities and differences between affinities, and on the number of dye-binding sites. Now imagine staining two substrates whose affinities for a given dye are equal. Excluding density differences (see Section 4.8) no selective staining will occur at equilibrium. However, if the two substrates stain or destain at different rates then selective staining can still be achieved, as long as equilibrium is not approached too closely. This is shown diagrammatically in Fig. 34.

In real situations the affinities of the different sites will usually differ. So selectivity due to rate of staining differences may in practice assist or oppose differences due to affinity. A further complication of practical procedures is that one stage of a multi-step staining procedure may be affinity controlled and the next rate controlled. For the sake of simplicity of exposition such issues will be largely avoided in this and the following Section.

4.11 What are the causes of differential staining rates?

Though the concept of selective staining due to differential staining rates has a long history in the biological staining literature (for instance EHRLICH, 1879; MANN, 1902; HEIDENHAIN, 1903; PAPPENHEIM, 1918; MÖLLENDORFF and MÖLLENDORFF, 1924; SEKI, 1932, 1933; SINGER, 1952; BAKER, 1958; GOLDSTEIN, 1963) little quantitative work has been carried out. Consequently the following remarks will be largely based on the many studies carried out in the textile dyeing and ion-exchange fields. General accounts include BIRD and BOSTON (1975), RATTEE and BREUER (1974, RYS and ZOLLINGER (1972), VICKERSTAFF (1954), HELFFERICH (1962) and KITCHENER (1961), the latter two considering ion-exchange chromatography.

Consider a dye molecule moving from solution into a solid substrate, perhaps tissue section, or textile fibre or ion-exchange resin particle. Any of the following steps could occur, and could take place *slowly* compared to the others:

1. Diffusion through the bulk solution of the dyebath,
2. Diffusion through the thin boundary layer surrounding the solid substrate,
3. Adsorption onto the surface of the substrate,
4. Diffusion through solvent-filled micropores inside the substrate,
5. Adsorption onto the surface of the pores,
6. Diffusion through solvent-free substrate,
7. Adsorption onto the substrate.

This (or these) rate controlling step(s) would then greatly influence the overall dyeing rate. Of course it will not be necessary to apply all the steps of this solvent-filled micropore model to any particular staining system. For example several textile dyeing systems, including large acid dyes being adsorbed onto cellulose and small acid dyes onto polyamides, have been shown to be satisfactorily described by steps 1 to 5 (WEISZ and ZOLLINGER, 1968). Note also that when staining with acid and basic dyes the flow of dye ions onto the substrate must be balanced by the flow of gegenions out, or *vice versa* if a destaining process is occurring. So the slowest step could be one not involving a coloured species.

A note on the concept of micropores is pertinent. In biological tissues at least these will not comprise sharp sided holes, filled with solvent, passing through a completely solvent-free solid substrate. However, the reality, gaps in a flexible, hydrated, three dimensional molecular mesh, is adequately described by the micropore model for our purpose.

In many systems, though see below, it is the rate of diffusion through the micropores which is the slow step. Fick's diffusion law, below, can then be applied to this step:

$$ds/dt = D\, dc/dt$$

(where ds/dt is the rate of diffusion of dye across unit area of substrate, D is the diffusivity and dc/dt is the concentration gradient inside the substrate).

The diffusivity, D, is numerically equal to the amount of dye diffusing in unit time across unit area of the substrate under a unit concentration gradient, and so offers a convenient comparative measure of the behaviour of different dyes and substrates. Various practical factors influence the diffusivity. For instance if other factors are equal, then the more numerous, the greater in diameter, and the less tortuous the micropores the greater the value of D. Similarly the smaller the size and lower the affinity of a dye the greater diffusivity in a given substrate. An increase in the temperature also usually increases diffusivity.

Although a high dye-tissue affinity will be expected to lower the diffusivity there are circumstances where such an effect may be masked. Thus if the affinity of a dye for a substrate is very high then the surface layer of that substrate will rapidly acquire an intense colouration. This effect, known to the textile dyers as «strike», has also been seen in biological staining systems, for instance the staining of cartilage matrix with Alcian Blue. Though the staining rate of the bulk of the cartilage matrix has been found to be slow, the surface rapidly acquires a heavy dye deposit (GOLDSTEIN and HOROBIN, 1974 b). Consequently cartilage matrix is one of the first structures to be strongly stained with Alcian Blue.

The surface area:volume ratio of the substrate is another real world factor complicating the diffusion-in-micropores model. For example the rates of staining of textile fibres obtained from different sources may differ. When allowance is made for the differing surface area: volume ratios of the various fibres however the differences in rates are diminished, as described for cellulose and keratin fibres by VICKERSTAFF (1954, p. 277, 416).

Staining systems where diffusion in pores is not the important determinant of overall staining rate are also known. A common case is if the diffusion across the boundary layer is slow, due to depletion of dye from the immediate vicinity of the substrate. Thus both with textile fibres (PETERS, R. H., 1968) and tissue sections (GOLDSTEIN, 1963) vigorous stirring of the dyebath has been found to increase staining rate. Such effects, currently largely ignored by histochemists, may be of considerable significance in biological staining where unstirred dyebaths, low temperatures and the complete enclosure of certain substrates inside others are commonplace. Another probable non-pore system is the staining of lipid droplets with non-ionic dyes. Here diffusion of dye presumably occurs through the, usually liquid, substrate itself.

In practice therefore the dyeing rate of any particular substrate could depend on its accessibility and surface area as well as the diffusivity of dye inside it. Consequently we require a straightforward descriptive measure of the rate of staining of various substrates with various dyes, not involving commitment to any particular theoretical model. Plots of dye taken up onto the tissue against time, as shown in Fig. 35, do provide a complete description of the rate of adsorption of a dye under particular conditions of temperature, concentration and so on. Such curves may be used to gain an idea of the staining properties of different dyes and substrates. The use of curves is rather inconvenient, and some more concise mode of description and characterisation is desirable.

Staining curves of the shape illustrated, namely with an initial fast reaction slowing as equilibrium is approached, are typical and may be described in terms of two factors, namely the equilibrium adsorption and the rate at which equilibrium is reached. A common method of giving a numerical value to this rate, involving no mechanistic assumptions, is to determine the time of half-staining ($t_{1/2}$). This is the time required for a substrate to absorb half as much dye as it will contain in the equilibrium state, see Fig. 35. Practical methods of determining half staining times when dyeing tissue sections are sketched in Supplement 9.

A clarification of the term «rate of dyeing» may be useful here. Dyes X and Y, in Fig. 35, are said to dye at equal rates but to have unequal equilibrium adsorptions. The practical histologist examining sections stained with these two dyes at an early stage in the dyeing is however likely to say that dye X stains more rapidly than dye Y. Of course if dyeing rate is defined as the amount of dye adsorbed in the first t minutes this is quite true; we need to be very aware of definitions.

Rates of dyeing are markedly dependent on temperature. Many complications exist here, the commonist being that affinity is also dependent on temperature. As shown for an idealised case in Fig. 36, a temperature rise commonly causes an increase in staining rate simultaneously with a decrease in affinity.

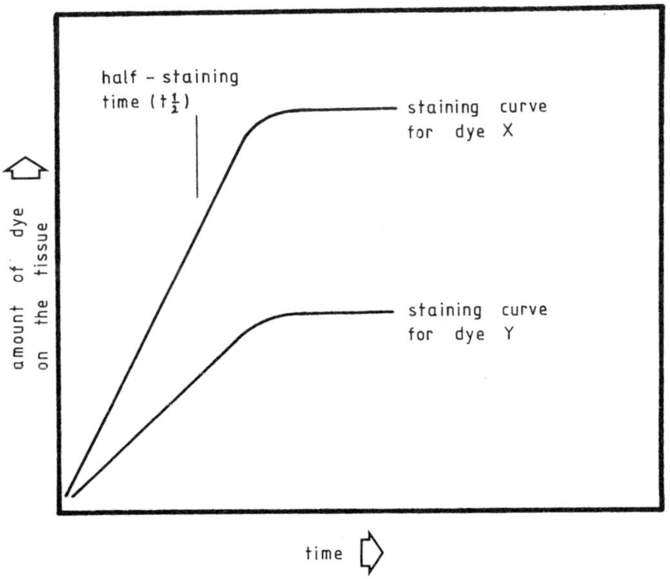

Fig. 35. Adsorption-time plots for dyes having equal half staining times ($t_{1/2}$ values) but differing equilibrium adsorptions. For an example of this effect see GOLDSTEIN (1963, Figs. 1–3)

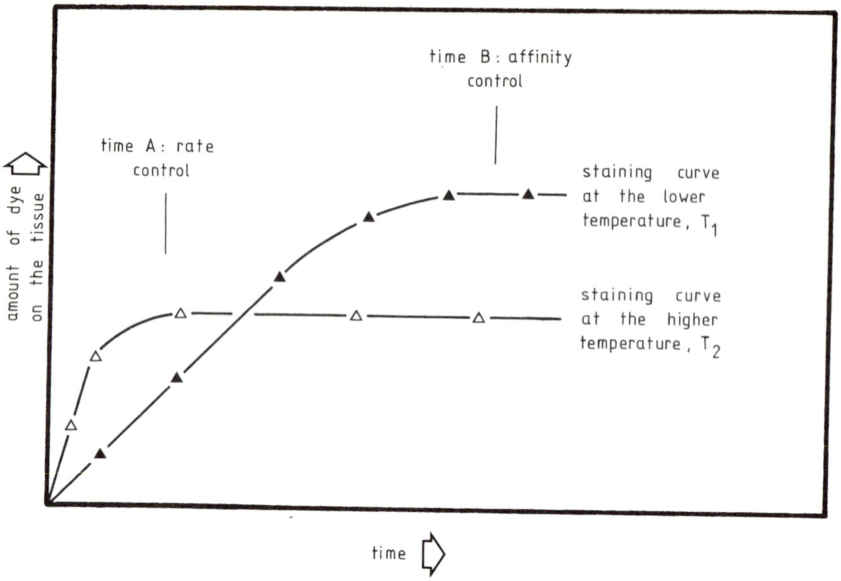

Fig. 36. Some common effects of temperature on dyeing rates and equilibria

The potential practical significance of this arises from the fact that biological stainings are carried out for finite times. So when using a slow staining dye the process may proceed, at the different temperatures, as far as line A in Fig. 36. Maximum dye uptake is at the higher temperature. In this case staining is under rate control. But when using a fast staining dye, for the same time period, staining may proceed as far as line B. So in this case maximum dye uptake will be at the lower temperature, and this staining process is affinity controlled.

A quantitative measure of the effect of temperature on dyeing rate may be obtained by determining the *activation energy* (E) of the process. This quantity is defined by the Arrhenius equation so:

$$\ln k = \ln A - E/RT$$

(where k is the specific rate of the process, A is a reaction constant, R the general gas constant, T the absolute temperature and E the activation energy).

The activation energy is a reaction constant which may be regarded as the energy required by a dye molecule to enter a transient transition state necessary for the reaction to occur. But what reaction? Although a rise in temperature might affect the staining rate in various ways (for example by altering the dye-tissue affinity, the swelling of the tissues, the state of aggregation of the dye or even the chemical content of the tissue and chemical nature of the dye) it has been often concluded that the dominant contribution to the activation energy is the process of diffusion inside the substrate (VICKERSTAFF, 1954; GOLDSTEIN, 1963; PETERS, R. H., 1968). The determination of activation energies for the dyeing of tissue sections is discussed in Supplement 9.

4.12 How is the rate of staining affected by the dye used?

In the absence of complicating factors the rate of diffusion of a dye molecule will be directly related to its size. Thus imagining for the moment that dyes are non-aggregating, spherical particles having no affinity for the medium through which they move, then their rates of diffusion would be proportional to (dye molecular weight) ½. The molecular weights of dyes used as biological stains do indeed vary quite widely. Commonly used acid dyes range from picric acid (with a molecular weight of 229) to Sirius Red F3B (1373), and basic dyes from Neutral Red (289) to Alcian Blue 8GS (1346). For a visual demonstration of the range of dye sizes see Fig. 16 and 17. As some indication of the effects of size differences consider Table 21. This shows the ease of penetration of certain acid dyes into agar gel, as measured by SEKI (1932).

Table 21. Penetration of certain acid dyes into agar gel

Dye	Approximate molecular weight	Penetration (mm) in fifteen hours
Picric acid	229	25
Acid Fuchsin	586	10
Nigrosin W	646	7
Methyl Blue	800	5

Though in histochemical practice the imaginary case noted in the opening paragraph above is too simple, the complications which do occur often only serve to accentuate the little dye-fast staining, big dye-slow staining trend. The value of such simple quantities as molecular weight is discussed in Section 11.14.

As discussed in Section 4.11 the rate of diffusion will tend to decrease as the dye-tissue affinity rises. Since in general affinity rises with increasing dye size, see Sections 4.4 and 11.14, the larger dyes will also tend to stain more slowly due to this effect.

Another factor affecting staining rate, again often related to dye size, is the aggregation of dyes in solution. An outline of the physical chemistry of this process is given in Supplement 10. Since diffusion rate is so dependent on dye size, aggregation of an appreciable proportion of the dye even to dimeric form will be expected to diminish the rate of staining. When the aggregates formed are large micelles few substrates will be stainable by them. Staining occurring will be due to the remaining non-micellised dye, so the dyeing rate will fall markedly.

As discussed in Supplement 10 and illustrated, in part, in Table 22, aggregation is favoured by large dye size; high dye concentrations; high ionic strengths; the absence of structure breaking co-solutes (such as urea or ethanol) or absence of an excess of structure making co-solutes (such as aniline or phenol); and low temperatures. However if conditions are suitable most types of dye *can* aggregate, regardless of charge, size and structural class. Biological staining solutions probably often contain highly aggregated dye micelles. It is consequently of interest to note the nature of «penetrating» stains, such as those traditionally used for mitochondria. These stains use hot dyebaths, which are either non-aqueous or contain co-solutes such as aniline or phenol.

A final point concerns the highly variable rates of the aggregation and disaggregation reactions. Sometimes they are extremely fast. For instance consider diluting a concentrated metachromatic solution, containing a high proportion of dimers, of a dye

Table 22. The aggregation, under various conditions, of typical dyes used as biological stains

Acridine Red and Rhodamine 6G are both cationic xanthene dyes, Rhodamine 6G being the larger by an extra $-C_6H_4COOC_2H_5$ grouping. The dyes were studied spectroscopically over a wide range of solvent, temperature and dye concentration by SELWYN and STEINFELT (1972). The following conclusions were reached concerning the situation at room temperature:

approximate concentration of the dye	solvent	dye species present
Acridine Red		
up to 3×10^{-5} M	water	monomer only
Rhodamine 6G		
up to 7×10^{-4} M	water	monomer and dimer only
up to 2×10^{-4} M	ethanol	monomer only

As the temperature was lowered both dyes became predominantly dimeric

Benzopurpurine 4B is a disulphonated, bisazo dye. The aggregation of this dye was studied by light scattering (ALEXANDER and STACEY, 1952). The following information, obtained at a dye concentration of ca. 7×10^{-5} M in water, is typical:

concentration of NaCl	Average number of dye ions per dye particle At 20 C	At 40 C	At 60 C
0.000 N	7	7	7
0.010 N	40	7	7
0.025 N	70	33	7
0.050 N	1400	40	7
0.075 N	precipitated		11

such as Toluidine Blue or Methylene Blue. The orthochromatic colour, indicative of the presence of a high proportion of monomers, is produced immediately. However in other situations the rate of formation of aggregates from the monomeric form and *vice versa* has been found to take hours or even days to reach equilibrium (COATES, 1969). A study on the aggregation of various biological stains is pertinent here (BENNION and HOROBIN, 1974b). Dyes were passed down columns of gel filtration media, through which monomeric or dimeric dyes would pass slowly but micellar aggregates would travel rapidly. Many small dyes, such as Azur A and Eosin Y, were found to move slowly down the column. Larger dyes (such as Alcian Blue 8GS, Crystal Violet, Supranol Bordeaux B) and some smaller ones (for instance Gallocyanin and Pyronin Y), especially if at high concentration, contained a second fast moving fraction, totally excluded from the gel. However as these micelles, probably containing many dozens of dye molecules, passed down the column they dissaggregated within a few minutes. No dyes forming kinetically stable micelles were observed. Such kinetic lability means that dye-tissue affinities will not be greatly reduced by such aggregation. However rates of staining, highly dependent on the instantaneous concentration of small dye particles, will be depressed. If dye micelles are formed whose rate of formation or dissociation is slow, then complications may arise from the consequent changing nature of the staining solution. Probably examples of such include the need to let Methyl Green-Pyronin staining solutions age before use if diluting from concentrated, and highly aggregated, stock solutions (MITCHELL, cited by GURR, 1973), and the increased blue (i. e. the colour of the aggregated dye) background staining occurring with old solutions of the Direct Blue 152 elastic fibre stain (HOROBIN and JAMES, 1970).

The comments so far have mostly related directly to dye size, or to the size of the dye's conjugated electronic sytem. *The charge on a dye also influences staining rate.* A high dye charge will tend to reduce the affinity of that dye for a substrate compared to a similar dye carrying a lower charge if ion-exchange is occurring, and also by reducing the hydrophobic character. Affinities between dye molecules will also be reduced, so reducing aggregation. So in general one can expect dyes of high charge to stain faster than dyes of similar size but lower charge.

4.13 How do the various tissue substrates influence staining rates?

Initially we will consider what appear to be the only major quantitative studies, the measurement of half staining times and activation energies for a variety of basophilic substrates when staining with Azur A and with Orcein (GOLDSTEIN, 1963; FRIEDBERG and GOLDSTEIN, 1969). The activation energies obtained from these studies are shown in Table 23, together with information pertaining to the porosity of the substrates obtained in another study by the same author (GOLDSTEIN, 1965).

Table 23. A listing of the activation energies of staining and of two porosity indicators for various substrates, obtained when staining with two basic dyes

	decreasing order of		Activation energies (kJ mole^{-1})	
Substrate	refractive index	effective thickness	Azur A (0.008% w/v in pH 4 aqueous solution)	Orcein (0.2% w/v in 70:30 v/v ethanol-water plus 1 v concentrated HCl)
Elastic fibres		3		98.7
Cartilage matrix (interstitial)	1	1	37.8	54.2
RNA-rich cytoplasm (pancreatic)	2	4	27.7	
Chromatin	3		26.0	37.8
Cartilage matrix (perilacunar)	3	2		18.9
Goblet cell mucin (gut)	4	5	5.9	

The two measures of porosity adopted were refractive index and effective thickness. The refractive index of fixed tissue components has been considered to be a function of their physical density (BARER and JOSEPH, 1954; DAVIES et al., 1954) and the effective thickness is the thickness of the solid material in the substrate when all the interstices filled with immersion medium are eliminated. If staining rate is determined either by diffusion inside the substrate, or by the cruder porosity corresponding to larger ultrastructural pores, then these measures of porosity will be expected to parallel the staining rates. Table 23 shows that structures with high refractive indices and effective thicknesses indeed do tend to stain slowly. Contrariwise, low refractive indices and effective thicknesses correlate with faster staining rates. Note also that calcified cartilage has an extremely high refractive index, whereas those of readily hydrated and so swollen polysaccharide-rich substrates such as cellulose, glycogen and mucins are low.

No comparably detailed and rigorous studies have been reported for the staining of acidophilic substrates. A few preliminary half staining times (HOROBIN, 1974) have been reported for various acidophilic substrates when staining with Biebrich Scarlet. For these see Table 24 in which are also displayed relevant porosity data. As with the staining of basophilic substrates the staining rates are paralleled by the porosity indicator. It is reassuring that the relative staining rates (namely collagen fibres > smooth muscle > red blood cells) are as expected from qualitative observations with other acid dyes, for example as discussed by BAKER (1958, p. 234).

Table 24. A listing of the half staining times and of a porosity indicator for various acidophilic substrates

Substrate	decreasing order of... refractive index[a]	half staining times[b]
red blood cells	1	1
pancreatic zymogen granules	1	2
smooth muscle	2	3
collagen fibres	3	4

[a] taken from GOLDSTEIN (1965)

[b] Results obtained using a 0.01% w/v solution of Biebrich Scarlet in aqueous solution buffered to pH 4 by acetate-acetic acid of ionic strength 0.05, with tissue fixed in neutral formalin (HOROBIN, 1974)

Most of the rate data considered so far has concerned Coulombically favoured situations, acid dyes onto acidophils and basic dyes onto basophils. When Coulombically unfavourable staining occurs the rates are apparently much slower. This has been seen with the staining of acidophilic substrates in tissue sections by salty solutions of cationic metal complex dyes (MARSHALL and HOROBIN, 1973c); and also with the staining of keratin with acid dyes. In the latter instance the half staining times using acidic and neutral baths were hours and days respectively (MEYBECK and GALAFASSI, 1971).

Since differentiation, that is controlled destaining, is such a widely used technique *the variation of destaining rates with different substrates* is an important variable. Once again GOLDSTEIN (1963) appears to have carried out the only quantitative study, using tissue sections stained in Azur A and destained in pH 4.0 buffer solution. The relative destaining rates of various substrates in fact followed the same sequence as the staining rates: the slowest was interstitial cartilage matrix, then RNA-rich cytoplasm, then chromatin, and the fastest destaining was of mucin. This first in-first out pattern is what is expected if diffusion through the substrate or across the Nernst boundary layer are rate-determining processes; or of course if surface area is critical.

4.14 How does substrate geometry influence staining rate?

It was noted earlier that in textile dyeing the surface area of the substrate had an important influence on staining rates. Other things being equal the greater the ratio of surface area to volume, the faster was the staining. The only definite evidence from biological staining for this kind of effect arises from fixation studies, see Section 4.16 below. Apart from this, GOLDSTEIN (1980) has plotted staining rate data on the basis of the substrate being either rod or plate shaped using the method of Crank and found an imprecise fit with either geometry. However it is at least suggestive that the fast staining basophilic and acidophilic substrates, goblet-cell mucins and collagen fibres, are both ultrastructurally finely divided, into spherical particles and fibrils respectively.

Another important geometrical factor, albeit an artifactual one, is *section thickness.* For slow staining-slow destaining substrates direct accessibility to the dyebath may markedly influence staining rate. This is most clearly apparent when staining small structures capable of being completely enclosed inside a tissue section. Thus LISON (1955) clearly showed that, in tissue sections, those cell nuclei slow to destain after iron-haematoxylin treatment, and also those slow to stain with a large dye, were in fact those nuclei completely inside the tissue section. Nuclei split by the microtome stained and destained more rapidly, and the thinner the section the more such nuclei there are. In this way were generated the alleged «two types of nuclei» reported by earlier (and later? see BLOCK, 1966; KHAN, 1978) workers. An analogous explantation could lie behind the two types of erythrocyte reported by BJERRUM and JOHNSEN (1971) and the two types of mitochondria reported by NOVELLI (1962). A similar effect probably underlies the marked influence of section thickness when staining renal basement membranes by the Luxol Fast Blue-Phloxine method of TAKAYA (1967).

4.15 How do solvent and co-solute affect staining rates?

Additives to dyebaths mentioned in the biological staining literature are legion. Those commonly used divide conveniently into organic solvents (usually ethanol but sometimes cellosolve, dimethylformamide, acetic and formic acids, and so on), organic co-solutes (small and aromatic, such as aniline and phenol; or large biopolymers such as

Table 25. Effects of some dyebath additives on the degree of aggregation of dyes

Type of additive	Dye(s) studied	Reference
Organic solvents which decrease dye aggregation		
Dimethylformamide, dimethylsulphoxide and pyridine	Alcian Blue 8GX	SCOTT (1970)
Ethanol	Acridine Red, Rhodamine B, Rhodamine 6G	SELWYN and STEINFELD (1972)
Formamide	Congo Red, Methylene Blue, Rhodamine B, Thionin	WAGNER (1969)
Organic co-solutes which decrease dye aggregation		
Phenol, urea	Benzopurpurin 4B	ALEXANDER and STACEY (1952)
Urea	Methylene Blue	MUCKERJEE and GHOSH (1963)
Electrolytes, if not of large size or of surfactant properties, increase dye aggregation, see Table 21 and below		
Sodium chloride	Methylene Blue	MUCKERJEE and GHOSH (1970)
Sodium chloride	Chicago Blue 6B	VALKO (1939)
Magnesium chloride	Alcian Blue 8GX	SCOTT (1970)
Surfactants, ionic and non-ionic, do tend to decrease dye-dye aggregation but simultaneously cause dye-surfactant aggregation		
For a review of this topic, discussing several dye-surfactant systems, see RATTEE and BREUER (1974, p. 133)		

dextrin and polyvinylalcohol), electrolytes (neutral salts, acids and bases, heteropoly-acids), and occasionally surfactants.

No direct study on the effects of such additives on rates of staining appears to have been carried out in the biological staining context. From considering Section 4.12 and 4.13 however it can be seen that if the additives either cause dye aggregation or shrinkage of tissue then staining rates will tend to fall, and vice versa. Some guidelines are offered in the form of Tables 25 and 26, where examples of the effects of additives on dye aggregation and also on tissue swelling and shrinkage are shown. Once again the information available does not in general refer directly to biological staining solutions or substrates. Furthermore the influences of a solvent or co-solute on aggregation and swelling are complex. Thus phenols swell tissues and disaggregate dyes, whereas alcohols also disaggregate dyes but shrink tissues, and so on. Rate measurements on biological materials would be very useful here.

The dye-tissue affinity is another factor of kinetic significance which may be altered by dyebath additives. In general, as discussed in Chapter 4–11, an increase in affinity will be expected to decrease the rate of diffusion through the substrate. The overall rate of staining may nevertheless be increased. In the absence of direct studies on biological staining systems such complications block further comment.

Table 26. Effects of some dyebath additives on the swelling and shrinking of tissues and tissue components

Type of additive[a]	Observations	Reference
Organic solvents		
Acetic and formic acids	swell unfixed collagen and elastic fibres, silk fibroin	JORDAN LLOYD et al. (1946), GUSTAVSON (1949)
Formic acid	swells fixed tissues	HOROBIN (unpublished observations)
Ethanol, methanol	shrink unfixed collagen and elastic fibres	JORDAN LLOYD et al. (1946)
Formamide	swells unfixed collagen and elastic fibres	JORDAN LLOYD et al. (1946)
Organic co-solutes		
Phenol and cresol	swell unfixed collagen and elastic fibres	GUSTAVSON (1949), JORDAN LLOYD et al. (1946)
Urea	swells unfixed collagen fibres	GUSTAVSON (1949)
Electrolytes		
Various neutral salts	complex results with unfixed collagen fibres: salts often initially produce swelling but, at high enough concentration, shrinkage	Reviewed by VEIS (1967)
extremes of pH	cause swelling of unfixed collagen fibres, and of fixed liver	GUSTAVSON (1949), HOROBIN (unpublished results) respectively
Surfactants		
Anionic	shrink cellulose fibres	EVANS (1958)

[a] Note that changes in temperature can cause volume changes, hot water for instance swells collagen fibres and shrinks reticular membranes (JORDAN LLOYD et al., 1946)

4.16 How do pre-staining treatments of the tissue affect staining rates?

Under this heading are discussed effects of fixatives, histochemical blocking agents, decalcification solutions, «step one» in multistep methods, and so on. Some information presented here derives from direct rate studies of biological staining systems, some from direct studies of relevant textile systems. Some indirect information derived from observations made on staining systems involving mixtures of large and small dyes is also useful, see Section 4.14. Since the larger dye of such a pair will often diffuse more slowly, processes giving rise to increases in staining rates will tend to produce excessive colouration by the larger dye, and *vice versa.* Some relevant background information on the effects of various reagents on the swelling or shrinkage of tissues and tissue components is provided, under usage headings starting with:

Fixatives and staining rates. A direct study of the staining rates of tissues with Methylene Blue after fixation with a variety of fluids has been carried out (HOROBIN and FLEMMING, previously unpublished results). The indirect assessments of rate by the big dye-little dye criterion has also been made for the same fixatives, using acid dye and

Table 27. Some examples of the influences of fixation on staining rates[a]

Fixative fluid used	Rate of staining
Based on half staining times for various tissues with Methylene Blue	
Bouin	fast
Ethanol	
Carnoy	intermediate
Neutral formalin	
Susa	
Zenker	slow
Based on big dye-little dye staining with Light Green SF-Chromotrope 2R	
Acetone	fast
Carnoy	
Ethanol	
Bouin	intermediate
Neutral formalin	
Gendre	
Susa	
Zenker	slow
Based on big dye-little dye staining with Methyl Green-Pyronin Y	
Bouin	fast
Carnoy	
Ethanol	
Acetone	intermediate
Neutral formalin	
Gendre	
Susa	
Zenker	slow

[a] Previously unpublished results of the present author, plus observations of GOMORI (1950), and of SWEAT et al. (1968)

Table 28. Examples of increases in staining rates following treatment with blocking or extracting reagents

Procedure	Observation	
Treatment of fixed tissues by HCl-methanol methylation and also alkaline desaponification	After both operations, an Aniline Blue-Orange G combination produces an over-blue colouration	BULMER (1962)
Treatment of keratin with alkali (c. f. saponification methods) or with chlorine (c. f. hypochlorite deamination procedures)	The staining rates of acid dyes are increased	LEMIN and VICKERSTAFF (1946)
Treatment of fixed tissues with β-glucuronidase and also with hyaluronidase	Staining of nuclei with Alcian Blue occurs rapidly	SALTHOUSE (1961)
Treatment of skin with collagenase	Staining with van Gieson produces a generalised over-red colouration	MURGATROYD (personal communication)

basic dye pairs (HOROBIN, unpublished observations). The conclusions are outlined in Table 27. The overall similarity in these three sequences is remarkable, the more so since acid and basic dyes will be binding to different molecular substrates. On comparing the sequences to the information concerning the ultrastructural morphology produced by these fixatives (see Sections 3.2 and 3.8) it seems possible that the ultrastructural porosity, effectively the surface area to volume ratio of the tissue, is an important rate controlling factor.

As some indication of possible additional complications, consider that the treatment of keratin with dichromates (WHEWELL et al., 1971) decreased the rate of basic dyeing, compare Table 27, but increased the rate of acid dyeing.

Reagents used for histochemical blocking and extraction procedures can also alter rates of staining, as is indicated by scattered observations on big dye-little dye staining systems, examples of which are given in Table 28. Only effects explicable on the basis of increases in rate are seen, perhaps due to non-specific effects of the various regents causing swelling, damage, and gross loss of tissue components. In keeping with this are the observed swellings produced by various reagents, see Table 29.

In multi-stage staining methods the *early steps* may influence the rate of subsequent staining, sometimes in a complex manner. For instance various oxidants in histochemical usage, such as chlorine and hydrogen peroxide, increase the rate of staining of keratin by acid and basic dyes (LEMIN and VICKERSTAFF, 1946; WHEWELL et al., 1971). On the other hand periodic acid was found to decrease the acid dyeing rate of keratin

Table 29. Examples of swelling of tissues or tissue components caused by reagents used in histochemical blocking procedures

Reagents	Observations	
Acetic acid (for instance used as the solvent in aldehyde blockades) and also thioglycollate (used as a thiol blockade reagent)	Both reagents cause swelling of unfixed collagen and elastic fibres	JORDAN LLOYD et al. (1946)
Alkali (used in saponification), and HCl-methanol	Causes swelling of fixed collagen fibres	VEIS (1967)
Alkali, and also HCl-methanol	Both cause swelling of fixed liver	HOROBIN (unpublished results)

whilst still increasing the basic dyeing rate, and potassium permanganate had the opposite results (WHEWELL et al., 1971).

Various other reagents, such as those used for acid decalcification and removal of mercury pigments, swell tissue components (see GUSTAVSON, 1949; WHEWELL et al., 1971). Furthermore a pre-staining treatment of fixed tissue sections with formic acid results in an increased rate of acid dyeing and an over blue colouration when staining by the picro-Sirius Blue technique compared to untreated sections (HOROBIN, unpublished observations). As a final example note that urea, present during certain staining procedures, can swell proteins and accelerate rates of dyeing (KILPATRICK and RATTEE, 1977; ASQUITH et al., 1978).

4.17 A summary of the causes of differential rates of staining

Inspection of Fig. 37 emphasises that practical staining variables such as fixation, microtomy, dyebath character, various pre-staining treatments of tissue may be regarded as influencing a limited number of dyebath and/or substrate properties. The wide variety of rate dependant staining patterns which are seen can be illuminated by consideration of these properties.

4.18 Various examples of rate-controlled selective staining procedures

Using the information, both concrete and tentative, described previously, a series of sketches of various common rate-controlled staining patterns will be given. These rate models start with staining solutions containing single dyes, and then consider dye mixtures and simple dye sequences. Both progressive staining and differentiation will be discussed, using examples involving both acid and basic dyes. Though some practical staining methods link several such unit steps together such added complications will not be considered here, except for the brief sketch of sequence staining in Rate Model 4.

Rate Model 1:
Use of a single dye, with progressive staining.

When a tissue section is stained from a dyebath containing only a single dye, one could expect (density factors aside, see Section 4.8) initial staining to be predominantly of the fast staining substrates. This is shown diagrammatically in Fig. 34. In such a way one might hope to achieve, for example, selective staining of collagen fibres and mucin

Table 30. Sizes of some traditional stains for mucins and collagen fibres

Stain	Stain ionic weight
Mucin stains	
Alcian Blue	1204
Southgate's mucicarmine	1009
Collagen stains	
Mallory's phosphomolybdic acid-hematein	>696
Sirius Red F3B	1235

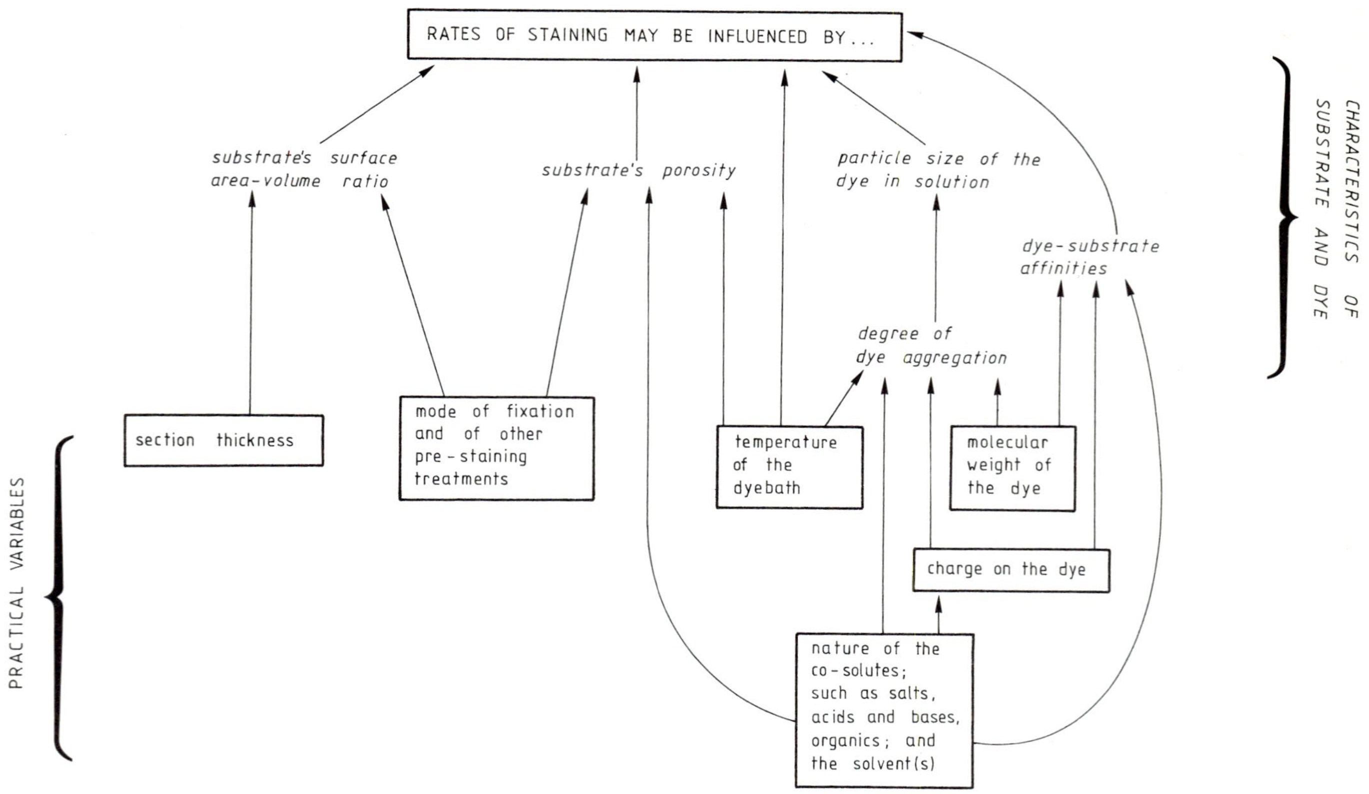

Fig. 37. Some origins of differential staining rates

granules (see Tables 23 and 24), the fastest staining acidophilic and basophilic substrates. This differential effect should be most conveniently exploited with slow diffusing dyes, that is those of large size and/or high degrees of aggregation. That this is indeed the case is shown in Table 30.

If tissues are stained with smaller acid and basic dyes then such differential staining is not easily achieved and a wide variety of substrates are quickly coloured. These effects have been discussed in some detail for the case of mucin stains by Goldstein (1962b), and are further discussed, in terms of structure-staining correlations, in Section 11.14. If long staining times are adopted then additional structures will stain even when using large dyes. For instance, although with short staining times an acid solution of Alcian Blue is quite a selective stain for mucins, if longer times are used additional structures such as cell nuclei and basophilic cytoplasms also stain well (see also Goldstein and Horobin, 1974a; Scott and Dorling, 1965).

Rate Model 2:
Use of a single dye, with controlled destaining.

Controlled destaining or *differentiation*, a valuable and indeed time honoured method of obtaining selective staining, is illustrated diagrammatically in Fig. 34. Considering the discussion in Section 4.13 one may expect that the rates of destaining from various substrates will parallel the rates of staining of the same substrates. Consequently (see Tables 23 and 24) if a preparation is overstained and then differentiated in a controlled manner selective colouration will be expected of slow-destaining structures such as elastic fibres and red blood cells, basophilic cytoplasms and nucleoli. These effects will be most marked when using large, and usually slow diffusing dyes. With such reagents the achievement of the initial overstaining may however require the use of agents such as heat or organic solvents to facilitate disaggregation of dye and swelling of tissue. In keeping with such expectations the traditional differentiation-based staining methods for structures such as mitochondria, elastic fibres and myelin sheaths often give background staining of the other slow staining substrates, for instance of keratin, nucleoli, red blood cells and zymogen granules. More detailed discussions of some of these methods may be found in Baker (1958, p. 241), Friedberg and Goldstein (1969), and Thompson (1966, p. 849).

Rate Model 3:
Use of a mixture of dyes, with progressive staining.

Consider preparations stained with a mixture of dyes of the same charge but of markedly different sizes, and so probably of different staining rates. In the early stages of staining at least it would be expected that whilst the fast staining tissue substrates would take up both the large and the small dyes, slow staining substrates would be predominantly coloured by the small fast staining dye. Various biological staining methods involving both acid dyes (for instance of the trichrome type) and basic dyes (the Methyl Green-Pyronin stain and its variants for DNA-RNA) may be rationalised by Rate Model 3 and will be discussed.

Of the many polychrome mixtures used to give selective staining of collagenous connective tissue fibres and other structures, one large group makes use of a single dyebath containing two or more acid dyes of different molecular weights. Taking the information in Table 24 into account (see also Baker, 1958, p. 241) one might expect that fast staining components such as collagen fibres would be coloured by the larger dye of the pair whereas slow staining substrates such as red blood cells will be coloured predominantly be the smaller dye. As Table 31 shows this is commonly so. A dramatic diagrammatic demonstration of this effect, and also of various complications and limits to Rate Model 3, is provided in the discussion of structure-staining correlations in Section 11.14.

As shown in Table 23, when staining basophilic substrates with a basic dye those structures containing ribonuclear protein (RNP) particles are slow to stain whereas chromatin stains relatively quickly. Consequently the use of Methyl Green-Pyronin and other big basic dye-little basic dye pairs to achieve the differential staining of chromatin-RNP particles (approximating to DNA-RNA) also fit into Rate Model 3. An indication of the generality of such effects is given by the example of Table 31. An account of this topic has been given by GOLDSTEIN (1961, 1962a). A diagrammatic demonstration of this is provided in Section 11.14.

Table 31. Examples of selective stains based on progressive dyeing with big dye-little dye pairs: the Rate Model 3 situation

Big dye	Little dye	Colour of	
Differential stains for chromatin-RNP particles			
		Chromatin («DNA»)	RNP particles («RNA»)
Janus Green, or Malachite Green, or Methyl Green	Pyronin	greenish-purple	red
Malachite Green	Acridine Red	greenish-purple	red
Janus Green	Safronin	greenish-purple	red
Celestin Blue or Victoria Blue 4R	Pyronin	blue	red
Bismark Brown R	Azur A or Toluidine Blue	brown	blue
Differential stains for collagenous connective tissue			
		Collagen fibres	Red blood cells
Indigocarmine, or Methyl Blue, or Nigrosine, or Sirius Light Blue GL	picric acid	blue	yellow
Acid Fuchsin, or Ponceau S, or Sirius Red F3B or Violamine R	picric acid	red	yellow
Methyl Blue	Orange G	blue	orange

It should be noted that dye-tissue affinities do in fact sometimes play a role in rate controlled processes, such as those involved in Rate Model 3 cases. Thus recent quantitative measurements of the affinities of a variety of basic dyes for tissue sections showed that the action of some of the big dye-little dye pairs used for differential staining of DNA-RNA will be aided by the relative affinities, though other pairs will actually be hindered (MARSHALL and HOROBIN, 1973b). For another approach (namely via structure-staining correlations) see Section 11.14.

Rate Model 4:
Using sequences of dyes.

Consider preparations stained first with one dye, and then stained again from a dyebath containing a second dye of the same charge sign as the first. When rate effects are the controlling factors the first dye applied will be initially replaced by the second from the faster staining and destaining substrates. However the first dye will still be present in the sites slow to stain or destain. Consider staining protein-rich sites with anionic (acid) dyes, from acid solutions. The information on the relative staining rates of such sites, given in Table 24, allows us to predict that it is structures such as red blood cells and muscle cytoplasms which will remain coloured by the first dye, whilst collagen fibres will be coloured by the second dye of the sequence. There are numerous methods which this model illuminates, though in fact most of them make use of dye mixtures in the first or second, or both, steps of the sequence. Such methods are typically designed to produce the multicoloured effects valued by histopathologists. A number of these methods have been assembled and discussed by LILLIE (1965, Chapter 15). This compilation indicates that sequential application of these dyes causes rates to be predominantly influenced by tissue characteristics. So either the first or the second dye can have the larger size, or the higher charge, without the staining pattern being significantly altered. It does seem possible however that a gross difference in size, and so of staining rates, might overwhelm tissue differences and prevent selective staining. Perhaps this why the extremely small picric acid never appears to be applied as the first dye in a sequence? Of course when dye mixtures are used then staining effects of the type discussed under the Rate Model 3 heading will also occur and will be superimposed onto the grosser effects we are primarily concerned with in Model 4. For example when the first step of the sequence applies the red dye Phloxine and the second step a mixture of Orange G (452 daltons) and Methyl Blue (738 daltons), the slow staining red blood cells retain the red dye and the fast staining collagen fibres become blue, whilst the muscle cytoplasm, which stains at an intermediate rate, becomes orange. Note that the use of «colourless dyes» in sequence stains, involving the same effects as are discussed here, is considered later in Section 4.19. The dramatic use of sequence staining with plastic sections is discussed in Section 10.6.

4.19 Rate-controlled selective staining involving «colourless dyes»

A variety of biological staining procedures utilise colourless dyes, to borrow JOHN R. BAKER's phrase (1958). That is, colourless compounds having a high affinity for tissue components. A variety of colourless dyes are in use, for instance the heteropoly acids (usually phosphotungstic or phosphomolybdic), various surfactants, and the complicated mixture of acid aromatic glycosides termed tannic acid.

Such methods mostly fall into one of three categories: procedures where the use of a heteropolyacid is merely an expensive way of achieving a low pH; methods where the colourless dye is applied in a separate step before or after the dye proper; and methods where colourless dye is applied simultaneously with the coloured species from a single bath. Only the latter two require comment.

The Masson trichrome method, with its many variants, provides an example of *selective differentiation using a colourless dye.* In the Masson procedure an initial staining with an acid dye colours all acidophilic substrates, whether slow or fast staining. The section is then treated with a phosphomolybdic acid solution. After a short period the very large phosphomolybdate anions have replaced the dye anions only in the fastest staining/destaining substrate, namely collagen fibres. Before exchange with the dye in slow staining/destaining substrates can occur the differentiation process is terminated. The tissue section is then stained with a large (slow diffusing) acid dye. This is usually Aniline Blue, though dyes such as Fast Green, Light Green and Wool Green are amongst those successfully substituted. Once again the period of staining is restricted so that the dye has time to exchange only with phosphomolybdate anions in the fast staining/destaining collagen, again leaving the dye inside the slower staining components unchanged. Such a sequence has been discussed by BAKER (1958, p. 239).

The other class of methods utilises a *colourless dye in the dyebath* when direct competition can occur between the coloured and the colourless compound. One example is the use of a mixture of Azur A and a cationic surfactant to selectively stain RNA-rich sites (BENNION and HOROBIN, 1975). Since the amount of surfactant present exceeded the critical micelle concentration, the species present in the dyebath included the small cationic dye molecules and very much larger cationic surfactant micelles. In fast staining sites the micelles could compete with the dye but the slow staining sites (such as those rich in RNP particles, see Table 22) would initially be stained only by the small, fast staining, dye molecules. This is effectively the Rate Model 3 situation from Section 4.18 but with the large dye replaced by a large colourless dye.

As a tailpiece it may be added that certain procedures use more than one colourless dye. An example is the tannic acid-phosphotungstic acid-Amidoblack sequence for the selective staining of haemoglobin (PUCHTLER and SWEAT, 1962). One can speculate that such procedures act by the different anionic colourless dyes being, due to selective staining and destaining, present in most substrates other than haemoglobin, so excluding the acid dye.

4.20 Everything may stain, and stain at the same rate – but in a variety of colours

This is the phenomenon of metachromasia, the physical-chemical basis of which is discussed in Section 4.2. No comment is needed here except to say that the vast histochemical literature on the subject, well reviewed by SYLVÉN (1956) and PEARSE (1968), does demonstrate that it is rare that histo*chemical* conclusions can be directly drawn from the occurrence of metachromatic staining: the technical advantages of exploiting metachromasia are of course appreciable.

5 Reactive Staining Methods

5.1 Some definitions, generalisations and cautions

What does reactive mean? In practice the term is applied to such stains as the Schiff reagent and Mercury Orange. In other words reactive methods are those where the staining process can, in part, be represented by a classical stoichiometric equation describing the making and breaking of covalent bonds or electron transfer, as exemplified by the examples shown in Fig. 38. Admittedly the choice of methods considered under this heading may seem arbitrary, since staining methods discussed in other Chapters may be stoichiometric, and may be capable of being represented explicitly; such as labelled antibody procedures (see Chapter 8), enzyme histochemistry (see Chapter 7), and at least certain methods involving the uptake of metals (see Chapter 6). However this definition of reactive (or *chromogenic* to give a perhaps even more misleading alternative) methods is time-honoured and not some novelty of mine.

Reactive methods have been applied to a wide range of biological staining problems. Thus general classes of compounds may be detected by utilizing marker groupings, such as the 1,2-diol structure present in most polysaccharides, or the aliphatic amino groups of proteins. Use of selective versions of such procedures reveals the distribution of particular groups. Such methods can for example distinguish between the side chain and terminal amino groups of proteins, or detect Ca(II) ions in structures undergoing pathological calcification, regardless of the chemical or ultrastructural environment of the grouping. At the other extreme methods may be sufficiently selective to restrict staining to a particular combination of reactive groups. For instance DNA may be stained without colouration of RNA; or catecholamines, but not other phenols, may be converted to coloured derivatives.

The variety of chemical mechanisms involved even in common procedures are very varied, as emphasised in Fig. 39. The technical problems are correspondingly complicated, see Table 33 for some examples.

The mechanistic simplicity implied by explicit formulations such as those shown in Fig. 38 is largely an illusion. The widely held belief that methods such as the Feulgen nucleal and the periodic acid-Schiff procedures avoid the physico-chemical complexities of staining methods involving dyestuffs is unjustifiably optimistic. Reaction, in the usuage of this Chapter, cannot occur until reagents are within the substrate. Hence reactive procedures in fact *add* the complexities of chemical reactivity to those of the rate of uptake of reagents into, and their affinity for, the substrates. Consequently this Chapter will be arranged rather like Chapter 4 and will first consider affinity and rate factors for uptake of reagents before proceeding to the reagents reactivities.

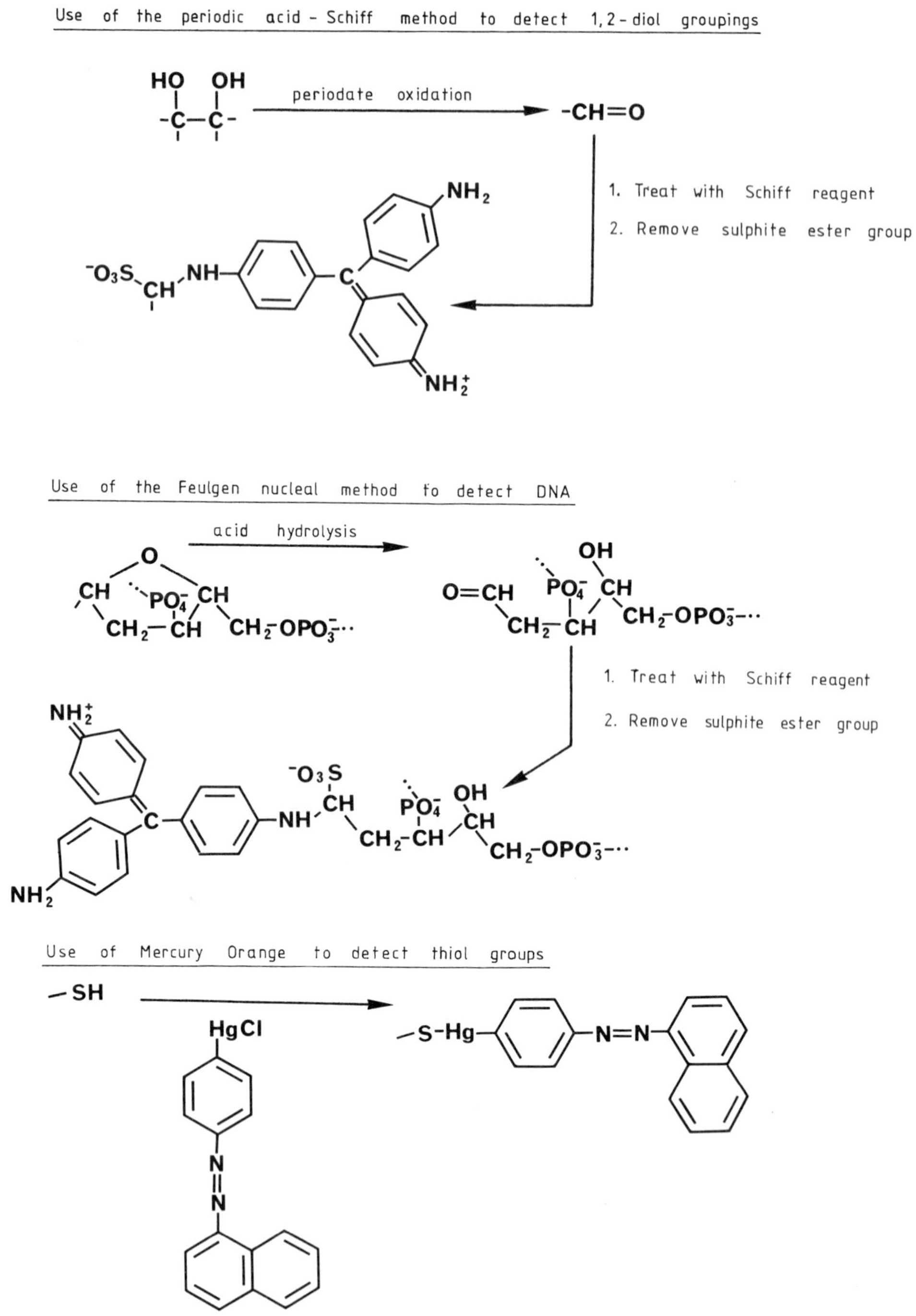

Fig. 38. Equations representing the reactive steps of some reactive biological staining procedures

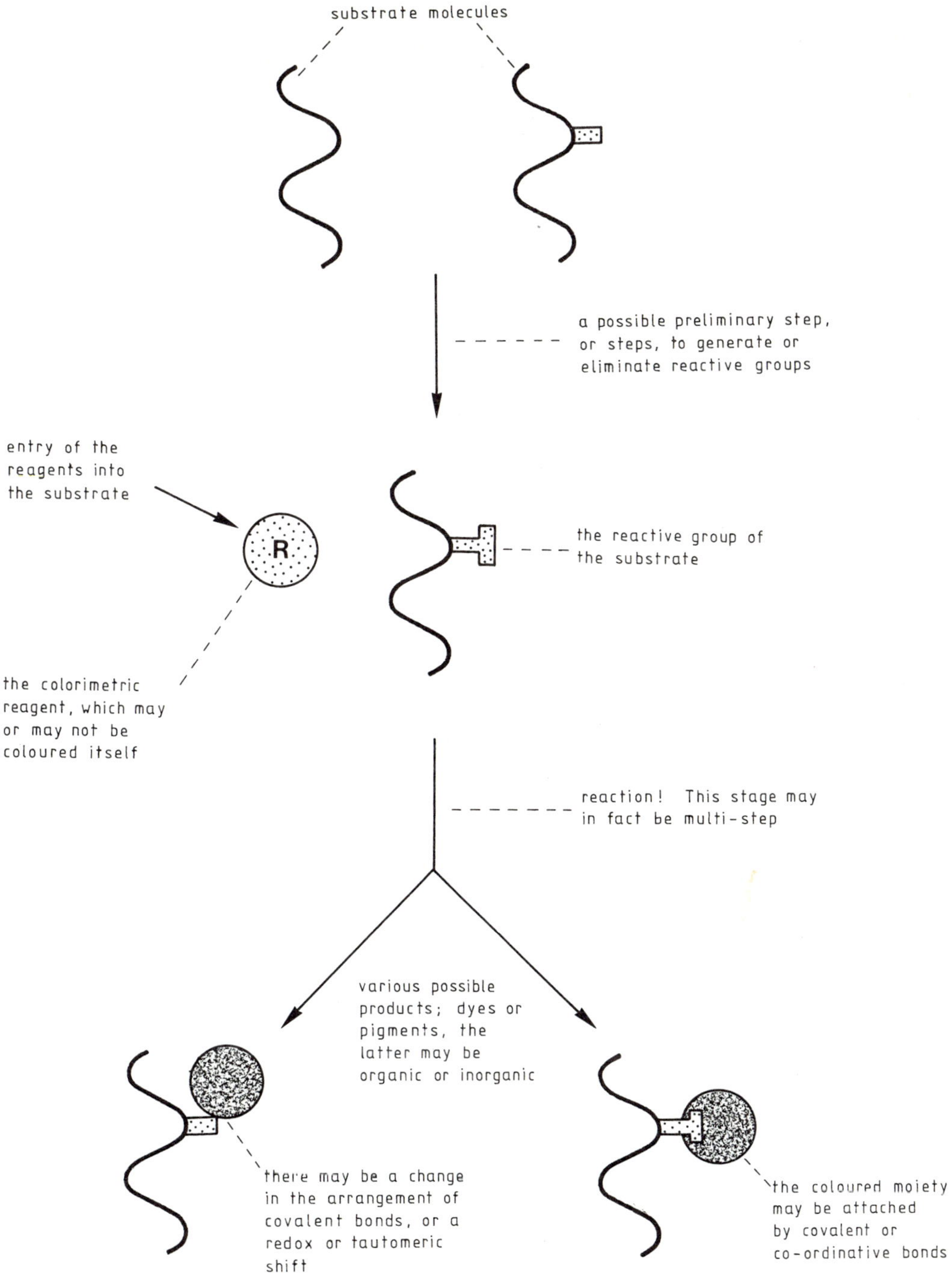

Fig. 39. An illustration of some potential chemical complications of the reactive stage of reactive staining methods

In the following discussion a large proportion of the examples involve either the Feulgen nuclear or the periodic acid-Schiff procedures. Because of the overwhelming importance of these two reactive methods more work has been done on these systems than on all other reactive staining procedures put together. There is however no reason to suppose that other, less intensely studied, reactive systems are any less complex.

5.2 The influence of reagent uptake on the selectivity of staining

As this topic has received little study, some of the evidence adduced is circumstantial, and distinctions between affinity and rate control of reagent uptake are not always clearcut. A further complication is that factors affecting reagent uptake, such as the pH or salt content of the staining solutions, may also influence reactivity. These possibilities will be returned to later.

Initially *affinity control* of staining will be considered, clear examples of which are found when Coulombic effects dominate. Thus when textile dyes of the Remazol class (which carry anionic nucleophilic reactive groupings) were briefly applied to tissue sections, the strongest staining occurred at low pH. Though covalent binding of the dye occurred most effectively at high pH, only at low pH was dye readily taken up into the tissue, by an acid dyeing process. The overall effect was to make dye binding most effective at low pH (SALTHOUSE et al., 1971). Another example is provided by the periodate oxidation step of the periodic acid-Schiff procedure. Studies on model systems as well as tissue sections have indicated that the periodate anion is repelled by anionic polysaccharides, reducing their rates of oxidation relative to those of non-ionic polymers (SCOTT and HARBINSON, 1968). This effect has given rise to the widely held, but erroneous, view that sulphated polysaccharides are inevitably PAS negative, see for example KASTEN (1960, p. 75).

Such effects, involving the mobile and fixed ionic species of reagents and tissues, are analogous to those occurring with acid and basic dyeing. Hence one might expect analogous behaviour to the salting-on and -off of the ionic dyes, see Section 4.4. Such behaviour does in fact occur, for instance with the periodate oxidation step mentioned above. Thus in the test tube the addition of a neutral salt to the periodate greatly increases the rate of oxidation of anionic polysaccharides, but scarcely changes the oxidation rates of an un-ionised polymer. Similar effects were seen in tissue sections (SCOTT and HARBINSON, 1968; SCOTT and DORLING, 1969). This is analogous to the salting-on of acid dyes onto basophilic substrates. A salt effect occurring with the Feulgen nucleal procedure also provides an example of an analogue to the salting-on of basic dyes to acidophils (BEDI, 1976). If the acid hydrolysis step is carried out with HCl to which a neutral salt has been added, the final staining intensity is increased relative to a salt-free control. This effect may be explained by the nature of the substrate, namely chromatin, in which the DNA is intimately entwined with the protein components. Hence the immobile ions of the chromatin may well carry an overall positive charge under the strongly acid conditions prevailing, and the cationic mobile ions (H^+) will be salted-on. This would be an internal-pH effect of the sort discussed in Section 4.4.

Apart from such direct Coulombic influences on reagent-substrate affinity two more indirect but interesting complications can occur. Firstly Coulombic effects will be expected if ionic reagents are used, whether or not the substrate is ionic. As soon as an ionic reagent has been taken up, the substrate has become ionic. This is exemplified by the periodate oxidation of glycogen in the initial step of the periodic acid-Schiff procedure. Adding neutral salt to the oxidant increases the oxidative rate. Though glycogen itself is non-ionic the reaction intermediates involved are anionic, as is the oxidant (OVADIA and STOWARD, 1971).

As an example of *rate-controlled reagent uptake* consider the action of thiol reagents such as aminobromoanthraquinone sulphonic acid and Mercury Orange. Under their recommended conditions of use both reagents stain many structures present in rat skin, such as muscle cytoplasm and red blood cells, in a few hours. The extremely dense hair shafts however are much slower staining, and with Mercury Orange (the larger compound) they are scarcely stained even after 24 hours. A number of probable rate-of-entry effects can be found in a detailed study of azo-coupling maleimide stains for thiols (SIPPEL, 1978). Compact tissue sites (e.g. the lens of the eye, prekeratin, myelin) were slow to stain with several maleimides. After reaction with azo-coupling maleimides such sites were in turn slow to react with diazonium salts. Rate effects may be more complex than this. In studies of the Feulgen nucleal procedure reported by KASTEN (1971) optimum hydrolysis time is defined by the relative rates of the hydrolytic removal of purines from the DNA, to yield the Schiff-reactive apurinic acid, versus the hydrolytic fission of the apurinic acid into fragments small enough to be readily lost from the tissue.

The influence of fixation on reagent uptake has apparently not received direct study. Fixative effects have been observed adventitiously however and two such instances will be briefly considered. Firstly, as has long been known, the optimum hydrolysis time in the Feulgen nucleal technique varies with the mode of fixation adopted. The fixation procedure with perhaps the most marked effect is chromation, whose influence may extend the optimum hydrolysis time fivefold (LILLIE et al., 1967). Such chromation also dramatically reduces tissue basophilia, in other words reduces uptake of dye cations (see Section 3.7). Hence it is quite possible that the lengthened hydrolysis period reflects an analogous reduction in the uptake of hydrogen ion, that is of hydrogen cations. The second example is taken from the applications of the aminobromoanthraquinonesulphonic acid reagent for thiols (TURNER and HOROBIN, previously unpublished work). If this reagent was used on porous, thiol-rich tissue substrates, then fixation in neutral formalin allowed stronger staining than occurred when the sulphur-oxidising Zenker's solution was used. However fixation in the oxidising fixative, known to open up keratin to dyestuffs (see Section 3.8), allowed stronger staining of the dense hair shafts than did formalin.

5.3 Influence of reactivity on selectivity of staining

It is apparent from Section 5.2 that the belief, apparently widespread amongst histochemists, in the supposedly superior specificity of reactive methods is of doubtful validity. The previous Section discussed only the influence of reagent uptake on staining patterns, and so on selectivity. We will now consider the selectivity of the reactive steps themselves. This part of the process is in the first approximation straight forward bio-/inorganic/organic chemistry, so no theoretical background will be provided, or indeed could be provided in the space available. For those seeking additional information elsewhere, two bibliographical suggestions can be made. Firstly to use texts with an analytical slant whenever possible, and secondly not to rely on elementary texts for other than general orientation. Few reactions, even when of real use in analytical chemistry, are specific, and it is important to grasp the degree of selectivity and of the factors affecting this.

Wide *variations can occur in the specificity of the reaction* in such staining methods. Certain procedures, such as the Feulgen nucleal and the use of Mercury Orange as a thiol stain, are remarkably specific, but most procedures are merely selective. Examples of the latter are given in Tables 32 and 33. Often the different reagents available for a given chemical function differ in their selectivities, as exemplified by the reagents in Table 33. Note that the basis of reactive selectivity varies. Some reagents, such as Mercury Orange, will only react with a single grouping, at least under the mild conditions used with tissues. Other procedures, such as periodic acid-Schiff, can give rise to coloured derivatives with a wider variety of groupings, but short reaction times limit staining to the fast-reacting groupings. Ironically it has been suggested that this principle may effectively and usefully be reversed. Thus 9-O-acyl sialic acid residues of mucosubstances are oxidised extremely slowly by periodate. If the initial aldehydes resulting from periodate oxidation are blockaded, and oxidation continued, then Schiff reagent is held to selectively stain 9-O-acyl sialic acid derivatives (Reid et al., 1978). Yet

Table 32. Some examples of non-specific chromogenic reagents

Acrolein as used in the acrolein-Schiff method for proteins, reacts with various nucleophiles. Most vigorously with amino and thiol groups, but also with the -NH- group of imidazole residues

Periodic acid as used in the periodic acid-Schiff procedure, reacts most vigorously with 1.2-diols but also, though with decreasing vigour, with 1.2-hydroxyamines, 1.2-hydroxyketones, 1.2-diketones and 1.2-hydroxycarboxylic acids

Schiffs reagent used in both the above techniques, reacts rapidly with most non-aromatic aldehydes and also with many epoxides, but more slowly with ketones

Table 33. Details of some stains for thiols, illustrating the wide variation in chemical specificity and in technical difficulty in reactive methods

Reagent or method	Specificity	Relevant practicalities
Aminobromoanthraquinone sulphonic acid	Reacts with $-NH_2$ as well as -SH, but the $-NH_2$ (and any -OH) derivatives are blue, whereas those with -SH are orange	Simple and rapid
Dinaphthyldihydroxydisulphide	Apparently can react with -COOH as well as -SH. Also tends to be adsorbed onto hydrophobic substrates such as elastic fibres	Multi-step, but quite rapid
Ferric ferricyanide	Non-specific, reacts with many reducing agents	Simple and rapid
Maleimide reagents	Uncertain, probably react with imidazole residues as well as -SH	Often multi-step, rapid Rapid and Simple
Mercury Orange	Considered to be specific	Slow, multistep
Phenacyl methods	Uncertain, probably high	
Tetrazolium methods	Non-specific, colour produced by many reducing agents. Also by hydrophobic substrates such as elastic fibres	Simple, rapid

another possibility, seen with the thiol reagent aminobromanthraquinonesulphonic acid, is that the reaction is not specific, but the products derived from different groupings differ in colour.

Two reagents may have equivalent selectivities and yet still not give rise to the same staining pattern, since *reagents may have varying sensitivities.* For instance it has been suggested that replacement of the periodic acid in the periodic acid-Schiff procedure with lead tetraacetate gives more selective staining. Much of the effect however appears to derive from the reduced reactivity of the lead tetraacetate. So after a short time of oxidation with this latter reagent only those regions rich in polysaccharides may be stained, though after a longer time of oxidation more regions stain, giving the less selective picture (GOLDSTEIN, 1965). Indeed Schiff reagent itself comes in varieties of varying sensitivity, some allegedly being «super-sensitive» (KASTEN, 1960). Such variations in reagent sensitivity are presumably widespread even though not well documented. Some studies are available, such as work on reagents for calcium visualisation (KASHIWA and HOUSE, 1964).

Even considering a single reagent and disregarding differential uptake the *reactivity of native tissue groupings may be site dependent.* One reason for such variation arises when there is in reality *more* than one reactive grouping in the tissue. An example of this is the demonstration of ferric ion, by formation of Prussian Blue pigment during application of ferrocyanide methods. The acidity of the standard reagents is sufficient to liberate and allow the demonstration of Fe^{+++} from, for example, hemosiderin pigment. However to stain the Fe^{+++} of hemoglobin often requires stronger acid, or a peroxide solution; and to stain the Fe^{+++} of particles of hematite needs even stronger acid. The difference is due to the Fe^{+++} being co-ordinated by quite different ligands in each case, so strictly speaking three different iron containing compounds are being demonstrated. Other kinds of substrate variation could occur. One such, sometimes cited to rationalise failures to stain, is the stereochemical factor. This concept does not however seem to have been subjected to experimental study. Another un-analysed substrate variable is the possible difference in internal pH of different structures, see Section 4.4. Since reactions of even non-ionic reagents may be pH-dependent, this could be an important characteristic.

A different situation occurs when *the reactivity of the tissue groupings are modified before staining.* This may occur adventitiously during the process of fixation. Section 3.7 describes several examples of fixation increasing or decreasing reactivity of tissue groupings. Modification is often carried out in a more calculated manner however. Only inhibition of reactivity, namely blocking procedures, need to be discussed. Enhancement of reactivity is just another name for a multi-step staining

Table 34. Examples of selective modification of tissue reactivity prior to staining

1. The selective staining of tyrosine residues by acidified dinitrofluorobenzene is possible if the other nucleophilic group reactive at low pH, namely -SH, is first made unreactive. This is carried out by treatment with N-ethylmaleimide
2. Histidine residues may be selectively stained using the coupled tetrazonium method if other reactive groupings, such as tryptophane and tyrosine, are first made inactive by acetylation with acetic anhydride
3. Before using the peracetic acid-Schiff technique to demonstrate carbon-carbon double bonds, native aldehyde groups may be eliminated by conversion to non-reactive derivatives by treatment with an acetic acid solution of aniline

method. Table 34 gives some examples of the use of inhibition of reactivity to enhance the selectivity of reagents. Modification of staining may also be achieved by the total removal of certain tissue constituents by extraction. For instance if the periodic acid-Schiff procedure is to be used to stain mucosubstances then a prior treatment with an amylase will remove hydrolytically any, potentially confusing, glycogen. Both these techniques are discussed more fully elsewhere, in Sections 11.3 and 11.2.

6 Staining Methods Involving Uptake of Metals

6.1 A general model for staining methods involving uptake of metal

Terms such as metal impregnation, heavy metal staining, and metallation recur throughout the biological staining literature. But what is meant by the uptake of *metal*? And are there common rationalising threads underlying the wide variety of methods? The nomenclature unfortunately is imprecise, and although the terms subsumed by the heading uptake of metal do commonly refer to metal ions (cations such as Ag^{+}, Fe^{3+} and uranyl; as well as anions such as dichromate and phosphomolybdate or tungstate) the uptake of metal has also been used to describe staining involving neutral species, such as mercuric nitrate or osmium tetroxide. Fortunately though the present account considers this wide usage of the term metal, the major principles underlying the selective stainings achieved are familiar, namely differential affinity and differential rates. Admittedly my selection of topics and examples does admit to elements of the arbitrary. For instance procedures in which a metallic compound is used as a reagent but no metal is permamently bound, e. g. the plasmal or chromaffin reactions, will not usually be considered here. Again, the possibility of tissue-metal-dye covalent links, that is of mordanting, will not be discussed here, though see Section 4.4.

The variety of staining methods making use of the uptake of metals is illustrated in Fig. 40, which may be regarded as a general phenomenological model of such procedures. The details and significance of the various steps are considered in the subsequent parts of this Section.

At least some well known staining techniques are readily classified in terms of this model. Silver impregnation is either type 1 (if argyrophil) or type 2 (if argentaffin). Heavy metal staining for electron microscopy involving the direct application of metal ions is type 2a. Chromation methods for myelin such as the acid haematein procedure is type 3, as are the enhancement stains used in electron microscopic enzyme histochemistry.

6.2 How does the uptake of metal occur

Examples of the wide variety of metal uptake processes are shown in Table 35. Notice that when no change in covalencies or in oxidative state occurs the process is more likely

Table 35. Examples of various modes of metal-uptake occurring during biological staining

Mode of uptake		Leading Reference
Examples involving no changes in oxidation states or covalencies		
Solution	Dissolution of osmium tetroxide in saturated lipids	BAKER (1958, p. 125)
Ion-exchange	Uptake of phosphomolybdate or phosphotungstate anions by proteins under acid conditions	SILVERMAN and GLICK (1969)
	Uptake of uranyl cations by DNA and RNA	HAYAT (1970, p. 266)
Examples involving changes in oxidation and covalency states		
Formation of metal complexes	Reaction of Ag(I) cations with the sulphur of tissue thiols and the nitrogen of imidazole residues	PETERS, A. (1955a), VELICAN and VELICAN (1972)
	Reaction of Cr(II) cations, often derived from dichromate anions reduced by the tissue, with tissue carboxylates and atmospheric oxygen yielding Cr(III)carboxylate complexes	HARTLEY (1969)
	Reduction of Os(IIX) by the tissue, followed by the complexation of Os(IV) and Os(V) by tissue thiols and sulphides	HANKER et al. (1967)
Precipitation	Reduction of Ag(I) cations by phenolic and sulphide groups of the tissues to yield metallic silver; i. e. the argentaffin reaction	SWIFT (1975)

to be reversible than when such change does take place. An exception is the binding of metal in the form of co-ordination complexes, many though not all of which are labile, especially if the solvent or solute has chelating properties.

In practice metal uptake is often more complicated than Table 35 implies, since the process may be multistep. A case in point is chromation, which has been studied extensively for the case of the protein keratin (reviewed by HARTLEY, 1969) as well as in the biological staining context, for which a leading reference is LILLIE, BERGEAUX and HENDERSON (1967). Such work indicates that in a protein-rich environment, such as the myelin sheaths of nerves, it is probable that dichromate anions are intially taken up from solution by an ion-exhange process. Under neutral or acidic conditions the Cr(VI) ions can then oxidise such substrate groupings as carbon-carbon double bonds, disulphide bridges and thiols, tyrosine and serine residues. In the course of this the Cr(VI) may be reduced to Cr(II) which can rapidly react with carboxylate to give a labile coordination complex of the type $-CO_2Cr(II)$. Such a complex can be oxidised by atmospheric oxygen to yield a kinetically stable complex, $-CO_2Cr(III)$. An analogous process occurs during the process of osmication, see for instance HANKER, ANDERSON and BLOOM (1972).

6.3 How is the metal visualised after uptake?

Detection is commonly achieved by applying some visualisation agent. However *in the simplest situation*, type 2 of the general model sketched in Section 6.1, *no visualisation is required* since the uptake process itself results in the formation of

Table 36. Some varied methods of visualising tissue-bound metals

	Leading references
By reduction of a bound metallic species to the free metal; i. e. type 1	
For example the argyrophil variety of silver impregnation. Silver ions bound to tissue ligands are reduced to silver metal by externally applied reagents such as formaldehyde or aminophenols	PETERS, A. (1955a, 1955b), VELICAN and VELICAN (1972)
By formation of coloured derivatives from the bound metallic species; i. e. type 3	
One possibility is to generate coloured, insoluble salts containing the metal ion. For instance protein-bound phosphotungstate anions may be visualised by forming salts with cationic dyes such as Azur A	QUINTARELLI et al. (1971a, b)
A more common example of this type is the visualisation of lead ions by the formation of lead sulphide	
Another approach involves generation of coloured co-ordination complexes from the bound metal. The possible ligands are very varied. For example iron(III), bound to tissue sites rich in DNA and RNA, has been visualised by complexing with ferricyanide to form the pigment Prussian Blue. Ions of Cr(III) and Fe(III) bound to tissue amino and carboxylate groups may be visualised by reaction with hematein, yielding co-ordination complexes of various colours	SORVARI (1972a, b) LILLIE et al. (1967), LILLIE and PIZZOLATO (1972)
By making use of the bound metal to generate a coloured, metal-free derivative	
For example metal ions such as cobalt or manganese can be used as catalysts to promote oxidative polymerisation of diaminobenzidine to a black polymeric pigment	HANKER et al. (1972)
Sometimes no visualisation is required; the Type 2 situation	
The argentaffin methods for instance; in which crystals of silver metal are deposited following reaction of silver(I) cations with reducing agents in the tissues such as ascorbic acid, melanin or urates	LILLIE (1965)
Another example is the reaction of osmium tetroxide with tissue components such as aldehydes, unsaturated carbon, phenols and thiols to yield osmium blacks. These are co-ordination complexes of lower oxidation states of osmium with tissue groups such as thiols	HANKER et al. (1967), PEARLMUTTER and LEMBI (1978)

coloured metallic species. Several examples of this type 2 situation are shown in Table 36, and some beautiful illustrations of the outcome may be seen in some recent micrographs published by WILLIAMS (1978). In this circumstance, as in other situations involving metal impregnation, it is not always known just what molecular substrate is responsible for the staining produced. For example in Cajals gold sublimate stain for astrocytes the aurous complex ion is reduced, at least in part to the metal, but by what species? Thiols perhaps?

Quite another type 2 situation is the *heavy metal staining* used in electron microscopy. Here metal atoms, or indeed any high atomic number atoms, are detected by their large electron scattering power compared to that of the background atoms of carbon, hydrogen, nitrogen, oxygen, phosphorus and so on. Several accounts of this method of staining are available, for example HAYAT (1970) and PEARSE (1972, Chapter 32).

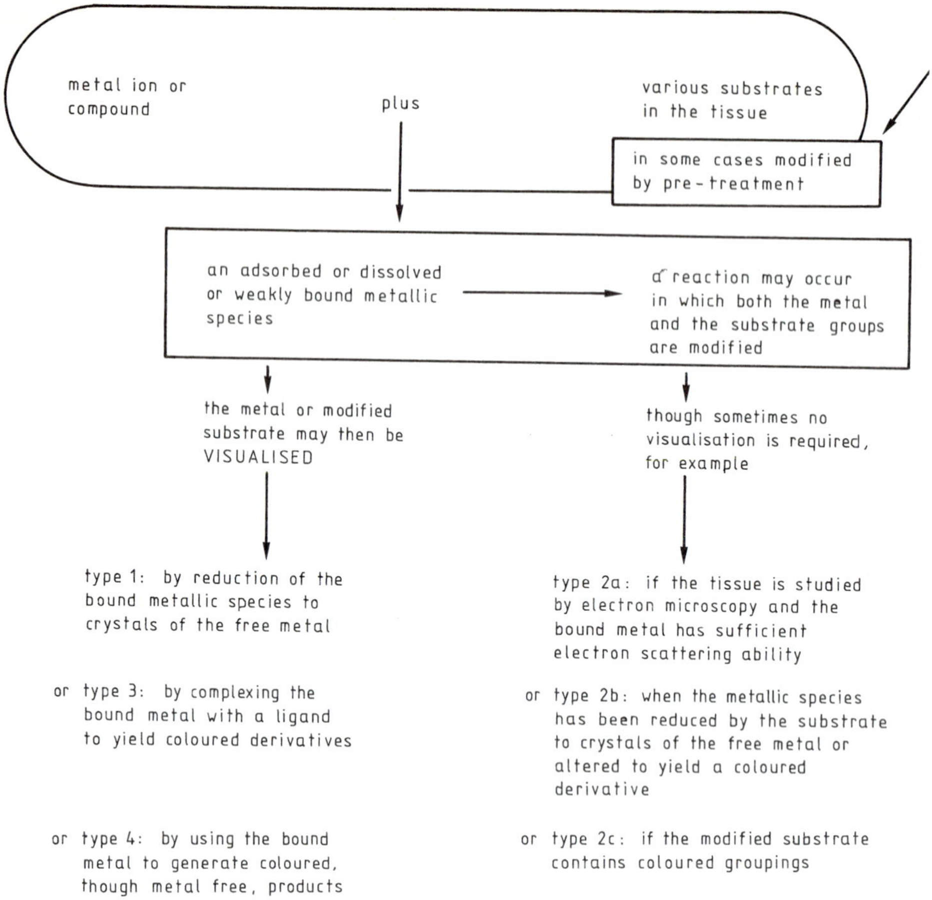

Fig. 40. A general model for biological staining methods involving the uptake of metals

A classical visualisation method, used in argyrophilic silver impregnation, involves the *reduction of a bound metallic species to the free metal*, the type 1 situation. Argyrophilic silver impregnation has been frequently studied and the following account leans heavily on the experimental and bibliographic efforts of PETERS, A. (1955 a, 1955 b), VELICAN and VELICAN (1972) and GALLYAS (1979 a, b). Certain aurophilic methods, such as Ranviers gold chloride procedure for peripheral nerve endings, could well be of the same type though no direct study seems to have been made of this.

The crucial factor in these methods is regarded as the binding of silver in two distinct forms. Firstly as Ag^+ attached to tissue ligands such as thiols and amines, and secondly as silver nuclei, so named by analogy to the photographic situation, which nuclei perhaps comprise small clusters of Ag(0) atoms. The adsorption of the Ag^+ is a fast process, as is its desorption into a silver-free solvent. The formation of the silver nuclei occurs very much more slowly but once formed they are not readily lost. The rate of uptake of both forms of silver depends on a variety of factors such as the pH, silver concentration, nature of co-solutes in the impregnation solution, and ambient light intensity.

Visualisation of the silver in such an impregnated tissue is achieved by treatment with a reducing agent, typically formaldehyde. Various phenols and arylamines may also be used. Though reduction occurs only at the sites of the silver nuclei, the crystals of metal produced are built up from the mobile Ag^+ from throughout the tissue. The efficacies and rates of the reduction processes vary with factors such as the hydrogen and sulphite ion contents, and the oxidation potentials of the reducing solutions.

The visualisation procedure can be modified by subsequent *toning*. This involves treating the silver stained section with aurous chloride so as to lay down gold at the site of the silver deposits. The procedure increases the size, and changes the colour, of the crystals. Suggested mechanisms for this include both ion-exchange [$Ag(0) + Au^+ \rightarrow Ag^+ + Au(0)$] and the catalytic role of the silver crystals in promoting the autoreduction of the gold chloride. It does seem possible that methods including a pre-impregnation treatment with metal salts amount to toning with the silver onto non-silver nuclei. Examples include the use of uranyl and cobalt nitrates of cobalt and barium chlorides in certain methods for demonstrating the Golgi apparatus, and also the use of so-called sensitisers such as ferric chloride or uranyl nitrate in certain stains for reticulin fibres.

Mention should be made of an arygrophil silver impregnation method which does not fit the pattern described so far, namely bromine-silver procedures for staining unsaturated lipids (e. g. NORTON et al., 1962). During this procedure a bromination step converts carbon-carbon double bonds to their saturated bromo derivatives. After this a silver impregnation is carried out in acidic solution of silver nitrate. Under these circumstances very little Ag^+ will be taken up in a reversible manner. However when the tissue is subsequently treated with a reducing agent a strong Ag(0) staining is still produced.

Another common visualisation procedure is the *formation of coloured derivatives from bound metallic species,* which is the type 3 case. As is shown in Table 36 such derivatives may be coloured insoluble salts, or may be co-ordination complexes. If the latter the complexing ligands can be organic or inorganic. In either case, for a given ligand to generate a derivative it must have a higher affinity for the bound metal than do the co-ordinating groups of the tissue. So different ligands, or different reaction conditions, would be expected to give different staining patterns. This has indeed been observed after osmication (ELLEDER and LOJDA, 1968) and also after chromation (for instance LILLIE, 1969b).

6.4 Why should such methods display selectivity?

Factors aiding selective staining can act at many points in the process, as is illustrated in Fig. 41. This figure takes the general model for biological staining methods involving the uptake of metals (see Fig. 40) and recasts it to emphasise the steps on which selectivity depends.

Taking first things first *the uptake of metal may be selective.* Such selectivity may be due to variations in either affinity or rate, or indeed to both as noted in Table 37. In fact all the influences on selective staining discussed in Sections 4.10 to 4.22 for the case of dyestuffs can be relevant here and in the subsequent steps. This being so, just the same

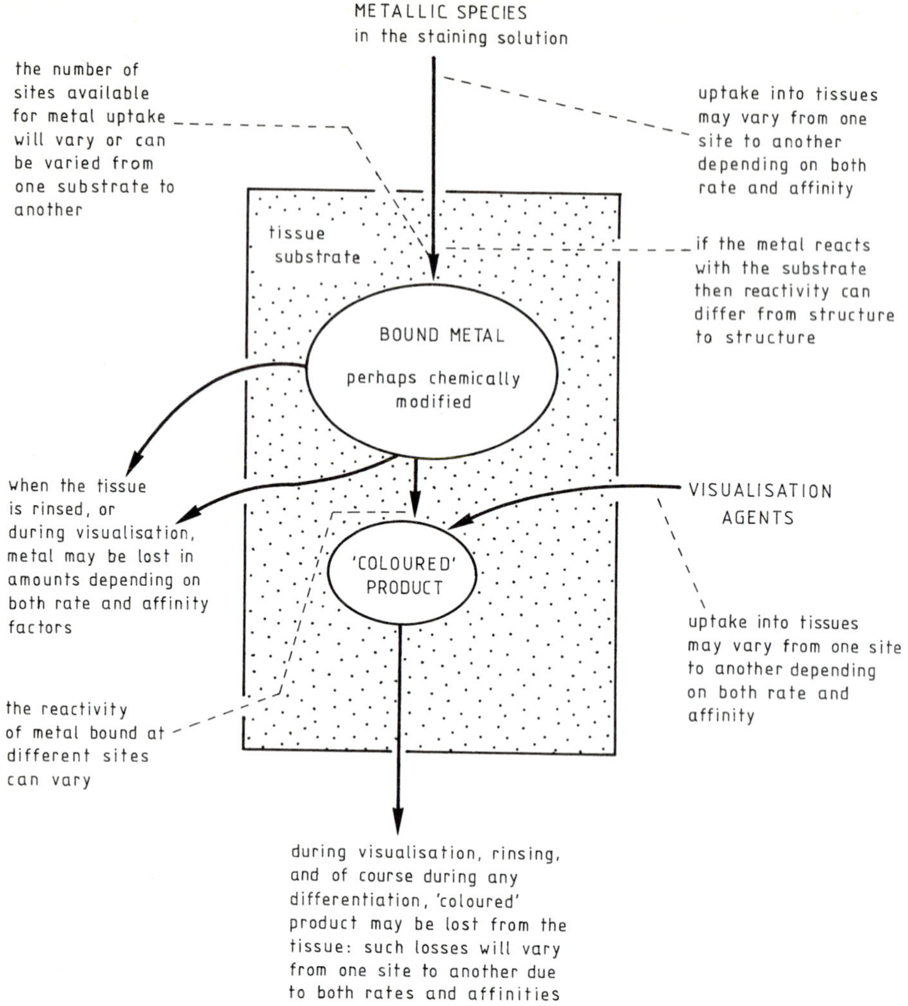

Fig. 41. A modification of the general model of the uptake of metallic species (compare to Fig. 40) illustrating factors facilitating selective staining

practical variables as those which controlled selective dyeing are important now: nature of co-solutes, the pH, ionic strength, temperature, stirring of the reagent solution and so on. For examples of this see SORVARI (1972a, b), LILLIE and PIZZOLATO (1972), and WILLIAMS, T. (1962).

It is convenient to consider the role of selective reactivity separately, though there is no sharp physical-chemical difference between affinity and reactivity. Reactivity is often loosely used to encompass effects of both affinity and rate. The reactivity of different tissue substrates with metals and metallic species vary vastly and no attempt will be made here to discuss this, though some illustrative examples are provided in Table 38.

Table 37. Examples of rate and affinity factors in the selective uptake of metals

Selectivity due to differential rate of uptake	
Fe^{+++} is taken into fine collagen fibres quite rapidly, but slowly into coarse fibres, and only very slowly into the cortex of hair sheaths	LILLIE and PIZZOLATO (1972)
Phosphomolybdic and phosphotungstic acids (whose ionic weights are of the order of 1800 or above) can bind to most acidophilic structures, but bind rapidly only to collagen fibres	BAKER (1958, p. 240), BULMER (1962)
Selectivity due to differential affinities	
The uptake of osmium tetroxide into hydrophobic lipids, and its exclusion from polar lipids and other structures, parallels the selective staining of lipids by the non-ionic fat stains	ADAMS (1969)
The uptake of some ionic species, e. g. of phosphomolybdate and phosphotungstate, resembles that of acid dyes; with coulombic forces aiding the staining of acidophils and preventing that of basophils	BAKER (1958, p. 240), QUINTARELLI et al. (1971 a, b)

Table 38. Examples of selective reactivity and the uptake of metals

Several metallic species (e. g. Cr(VI) and OsO_4) react readily with unsaturated carbon-carbon bonds, but less readily or trivially with saturated linkages	ADAMS (1969), WATERS (1958)
Some metal ions form stable co-ordination complexes only with a limited number of ligand groupings. E. g. Al^{+++} complexes with carboxylate and phosphate but not with sulphate: Hg^{+} reacts with thiols but not hydroxyls; and UO_2^{2+} reacts well with phosphate but poorly with carboxylate	HAYAT (1970, p. 266), PEARSON (1968)
Some metallic species are reduced only by certain tissue groups. E. g. Ag^{+} and OsO_4 are reduced by thiols but not alkyl hydroxyl (but c. f. Cr(VI) which is reduced by both types of groupings)	HARTLEY (1969), PETERS, A. (1955 a), WATERS (1958)

A practical variable which does call for specific comment is the variation in the *concentration of sites for binding or reaction.* As was implied above in Section 6.2, and also earlier in this Section, metals and metallic species bind to or react with a very wide variety of native sites. If these differ in concentration, selective staining may be achieved. For instance it is the high concentration of phenolic compounds in the granules of argentaffin cells which is responsible for the selective staining of these organelles by silver. However selectivity is sometimes achieved by modifying the tissue prior to metal uptake, with reagents which generate new binding/reaction sites in specific substrates. Best known is the use of oxidising agents such as chromic acid, periodic acid or potassium permanganate prior to silver impregnation in the reticulin stains. Such a step generates aldehyde groups, capable of reducing Ag^{+} to silver metal, by the oxidation of polysaccharides such as those of glycogen granules, reticulin fibres or basement membranes. In this context note the marked similarity in staining patterns, though not of intensities, of the periodic acid-Schiff procedure and the classical silver stains for reticulin fibres. A variety of final staining outcomes can thus be achieved by varying the mode of generation of the binding/reaction site. For example one study demonstrated that if, in a particular silver stain for reticulin, potassium permanganate was used as oxidant then most reticular fibres and basement membranes, excluding those of the kidney glomerulus, stained. When periodate oxidation was used no such selectivity occurred (Velican and Velican, 1972). Similarly, when short periods of oxidation with chromium trioxide were used then glycogen was stained but not reticular fibres, whereas with long periods of oxidation reticular fibres were stained but glycogen did so no longer (due to further oxidation of aldehyde to carboxylate?).

Other examples of this effect include the bromination of carbon-carbon double bonds prior to silver impregnation (e. g. Norton et al., 1962) and the unwanted case of the formation of S- and N-methylol groupings during prolonged formalin fixation (e. g. Table 11). A rather different mode of creating metal-binding sites is to first treat the tissue with a mobile agent capable of subsequent metal binding. In the tannic acid-silver methods for example the tissue is first treated with the «colourless acid dye» tannic acid, whose phenolic groupings are able to reduce Ag^+ subsequently applied to the tissue. Another example of this kind is the binding of phosphotungstic acid to basic dyes previously selectively adsorbed onto the tissue.

The *selective loss of bound metal* is undoubtedly most important in the case of silver impregnation methods of the argentaffin type, when the loss occurs during the visualisation process. For a full discussion see Peters, A. (1955 a, b) who concluded that in such methods silver is bound to most tissue structures during impregnation, but deposition of silver crystals is restricted to structures from which loss of silver ions is slow. An observation supporting Peters' proposal is that argentaffin structures include those slow to stain or destain with dyestuffs, e. g. calcified bone, myelin sheaths, nucleoli and red blood cells.

Selective loss of metal could occur during post-impregnation washing, though this does not seem to have been studied. A possible example of this can be seen by comparing staining results following iron-impregnation with different washing modes (Lillie and Pizzolato, 1972; Sorvari, 1972).

The *differential uptake and reactivity of the visualising agent* constitute other little studied critical steps. Published work is so sparse that it is not possible to distinguish between affinity, rate, and reactivity effects. A good example of one or more of these factors being important is found in the argentaffin procedure. If the reducing agent used to generate the deposit of metallic silver is too vigorous in action then most sites tend to be argentaffin, whereas a very slow acting reducing agent will give a completely negative picture (Peters, A. 1955). Other examples include the variable staining patterns resulting with different visualisation agents after treatment with osmium tetroxide (Elleder and Lojda, 1968).

Finally consider the *selective loss of the coloured product* formed by the visualisation agent acting on the bound metal. An example of this is provided by the so-called phospholipid stains of the Baker's acid haematein variety. Such methods typically involve chromation of the tissue followed by a treatment with hematein to yield fairly non-specific staining. Differentiation then yields a selective staining of mylein sheaths, along with other structures such as mitochondria, red blood cells and pancreatic zymogen granules. It is noteworthy that these differentiation-resistant structures are amongst those slow to stain and destain in staining methods involving dyestuffs. Hence differential staining is probably rate controlled, though affinity control could presumably occur in other circumstances. It is noteworthy that the model system studies supposedly validating such procedures (reviewed by Adams et al., 1965, and Chayen, 1968) are only likely to be modelling differences in reactivity, not rate.

6.5 Why are many metal impregnation methods so fickle?

Undoubtedly the most important reason is the sheer physico-chemical complication of the methods (as exemplified by Fig. 41) compared to many other biological staining systems. Many potentially critical steps occur in each method, and at each step variations in reaction time, temperature, pH, reagent concentration and so on can prove influential. In addition methods involving changes in oxidation state of metals can also be influenced by the intensity of the illumination.

Another practical variable which requires specific comment is *fixation.* The changes in the structure and chemical nature of the tissues induced by the fixative can sometimes be plausibly related to the final staining patterns. Thus with silver impregnation, prior fixation with formaldehyde or glutaraldehyde tends to produce false positive staining due to bound aldehyde or aldehyde-generating groups of the fixative molecules. If fixation in Zenker's fluid is adopted aldehydes may again be present, this time due to the oxidative action of the fixative on the tissues (VELICAN and VELICAN, 1972). Another effects is seen when the binding of Fe^{+++} to hair is studied (LILLIE and PIZZOLATO, 1972). Little metal is bound after fixation in formol or mercuric chloride at room temperature, but much more is bound after treatment with (the mechanically disruptive?) hot chloroform-methanol solution.

A final point is that of localisation. Since the visualisation step is separate from the binding step, and may indeed involve the growth of large crystals of metal or pigment,

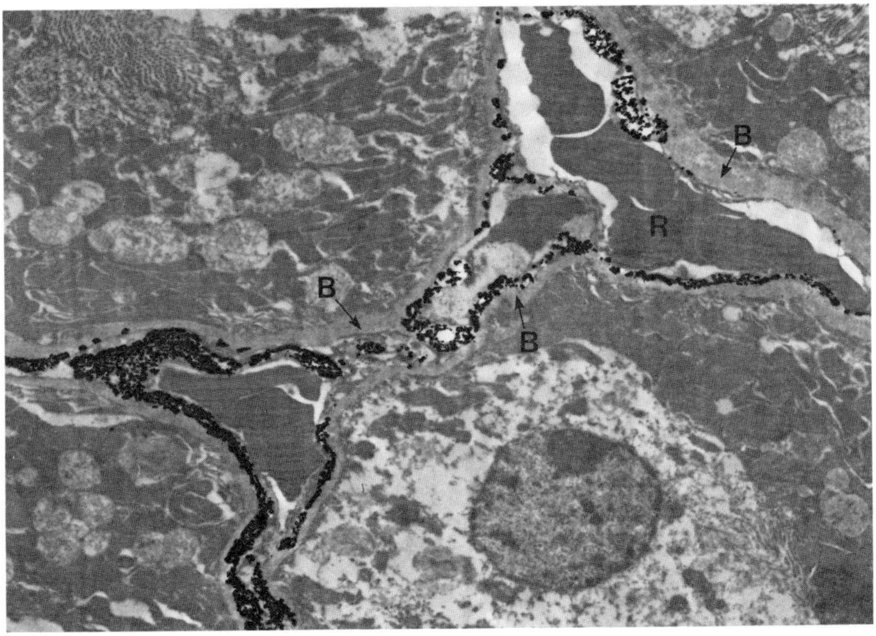

Fig. 42. An electron micrograph showing rat kidney cortex fixed in neutral formalin (only, hence loss of membranes) and stained by the Gordon and Sweet silver impregnation method after prior permanganate oxidation. Note that crystals of silver are deposited in the region of, but not always within, the basement membranes. B basement membrane, R red blood cell. ×6250

localisation can be poor. This is illustrated in the electron micrograph of tissue stained by the Gordon and Sweets reticulin procedure in Fig. 42. Note that the deposition of large crystals does have the advantage that since the crystals may extend beyond the boundaries of the structure giving rise to them, then the detection of small entities may actually be facilitated.

7 Enzyme Histochemistry

7.1 Essential problems and characteristics of enzyme histochemistry

The basic problem of enzyme histochemistry is that an enzyme is merely another protein, and cells are full of proteins. How can we distinguish one protein from all the others? The group of techniques usually called enzyme histochemistry does so by utilizing the enzymic properties of the protein which is to be demonstrated. Other strategies, such as using a labelled anti-enzyme antibody or a labelled inhibitor are mentioned in Chapter 8, and will not be discussed here.

The characteristic strategy of enzyme histochemistry, illustrated in Fig. 43, involves the enzyme-catalysed conversion of an applied substrate into a coloured final reaction product. This may be achieved directly, or via a colourless initial reaction product with the aid of capture and/or visualisation reagents. To aid the process the native substrate may be replaced by a synthetic compound. A modification of this is to detect the loss of

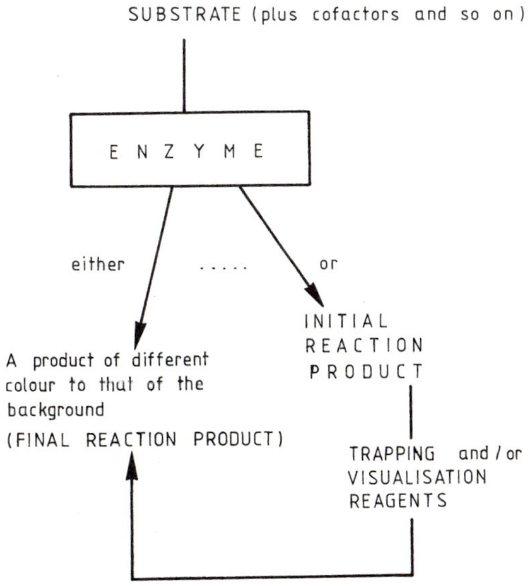

Fig. 43. The basic strategy of (most) enzyme histochemical staining methods.

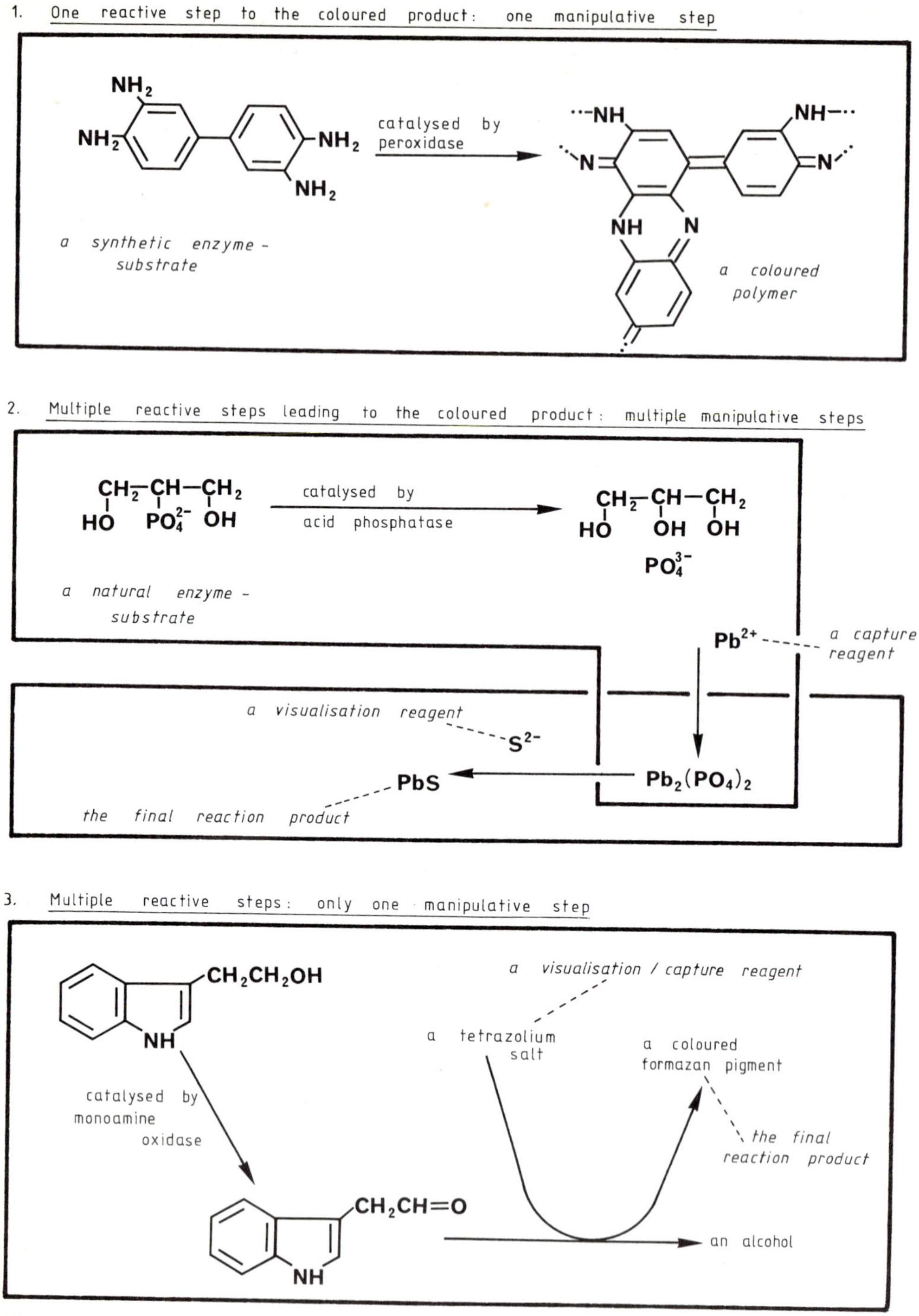

Fig. 44. Examples of various practical procedures used in enzyme histochemistry. Each box represents a separate manipulative step with a separate solution of reagents, each heavy arrow a separate reaction step

substrate, by using a coloured or colourable substrate. The chemical outlines of several common methods are shown in Fig. 44, to demonstrate tactical variants in current use and to illustrate some of the jargon used in this field.

By now you may be thinking how well defined and rational enzyme histochemistry seems compared to, say, staining with dyestuffs. Alas, a major admission must be made: most of the problems now to be discussed arise from the fact that enzyme histochemistry is *not* merely assorted biochemical assays carried out on non-homogenised tissue. All the complications involved in biochemical assays, such as the

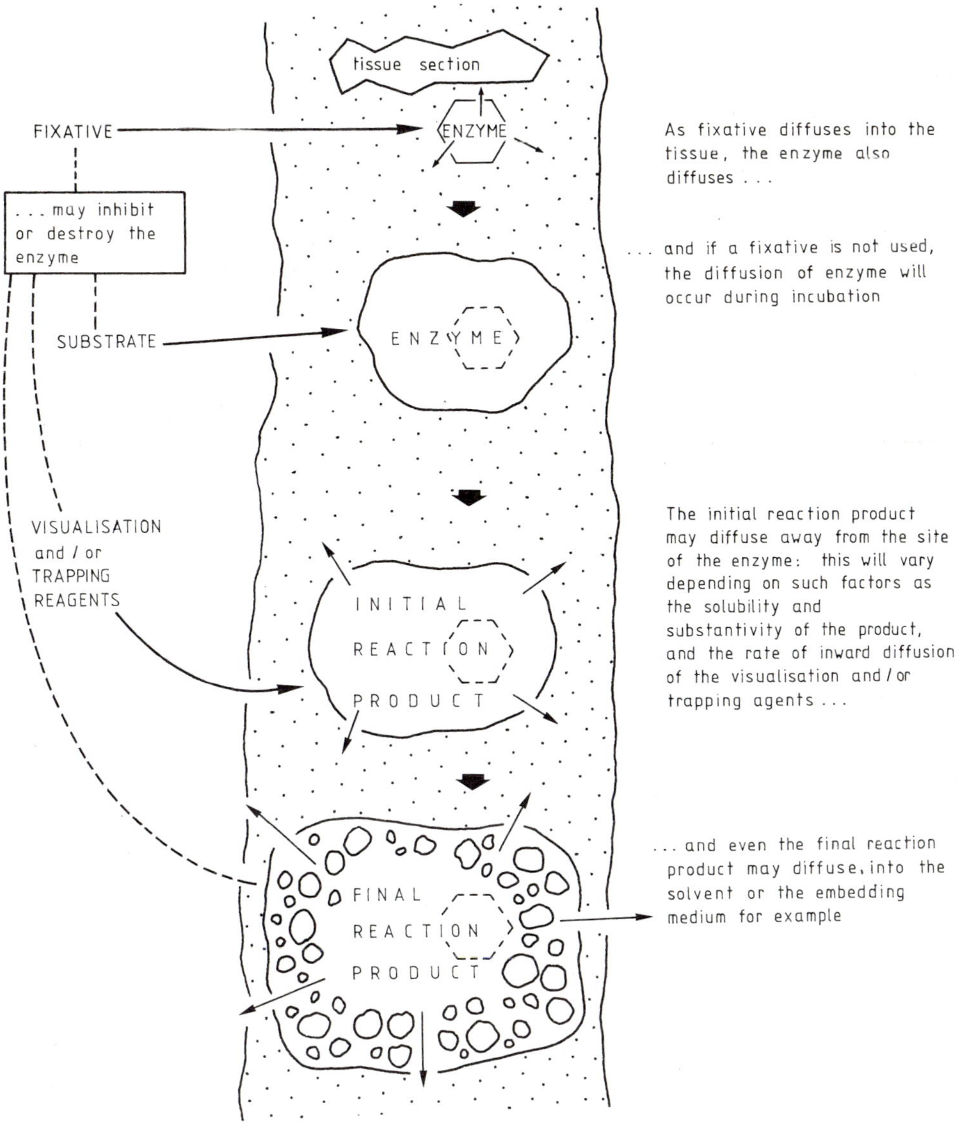

Fig. 45. An illustration of some of the complications of histochemical, as compared to biochemical, enzyme assay involving an example where the initial reaction product is not coloured

question of enzyme specificity and of inhibition by substrate, reagents and products, must be taken into account. The need to preserve the histological and intracellular structures causes additional complications. Furthermore there is the tendency of enzymes and of reaction products to diffuse through the tissues, and the ability of reagents and reaction products to become adsorbed onto certain tissue constituents. Such complications are illustrated diagrammatically in Fig. 45.

In this Chapter the various steps and procedures outlined above will be discussed in more detail. Emphasis will be placed on the characteristic complications of this methodology, namely enzyme localisation versus loss of enzyme integrity, and the addition of histochemical complexities to biochemical complications. The discussion will be centred around the practical problems and techniques. A more rigorous physico-chemical approach is provided by VAN DUIJN's (1974) lucid essay.

7.2 Ultrastructural localisation of enzymes versus preservation of their biochemical integrity

The essential, and conflicting, requirements for successful enzyme histochemistry are:

1. That morphology, both of tissues and cells, should be preserved: otherwise the precise site of the enzyme will remain uncertain.
2. That enzymes must be either insoluble or insolubilised: otherwise they will be lost into the reagent solutions, or will diffuse through the tissues.
3. That enzymes should retain their biochemical potency and specificities: or they cannot be demonstrated or recognised.

Preservation of morphology and retention of macromolecules in the tissue is traditionally achieved by fixation. However, this will almost inevitably reduce enzyme activity. In efforts to avoid this impasse many alternative strategies have been explored. When the final study uses light rather than electron microscopy, and so when morphological preservation is less critical, fixation is sometimes merely omitted. From the viewpoint of enzyme loss this is often satisfactory for those enzymes enclosed within membranes, such as the mitochondrial dehydrogenases, but is of little value for soluble enzymes. Quite different loss-limiting, though hardly morphology-preserving, methods do exist. For instance adding various polymers to the reagent solution, using gelled reagents, or applying a semipermeable membrane to the surface of the tissue. When good morphological preservation is essential and the tissue is embedded, as with electron microscopy, a post-staining fixation step is sometimes adopted. Of course the mere preservation of an enzyme, even if undenatured, is not synonymous with the preservation of enzymic activity. In addition the common need for cofactors or activators must be taken into account.

It is instructive to introduce the problems of enzyme fixation by noting some consequences of *not* utilizing a fixative. In one investigation (ALTMAN, 1972) sections of unfixed rat liver were incubated for 20 minutes, during which time some 40% of glucose-6-phosphate dehydrogenase and 90% of 6-phosphogluconate dehydrogenase were lost into the incubation medium. However, even after action of a fixative, diffusion of enzymes can still take place. For instance a study of catalase-containing peroxisomes showed that after fixation with glutaraldehyde the enzyme diffused from the peroxisomes during post fixation storage in buffer, and became adsorbed onto endoplasmic reticulum, mitochondria and ribosomes (Fahimi, 1973). In addition the ultrastructural morphology of cells can be grossly altered from the *in vivo* state by fixation. This is exemplified by the demonstration of non-specific esterase in acetone-fixed, rat hepatocytes. In this study the enzyme appears, artifactually, to be concentrated along the peripheries of cells (Nachlas et al., 1957, their Fig. 1).

The usual effect of a fixative is to grossly reduce enzymic activity. There are exceptions, alkaline phosphatase and ribonuclease for instance are both enzymically active even after prolonged treatment with fixative agents (e. g. Tomimatsu et al., 1971). However, the more commonplace loss of activity is to be expected since, as discussed in more detail in Chapter 3, fixation of tissues is essentially the fixation of protein, such as enzymes. This fixation is usually achieved by inducing gross structural changes in the protein molecules. Disruption of the enzymes' active centre, either by change in the protein's conformation or by chemical modification of the chemical groupings of the active centre, as illustrated by Fig. 46, will be commonplace.

To minimise such loss of activity all sorts of manœuvres have been adopted, as reviewed by Pearse (1968) and Hopwood (1972). The first damage-limiting step is to choose a fixative appropriate to a given enzyme. For instance when demonstrating alkaline phosphatase in tissue sections processed through paraffin, enzyme activity was greater after formalin than after acetone fixation. Another example is provided by glutaraldehyde, which was found to exert markedly different effects on different enzymes. For instance a 4% w/v solution at 4C acting for 1 hour preserved some 75% of the activity of cholinesterase but only 12% of that of creatinine phosphokinase. Such instances could be multiplied vastly, but to little purpose since the number of practical variables is so large that rigorous comparisons of published work are rarely possible. However, acetone and alcohol are often least damaging to enzymic activity. Amongst the aldehyde fixatives hydroxyadipaldehyde is often least damaging and glutaraldehyde the most, with formaldehyde occupying an intermediate position. Not surprisingly the fixatives least damaging to enzyme activity often give the worst preservation of morphology.

Several practical factors have been identified as important for a wide variety of enzymes and fixatives. Enzyme activity is often better preserved by use of short fixation times, even down to a few seconds; low fixation temperatures, but beware artifacts due to excessive cooling rates; low concentrations of fixative; and careful washing out of fixative before staining. Some workers claim that the use of a brief period of pre-incubation fixation achieves a reduction in enzyme lost from the tissue without gross enzyme inactivation, with the major period of fixation being after incubation (for example, Hanker et al., 1973). Another dodge which, on occasion at least, appears valuable is to fix in the presence of the enzyme's substrate. Perhaps this protects the active site? Thus substrate can be added to the fixative solution, for example β-glycerophosphate to alcohol when fixing alkaline phosphatase (Mowry, 1949), and ketoglutarate when fixing glutamate oxaloacetate transaminase (Papadimitriou and van Duijn, 1970). Alternatively fixative can be added to the staining medium, alcohol and formaldehyde for instance have been so used.

Certain studies involving biochemical enzyme-assay before and after fixative treatment have a potential source of error, since post-fixation activity may be due to both insoluble *and* still-soluble enzyme. For example after several hours in hydroxy-adipaldehyde solution more than half the remaining β-glucuronidase activity was found

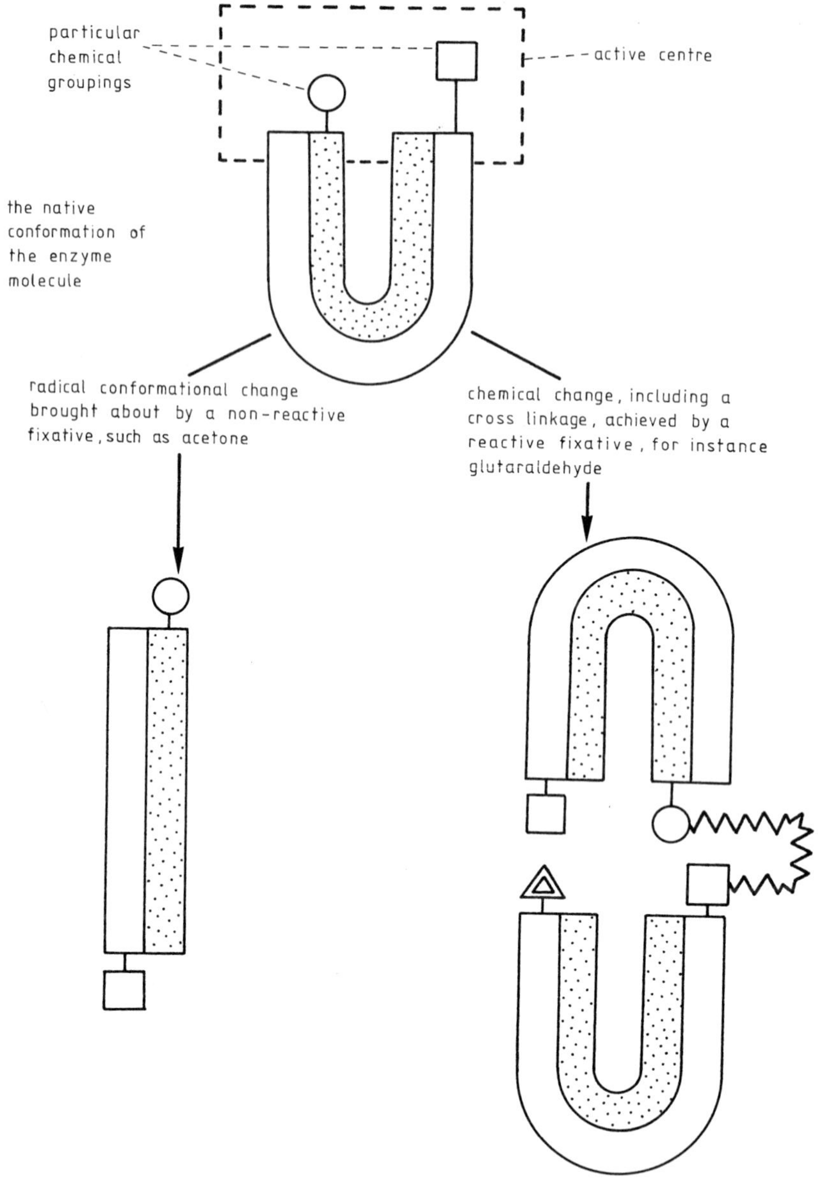

Fig. 46. A diagrammatic illustration of the destruction of an enzyme's active centre by fixatives. Compare to Figs. 7, 9 and 10

to be due to soluble enzyme, and 28% when formaldehyde was the fixative. Only with glutaraldehyde was all residual activity due to insolubilised enzyme.

The non-fixative procedures for limiting enzyme loss are various. Most involve addition of polymers to the incubation solutions, even to the extent of gelling these media. Another approach is to cover the tissue section with a semipermeable membrane, and yet another is to *allow* the enzyme to diffuse: into a solid film in which enzyme action may be detected. These varied approaches are illustrated diagrammatically in Fig. 47.

Perhaps the most direct approach is to separate the tissue section and incubating solution with a membrane permeable to the enzyme substrate and to the capture and visualisation reagents, but through which the much larger enzyme molecules cannot pass. This has been done deliberately, e.g. using a cellulose film for the demonstration of acid phosphatase, β-glucuronidase, non-specific esterase and lactate dehydrogenase (see Meijer, 1973 and cited references) and, more adventitiously apparently, using a celloidin membrane (Christie, 1973). These methods have been applied for example to oxidative and hydrolytic enzymes, and certainly provide increased staining intensities. Though there is neither inhibition by fixative nor loss of enzyme into the staining medium, the question of the validity of the localisations achieved remains open since diffusion behind the membrane could still occur. Hence one wonders what is the significance of the observation (Meijer, 1972, my emphasis) that «Compared with the

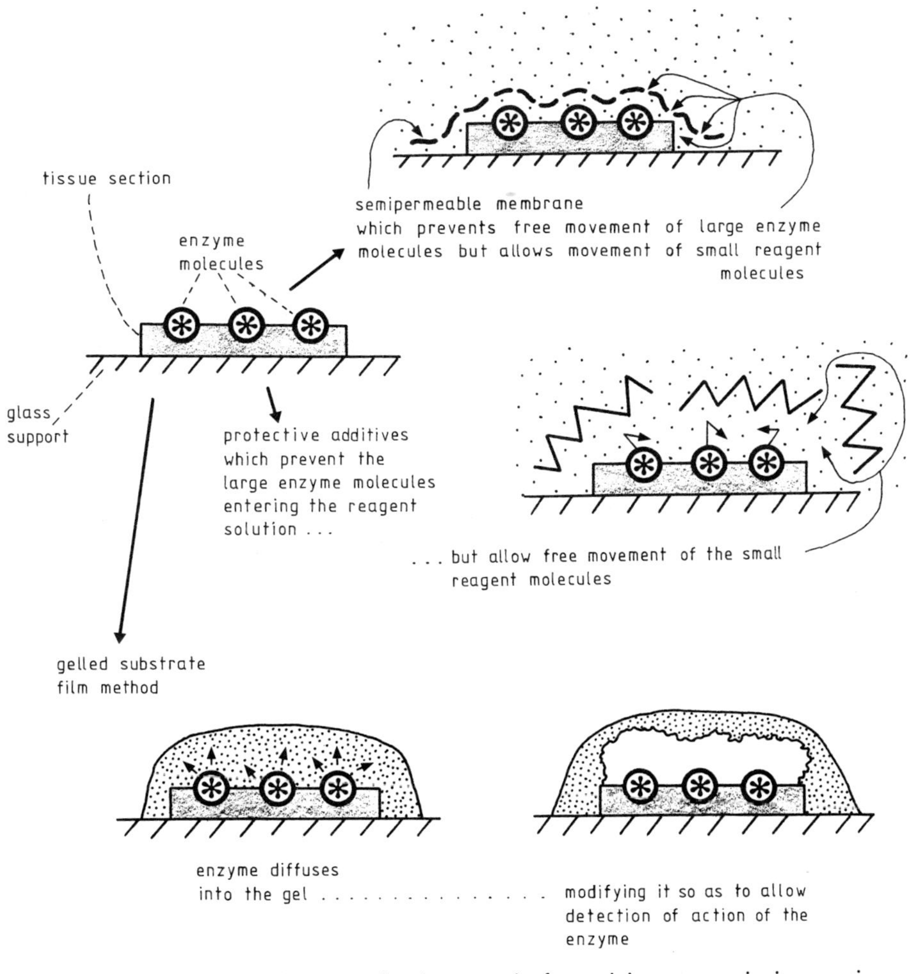

Fig. 47. A diagrammatic representation of two non-fixation strategies for retaining enzymes in tissue sections; also a method of localising enzyme even when diffusion occurs

results obtained with the conventional techniques the *diffuse* staining is much more intense»?

Addition of protective additives to incubation media has become more commonplace. A variety now exist and have been applied to the demonstration of a number of enzymes, some examples are given in Table 39.

Table 39. Examples of protective additives or stabilisers added to incubation media to prevent loss of enzyme from unfixed frozen sections

Additive or Stabiliser	Enzyme demonstrated	
Acrylamide	A wide variety of oxidative enzymes	Ruitenbeek and Scholte (1976)
Agar	enterokinase; thiamine pyrophosphatase	Lojda and Malis (1972), Lukas et al. (1978)
Collagen polypeptides	glucose-6-phosphate dehydrogenase	Butcher (1971)
Dextran	Acid phosphatase, ATPase, DPN diaphorase, β-glucuronidase, succinate dehydrogenase	Scarpelli and Pearse (1958)
Gelatin	creatin kinase	Khan et al. (1972)
Polyvinyl alcohol	glucose-6-phosphate dehydrogenase, 6-phosphogluconate dehydrogenase	Altman (1972)
Polyvinyl pyrrolidone	phosphorylase	Eränkö and Palkama (1961)

Some impressive claims for the efficacy of such additives have been made on the basis of quantitative studies. Either collagen polypeptides or polyvinyl pyrrolidone will apparently entirely prevent the loss from frozen sections into the incubation media of a soluble enzyme such as glucose-6-phosphate dehydrogenase (Altman, 1972; Butcher, 1971). However, other workers, studying the mitochondrial enzyme lactate dehydrogenase, found appreciable though reduced preservation (Ritter et al., 1971). A complicating factor is the possible inhibition of enzyme activity by such additives. In a study of brain homogenates, both polyvinyl pyrrolidone and sucrose diminished acetylcholine esterase, lactate and succinate dehydrogenase activities (Lisy et al., 1971). However, such effects generally seem to be regarded as unimportant in the histochemical context.

How do such additives work? They were orginally introduced as «osmotic protective agents», though most are polymeric and so have only a minor osmotic influence. A later suggestion was that loss of enzyme from tissue sections was reduced due to the additives raising the viscosity of the media. Acrylamide, agar and gelatin gels have been used, and polyvinyl pyrrolidone and some polyvinyl alcohol solutions are also extremely viscous. A more recent view, accounting for the effectiveness of agents which do not increase the gross viscosity of the reagent solutions, is that polymeric additives act as excluded volume to the bio-polymers of the tissue, though not to small substrate and other reagent molecules (Scott, 1974). This amounts to a solute-selective viscosity increase. Regardless of the details, such methods are conceptually very similar to the semipermeable membrane strategy: except that the membrane is now as thick as the incubation solution.

Admittedly somewhat anomalously we will now consider the strategy of *allowing* the enzyme to diffuse. Substrate may be applied to the tissue section or smear as a solid layer or film, hence the term substrate film methods. Enzyme is allowed to diffuse into and induce reaction in this layer. The altered region is then visualised in some way. For details of the varieties of substrate and visualisation available see Sections 7.3 and 7.5 below.

To round off this discussion we may note that the choice of appropriate fixative or additive is not the only practical decision of importance. Other, erratically, damaging operations can include the mode of killing of the experimental animal (Al-Azzawi and Stoward, 1973); the pre-fixation period of anoxia; the freeze-thaw cycles in cryostat sectioning; the desiccation of sections or smears; and osmotic imbalance between organelles and fixation or incubation media. Such operations influence staining largely by disrupting organelles and so causing losses of integrity of enzyme systems as discussed by Pearse (1968, p. 477) and Altman (1972). Loss of integrity may be due to dispersion of enzyme as already considered, or of a structure, or of cofactors and activators. Losses of low molecular weight compounds into pre-staining or staining solutions will not be readily prevented either by fixation (compare Section 3.6) or the use of additives (Ritter et al., 1971), though replacement may sometimes be possible. An example is the demonstration of alkaline phosphatase in hard tissues after decalcification by EDTA-HCl. Staining is only possible if Mg^{++} ions are applied in a pre-incubation step (Yoshika et al., 1972).

7.3 Substrates and incubation media

Discussion of some major biochemical topics is reserved until Section 7.7. We here limit the breadth of our biochemical gaze to consequences of the essential biochemical requirements of substrate and medium allowing demonstration of an enzyme. Simultaneously the characteristically histochemical needs can be considered. Overall the minimum requirements are:

1. The substrate and medium should be chosen keeping in mind the specificity and inhibition/activation characteristics of the enzyme under study.
2. The substrate should be such that the reaction product (Initial, Final or indeed both) is strongly coloured, immobile on the tissue, and is produced in amounts proportional to the activity of the enzyme.

Point (1) implies that the favoured substrate should be the natural one, not be present in excess or substrate inhibition could occur, and be free of extraneous inhibitors (or activators?). The proportionality of reaction product to enzymic activity, mentioned in Point (2), means that substrate should be present in excess at the site of enzyme action (Holt and O'Sullivan, 1958). This in turn requires that the substrate should diffuse into all parts of the tissue rapidly compared to the rate of subsequent reaction. And so on. These requirements only too clearly conflict in various ways. Some of these contradictions are shown dramatically in Fig. 48.

In practice one's compromises are largely empirical since there are just too many variables to permit rational predictions. However, substrates are usually chosen to have low molecular weights, aiding rapid diffusion, and are often highly soluble in water. If not then suitable organic solvents may be added to the medium. Such solvents can also actively assist the penetration of other reagents, an example being the use of

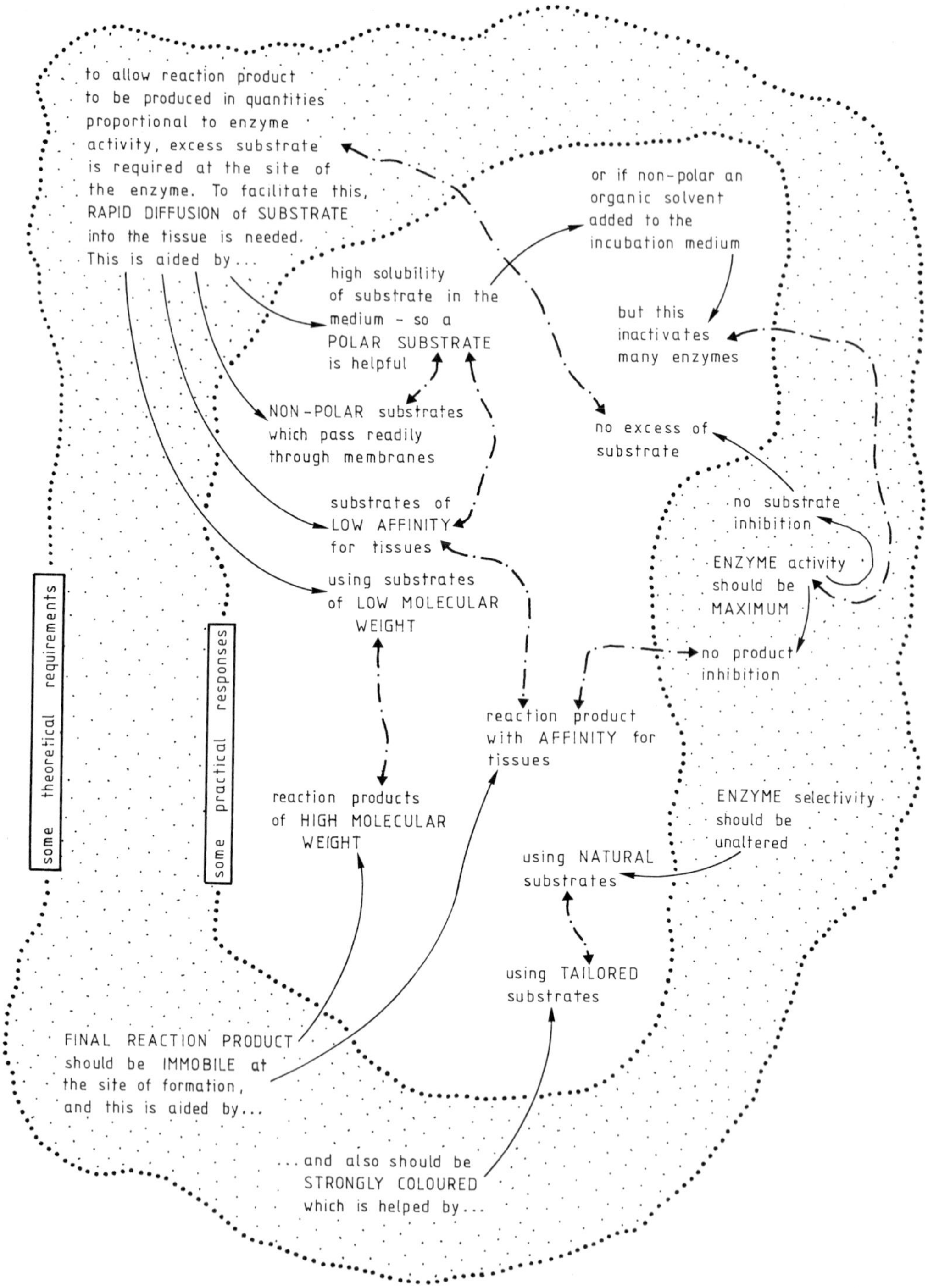

Fig. 48. A diagrammatic illustration of some conflicting requirements of an incubation medium used for enzyme histochemistry. Conflicts are emphasised by the heavy, double-ended arrows. Readers reminded of a snakes and ladders board are, unfortunately, only too correct

dimethylsulphoxide. Factors other than molecular size and polarity can influence the rate of reagent penetration of tetrazolium salts (CASCARANO and ZWEIFACH, 1961).

One topic requiring further comment is enzyme specificity. Consider the type of substrates used in enzyme histochemistry. Oxidoreductase demonstrations often require the use of the natural substrates, malate for malate dehydrogenase, succinate for succinate dehydrogenase, and so on. Here no direct worries concerning specificity arise, though for more subtle ambiguities see Section 7.7. Many hydrolases, however, require the use of synthetic substrates to facilitate development of a coloured final reaction product; such as arylphosphates and sulphates used for demonstrating phosphatases and sulphatases. Here pertinent questions are what does the «aryl sulphatase» demonstrated histochemically represent in biochemical terms? And is the substrate hydrolysed by more than one enzyme? An example of such ambiguity arises from the popularity of both β-glycerophosphate and p-nitrophenyl phosphate as substrates for the demonstration of «non-specific phosphatase». p-Nitrophenyl phosphate is sometimes regarded as the better substrate since heavier and faster staining can be achieved by its use (BUTTERWORTH, 1971). As biochemical studies indicate p-nitrophenyl phosphate is hydrolysed by more phosphatases than is β-glycerophosphate (NEIL and HORNER, 1962), the heavier staining may thus merely represent a less specific enzyme demonstration.

We will end this discussion with a ragbag of cautionary tales. For instance an important technical factor widely appreciated in biochemical assays, and acknowledged by some (for example LOJDA et al., 1964) in the histochemical literature, is substrate purity. If a naphthyl substrate contains a mixture of isomers these may well have mixtures of properties, for example different rates of reaction. If free naphthols are present, and in some commercial products this is still the case, these substantive compounds may be adsorbed onto the tissue and so give rise to non-specific staining, see Section 7.7. A related problem is substrate instability. 6-Bromo-2-naphthyl acetate for instance hydrolyses rapidly in alkaline aqueous solutions in the absence of enzyme, giving rise to the more substantive 6-bromo-2-naphthol (DEFENDI, 1957). Analogous is the non-enzymatic hydrolysis of ATP, to ADP and inorganic phosphate. This reaction is catalysed by lead ions, such as those present as part of Gomori-type phosphate-visualising systems. This has been considered a serious histochemical problem, for instance by MOSES et al. (1966), though this is not a universal viewpoint (MEIJER, 1970). Incidentally, the substrate may have unexpected interactions with other parts of the enzyme-demonstrating system. The ATP in certain ATPase demonstrating systems is not merely a substrate since it grossly alters, by complex formation, the nature of the lead species present (BERG et al., 1972). Similarly complexes with lead ions may be formed by simple species, such as fluoride ion, added to the incubation medium to increase specificity or activity (MOSES et al., 1966). Such simple species are sometimes essential, for example calcium ions in demonstrating myosin adenine triphosphatase (MEIJER, 1970). In other instances however activators necessary in biochemical assays are not required in the histochemical system, perhaps due to adsorption of sufficient ions such as Mg^{++} and Ca^{++} onto the tissues. Finally note that such a seemingly straightforward operation as washing out substrate and so on after incubation may in practice prove difficult, so allowing the generation of positive artifacts (HAYASHI, 1977).

7.4 Formation and properties of the initial reaction product

An objective of enzyme histochemistry is that the final, coloured, reaction product should be formed at the site of action of the enzyme. Consequently various requirements concerning any initial reaction product can be specified as follows:

1. The initial reaction product should preferably be immobile.
2. Immobility may be achieved either by:
   a) High product-tissue affinities, or
   b) insolubility in the incubation medium.
3. If however the product is mobile then it should be trapped or visualised very rapidly.
4. Both 2b and 3 will be facilitated by rapid formation of the product.

The various aspects of this situation will be discussed separately, as a narrative convenience.

Perhaps the most straightforward situation is when the initial reaction product is insoluble and precipitates at the site of enzymic activity. In this instance the essential problem is how to produce an insoluble initial reaction product from a substrate of high solubility. This is usually achieved by making the enzyme-modified portion of the substrate a solubilising group. For instance ionic naphthyl phosphates and sulphates are converted by phosphatases and sulphatases respectively to non-ionic, and so much less water-soluble, naphthols. Careful design or choice of substrate will enhance this effect. Thus the introduction of hydrophobic substituents into 2-naphthol may reduce its solubility in water several hundred times. This tactic is used in the preparation of substrates from the Brentamine or Naphthol AS series of naphthols. The need to transform soluble substrates to insoluble products can at times be avoided, by using substrates which are not truly water-soluble. In certain methods for demonstrating lipases the insoluble substrate is applied as an emulsion, producing as initial reaction product a precipitate of an insoluble fatty acid-calcium soap.

Solubility of the initial reaction product may be further reduced by careful consideration of the nature of the incubation medium. Here we must appreciate that distinguishing between insoluble initial reaction products and soluble ones requiring trapping is a mere convenience of classification and not one of rigid physicochemical significance. In this context pH is often an important factor. The water-solubility of initial reaction products such as naphthols and thiols rises due to ionisation as the pH is increased, whilst naphthylamines show the reverse effect. Consequently the optimum pH of the incubation medium is sometimes a compromise between insolubilising the reaction product and activating the enzyme, and on occasion also of activating the capture or visualisation reaction. Another stratagem involving the incubation medium is to saturate it with the initial reaction product. An example is found in the demonstration of choline esterase using thiocholine esters, when the incubation medium may be saturated with copper thiocholine. This procedure is of course limited to cases where reaction product-tissue affinities are not significant.

Note that *in*soluble only means very slightly soluble. A redistribution of the reaction product via the solvent is always possible. This may be a straightforward outward diffusion, giving rise to a region of reaction product adjacent to the enzyme sites. Alternatively the reaction product may be adsorbed onto specific sites of high affinity far from or near to the enzyme; and this is discussed further in Section 7.7. The rate of formation of the reaction product is consequently important. Since no precipitation, as distinct from adsorption, of the reaction product will take place until some critical concentration is reached, a high rate of formation will limit whatever diffusional spread does occur.

Quite another factor limiting mobility of the initial reaction product is reaction product-tissue affinity. This does seem of some significance, for instance naphthols substituted by halogen and other non-polar groupings seem to bind more effectively than the less substituted derivatives (see DEFENDI, 1957; HESS and PEARSE, 1961). Perhaps (see Section 4.4) because of enhanced van der Waals attractions and hydrophobic effects? With substituted naphthyl compounds as substrates however this aspect of immobilisation has largely been by-passed, due to the widespread use of substrates derived from the Brentamine/Naphthol AS series. An outline discussion of these issues from a structure-staining viewpoint may be found in Section 11.14.

Finally, a few comments on those initial reaction products which are soluble and are not attracted by the tissue. In such cases localisation depends on effective and fast acting capture-visualisation procedures. This approach, which is discussed in Section 7.5, is only by convenience of decription a new methodology. The classic example is found in the Gomori-type methods. Here the reaction product is phosphate ion and this is trapped by lead or calcium ions. In some procedures of this sort, effective intracellular localisation of the initial reaction product is not achieved. An example is the demonstration of disacharidases, where the reaction product is glucose, see LOJDA (1965).

7.5 Characteristics of the capture-visualisation steps

Sometimes it is the capture of the initial reaction product that is of paramount importance, as with the precipitation of phosphate ions by lead or calcium. Sometimes, with post-coupled azo-dye methods for instance, the initial product is fairly immobile and the important aspect is the generation of colour. Sometimes, as with simultaneous-capture azo-dye procedures, the capture and visualisation steps are one and the same. Alternatively the steps are multiple, as when phosphate is first trapped as lead phosphate, and this colourless material is then transformed to black lead sulphide. Sometimes no capture or visualisation is necessary as the initial reaction product is both immobile and coloured, an example being the coloured polymeric materials generated from benzidine and its derivatives by the action of peroxidases.

In spite of this diversity one can specify certain desirable characteristics of capture-visualisation reagents. To limit the diffusion of the initial reaction product from the site of the enzyme the reagents themselves should diffuse rapidly, and react rapidly with the initial products. Yet these reagents must generate final reaction products which are coloured and immobile. Such varied requirements are, as with the analogous case of substrates, in conflict. A chart could easily be constructed ressembling Fig. 48. Here we will merely consider how these factors are balanced in a variety of capture-visualisation procedures. The major practical approaches will be emphasised: azo-coupling of naphthols and naphthylamines to yield azo dyes; reduction of indoxyls to indigos, or of tetrazolium salts to formazans; and production of lead or transition metal sulphides from precipitated phosphates or carboxylates.

Firstly methods, other than redox reactions, in which an (organic) initial reaction product is converted to a coloured (organic) final reaction product. Examples of various strategies are shown in Fig. 49, though here azo-coupling methods are emphasised. The

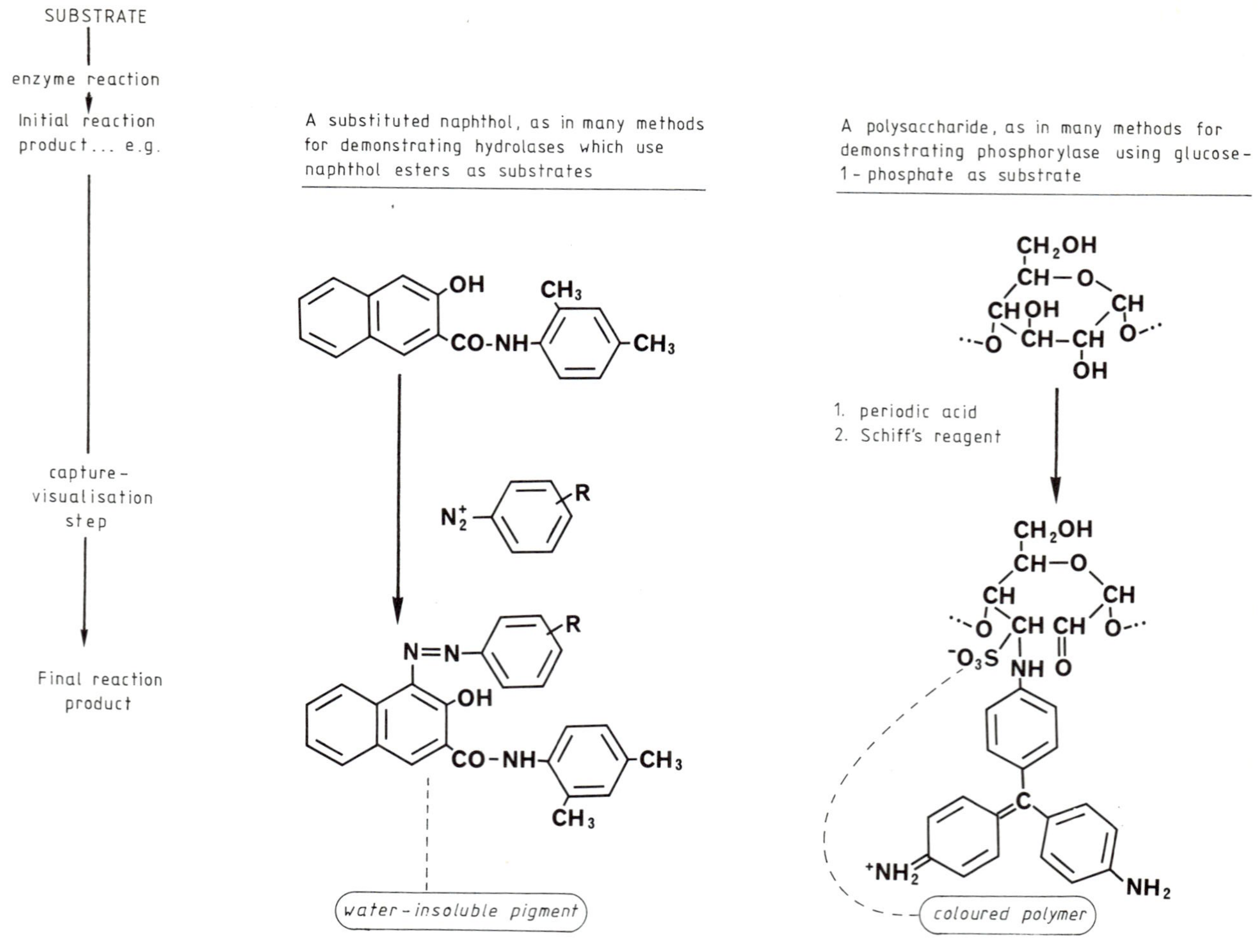

Fig. 49. Examples of capture-visualisation procedures in which an organic initial reaction product is converted to a coloured organic final reaction product

capture-visualisation procedure may involve a single reactive step, or several steps, and may use a single reagent solution or several.

The success of the capture-visualisation process, at least as far as localisation is concerned, is critically dependent on the balance of the rates of various processes. Not surprisingly these requirements can conflict. For instance in methods using azo-coupling as the capture-visualisation reaction, the rate of the azo-coupling can be the critical variable. With poorly soluble and so effectively poorly diffusing AS Naphthols, good enzyme localisations are achieved using naphthols which react rapidly *or* slowly with diazonium salts. However more soluble and so more diffusible naphthols give good final enzyme localisations only if the rate of azo-coupling are high (DEFENDI and PEARSE, 1955). Reaction of the diazonium salts with tissue groupings (for instance tyrosine, tryptophane or histidine residues) can occur but will probably give rise to azo-dyes of different colours to the final reaction products of the enzyme demonstration process. Loss of diazonium salt due to such reactions, and also due to reaction with the aqueous solvent, is usually trivial since the salt is present in great excess. Another factor is the optimum pH of azo-coupling. If the initial reaction product is a naphthol then its rate of azo-coupling will increase with pH, since the active species is the phenoxide anion. However phenoxide ions are much more soluble in water than the unionized phenols. Also, if a simultaneous-capture procedure is used, the pH optima for the azo-coupling and the enzyme-catalysed reactions may be very different. Such problems are normally resolved by trial and error.

Another important group of methods use as a capture-visualisation process *a redox reaction converting the organic initial reaction product into a coloured organic final reaction product.* Such a process is involved in the indoxyl methods. Here the enzyme substrate consists of a suitably substituted indoxyl derivative, such as an acetate or glucoside for demonstration of an esterase or glucosidase. From such substrates enzyme action generates as initial reaction products indoxyls with free enol-keto groupings. Oxidation (the visualisation step) converts this to the final reaction product, an insoluble, coloured indigo derivative. Such a process is shown in Fig. 50.

If the capture-visualisation step involves oxidation by atmospheric oxygen the reaction is slow compared to the rate of generation of the initial reaction product, and consequently there is poor localisation of the enzyme. However when the oxidative rate is accelerated by the addition to the system of catalytic amounts of copper or ferro-ferricyanide ions excellent localisations may be obtained.

Perhaps the best known capture-visualisation methods based on redox processes utilise the reduction of colourless, water-soluble tetrazolium salts to coloured, poorly soluble formazans. For an example see Fig. 50 where the initial reaction product is a reducing species, such as a reduced cytochrome, in the electron transport chain. The problem of identifying just what is the initial reaction product in this case is considered in Section 7.7. The rate of entry into and affinity for the tissues of tetrazolium salts has received some study. For instance the entry rate has been found to vary with tissue oxygenation (CASCARANO and ZWEIFACH, 1961). However the only physicochemical factor regarded as commonly significant is the ease of reduction of the tetrazolium compounds. For example in a demonstration of monoamine oxidase using tryptamine as substrate the initial reaction product is an aldehyde which will reduce tetrazolium salts. Best results are said to have been obtained using «easily reduced» tetrazolium salts, where ease of reduction could be predicted by considering the measured redox potentials of the salts. Since good correlations of histochemical utility with such

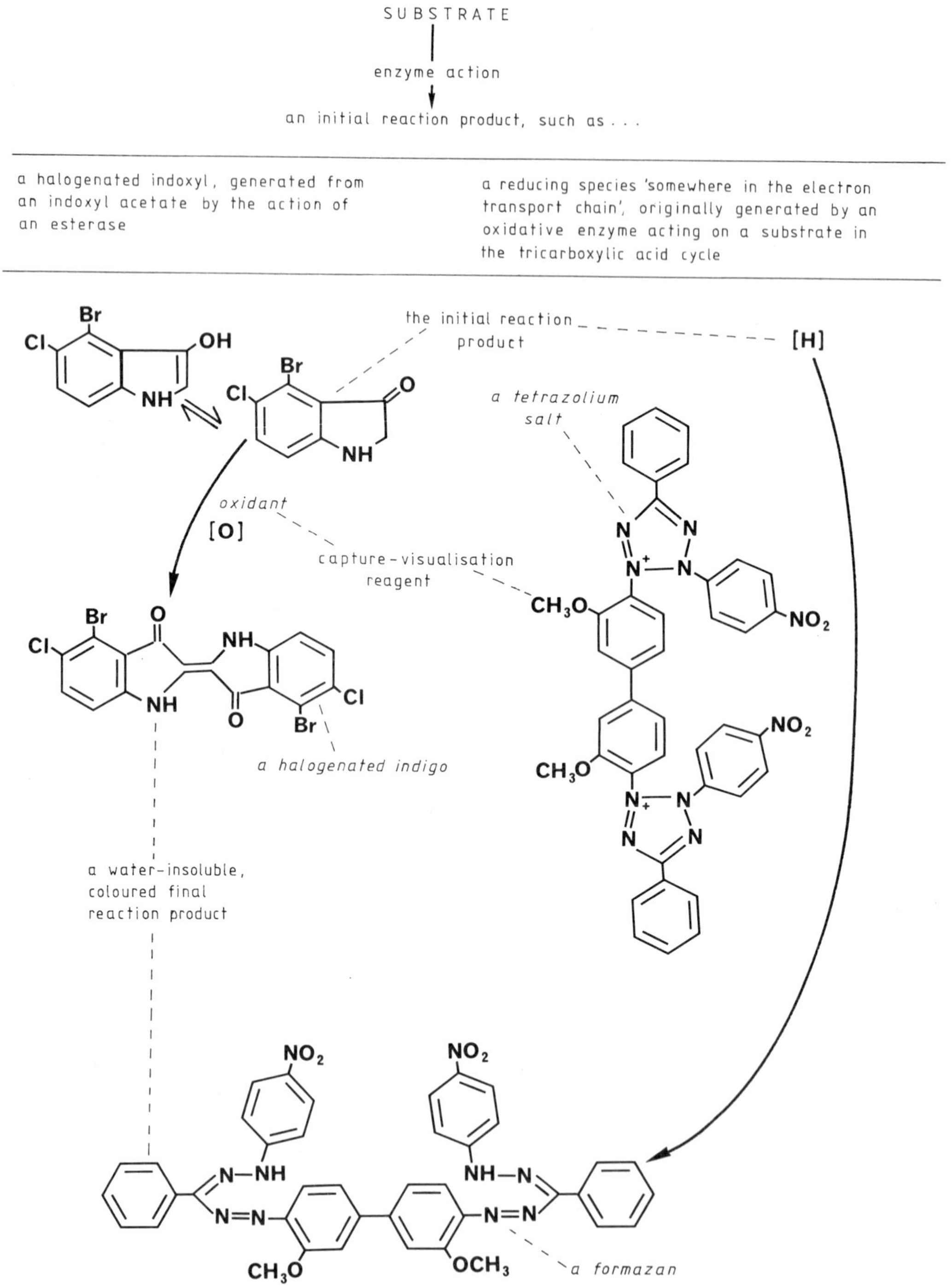

Fig. 50. Examples of capture-visualisation procedures in which an organic initial reaction product is converted to the final reaction product by a redox reaction

physical measurements are not always found, some authors (for example ALTMAN, 1972) are inclined to feel that factors other than or additional to redox potential must be important.

Other redox-based capture-visualisation methods are used. For example, cholinesterase and acetyl-choline esterase may be demonstrated with incubation medium containing a suitable acyl thiocholine as substrate, plus ferricyanide and citrate-chelated copper ions. From the substrate the esterase generates free thiocholine, whose thiol group reduces the ferricyanide to ferrocyanide. This is then precipitated as its copper salt, which is known as Hatchetts Brown. Though this is indeed a redox process it bears more resemblance to the metal-precipitation methods discussed below than to the tetrazolium or indoxyl procedures previously described.

A third widely used capture-visualisation strategy is to *precipitate the initial reaction product as an insoluble metal salt or co-ordination complex.* This may itself be coloured or may require conversion into the final reaction product. Several examples of this approach are shown in Fig. 51. A given initial reaction product may be captured in more than one way, and a given captured product may be visualised in more than one way. Each approach represents a different compromise. If lead ions are used to capture phosphate ions then enzyme inhibition will be appreciable; but if the less enzyme-inhibitory calcium ions are used then an additional pre-visualisation step becomes necessary. Similarly the payoff from introducing an additional silver intensification step in choline esterase demonstrations is that substrate incubation times may then be shortened, so reducing diffusion of the initial reaction product.

The precipitation of inorganic phosphate by lead ions is the best known and most studied capture process and will be here taken as a general case. The critical physico-chemical events in the precipitation of an insoluble salt such as lead phosphate are sketched in Supplement 11. For good visualisation the initial reaction product, namely the phosphate ions, must be efficiently precipitated near the site of the enzyme. As the Supplement indicates, this will entirely depend on the lead ions being super-saturated. If such supersaturation is not achieved, phosphate ions will move away from the enzyme site. Eventually these ions will be either lost into the incubation medium (reducing the sensitivity of the process), or precipitated onto some non-enzymic site (reducing the selectivity). Consequently as the concentration of lead ions falls, diffusion artifacts occur, commonly including nuclear staining. This has been attributed to heterogeneous nucleation aided by high concentration of the lead cations in the region of the polyanion (CORNELISSE and VAN DUIJN, 1973). It is of considerable practical importance, as numerous workers (BERG et al., 1972; HOLT, 1959; MOSES et al., 1966) have shown that the critical lead concentration is that of those lead complexes active in precipitating phosphate. Thus the critical lead concentration in the incubation medium can be lowered either by the presence of precipitants (such as glycerophosphate present as an enzyme substrate, which slowly gives rise to a lead-containing precipitate), or by ligands such as fluoride or ATP binding to lead ions to give soluble lead compounds that are ineffective as capture species.

The presence in the incubation medium of cations such as calcium or magnesium which compete with lead for ligands such as ATP or fluoride, increase the efficiency of the precipitation process. As do anions, such as chloride, which lower the solubility of lead. Furthermore the material precipitated in the capture process need not be lead phosphate only. When demonstrating ATPase for example a mixture of lead orthophosphate and a lead complex of ADP is precipitated (BERG et al., 1972).

Visualisation of a metal-containing precipitate is commonly achieved as illustrated in Fig. 51, by conversion of the initial metal salt to its sulphide. This may be direct when the metal present in the precipitate gives rise to coloured sulphides, as is the case with cobalt, copper and lead; or indirect as when an initial cation exchange procedure is used to replace calcium with cobalt or lead ions. A further modification is to exchange the metal cation initially present in the precipitate for silver cations, and then to reduce these to silver metal crystals. Occasionally the metal cations are visualised by reaction with an organic ligand yielding a coloured co-ordination complex. A new and extremely promising approach is to use the visualisation step as an amplifier. Thus certain demonstrations of lactate dehydrogenase or monoamine oxidase produce an initial reaction product of copper ferrocyanide. This compound may then be used to catalyse the oxidation of diaminobenzidine, yielding coloured indamine polymers as the final reaction product. Since the initial reaction product is not consumed in this process, only very small amounts of the ferrocyanide need be produced, so incubation times may be shortened (HANKER et al., 1973). In general there has been little study of the physical chemistry of any of these visualisation processes, in their histochemical setting.

Table 40. Some examples of visualisation procedures used with substrate-film techniques

Nature and function of the substrate film	Mode of visualisation adopted
RNA, contained in a gelatin film, is hydrolysed by tissue RNAase, and the fragments of degraded nucleic acid are lost from the film	Treatment with Toluidine Blue stains only those areas of film still containing the basophilic RNA, leaving colourless the areas over sites containing RNAase
A film of chitin is modified by the action of tissue lysozyme	The modified chitin, but not the original material, is intensely stained by Alcian Blue
A layer of bacteria contained in a film are lysed by tissue lysozyme	Staining by the Gram method fails to colour the lysed bacteria, leaving clear areas above lysozyme containing sites
A stack of coloured gelatin films are hydrolysed by the proteases of tissue sections placed on the uppermost film	No visualisation is required since hydrolysed regions display the colours of the underlying layers
A starch film is hydrolysed by tissue amylases	Unhydrolysed starch is stained by use of Lugols iodine or the periodic acid-Schiff procedure
A tissue section is mounted on a starch film and the whole then incubated in aqueous hydrogen peroxide solution. In the vicinity of tissue catalase the hydrogen peroxide is destroyed whilst elsewhere the film remains saturated with the reagent	On subsequent immersion of the film and section in iodide solution the retained peroxide converts iodide to iodine which then reacts with the starch yielding a blue complex. The peroxide-free regions around the sites of the catalase remain white

In the final paragraph of this discussion, visualisation methods used with the substrate-film procedures will be briefly discussed. The examples in Table 40 show that the staining of the enzymic sites may be positive when the altered substrate layer is itself coloured. However it is more usually negative, when it is the unmodified film which stains leaving the modified region colourless. The mechanisms of visualisation used are very varied. Sometimes none is needed, as when coloured gelatin films are digested away to reveal another coloured layer underneath. Sometimes basic dyeing or reactive staining of the substrate or of the modified substrate is used.

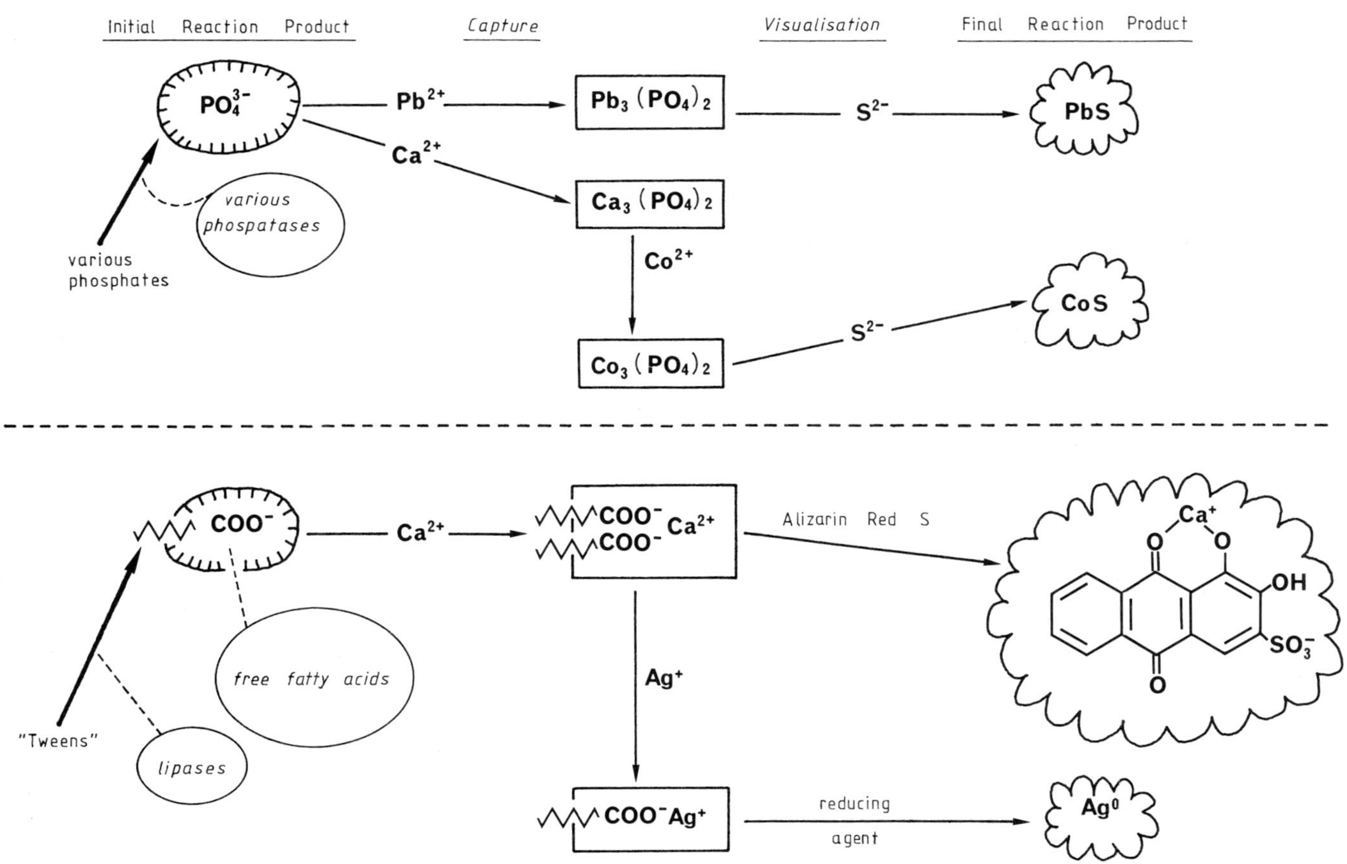

Fig. 51. Examples of capture-visualisation processes in which the initial reaction product is precipitated as an insoluble metallic salt or co-ordination complex

7.6 Properties of the final reaction product

Good enzyme localisations require the final reaction product to remain immobile in the tissue, and be of a non-particulate nature, or if particulate the particles should be very small. Consequently the final reaction product should preferably be chemically and photochemically stable; have a low solubility in reaction solvents, tissue lipids and mounting media; and possess high affinity for tissue proteins. The latter two properties influence both the mobility and the partical size of the final product, since both low solubility and high affinity will inhibit crystal growth. These various characteristics will now be discussed separately.

A variety of unwanted consequences follow from a mobile final reaction product. The diffusion of reaction product away from the enzymic site may be random, so merely decreasing the resolution obtained. Deterioration of resolution can also occur when mobilisation of the reaction product allows its transformation into large crystals. Nitro Blue-formazan for example, though initially amorphous (microcrystalline?) turns into large crystals after mounting the section in a synthetic resin. Crystals are not formed if the mountant is aqueous, in which media the formazan is less soluble. Alternatively, as described in Section 7.7, selective uptake of mobilised reaction product may occur, giving rise to positive artifacts.

A non-particulate stain is one in which the final reaction product is either molecularly dispersed or whose crystals are too small to be resolved by the microscope used. One factor which minimises crystal growth is the chemical nature of the reaction product. For instance the bis-formazans derived from bis-tetrazolium salts are typically non-crystalline, or at least microcrystalline; whereas mono-formazans formed from mono-tetrazolium salts are commonly coarsely crystalline. When demonstrating choline esterase using as substrates choline esters of higher fatty acids, the usual initial reaction products are fatty acid soaps of cobalt. In this instance the final reaction product is finely crystalline cobalt sulphide. If calcium, magnesium or manganese soaps are produced then much larger crystals form. Another influential factor is the nature of the tissue. If the final reaction product is lipid soluble the presence of lipids can facilitate crystal growth. This occurs when using certain indoxyls as substrates, and removal of tissue lipids by solvent extraction has been found to diminish crystal size. Other environmental variables may be important, in the previous example crystal growth is also enhanced if acidic reaction media are used. The crystal size of a final reaction product can also depend on when it is looked at. For example the intra-mitochondrial crystals of Indonaphthol Purple, the final reaction product of a Nadi-type method for demonstrating cytochrome oxidase, are small when first formed but coalesce on standing.

Various strategies are used to obtain *final reaction products of low solubility*. One is to arrange for the reaction product to be polymeric, as with the coloured glycogen derivative generated in certain methods demonstrating phosphorylase; and also the coloured indoaniline and azine polymers resulting from the cytochrome oxidase catalysed oxidation of various aromatic diamines. A more common strategy is to generate a reaction product whose molecules are small but non-ionic, such as formazans, azo-dyes, indophenols; see Figs. 51 and 52 for some examples. However compounds exhibiting low water solubilities may well be soluble in non-polar substances, such as tissue lipids and resinous mounting media (e.g. KUGLER and WROBEL, 1978). To minimise such effects metal co-ordination complexes are sometimes formed, for instance from azo-dyes or formazans, as *final* final reaction products. Examples may be seen in Fig. 52.

The formation of co-ordination complexes leads to final reaction products with high affinities for the tissues. The formation of a co-ordination complex often markedly increases the molecular size of the final product, see Fig. 52. This will typically increase product-tissue affinity due to increased van der Waals attractions, as discussed in Section 11.14 (see also HOROBIN and BENNION 1973; MARSHALL and HOROBIN, 1973 a). The consequent reduction in mobility will also inhibit crystal growth. Large, high

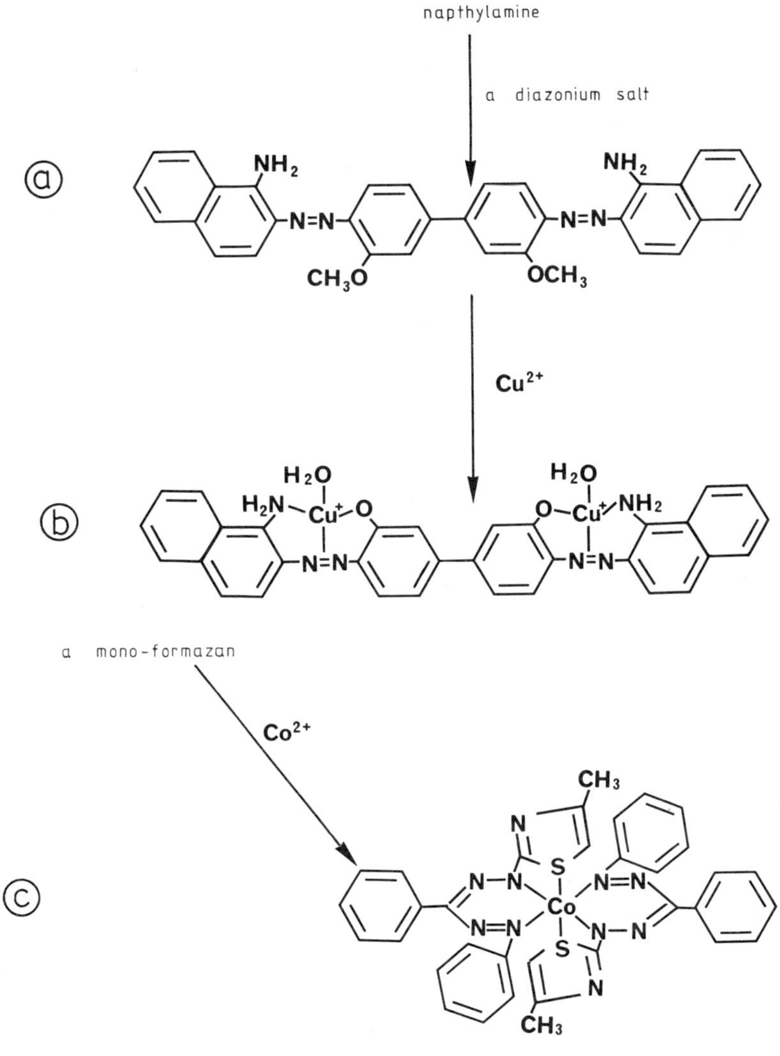

Fig. 52. Structural formulae of some final reaction products. The fairly substantive azo-dye (a), formed from a naphthylamine initial reaction product by reaction with a diazonium salt, may be transformed into an even less soluble and more substantive copper co-ordination complex (b) as a *final* final reaction product. The final reaction product (c) is a poorly soluble, substantive, cobalt co-ordination complex formed from a thiazoylmonoformazan precursor, generated from a tetrazolium salt by some tissue reducing system

affinity final reaction products with such desirable properties can also be obtained in other ways. For instance substantive, though metal-free, formazans are formed from the bis- rather than the mono-tetrazolium salts. This has been studied by PEARSE and HESS (1968) and SIEDLER and KUNDER (1968), and their results were further discussed by BENNION and HOROBIN (1973) and in Section 11.14 of the present book. Azo-dye final reaction products are also often large dyes, resulting for example from the azo-coupling of naphthol or naphthylamine initial reaction products with bis-tetrazonium salts. The affinity of the final product may also be raised by using halogenated substrates, as in certain indoxyls where the final reaction products are dibromodichloro indigos.

Another highly desirable characteristic for final reaction products is chemical stability. Where the final product is a metal co-ordination complex this must not readily decay back to the free ligand plus the metal ion. When a post-staining fixation step is used (with formaldehyde for example) the final product must not be degraded by the fixative agent.

7.7 How do we interpret the results of a staining procedure?

Here we consider problems of interpretation arising from the intrinsic biochemical and ultrastructural complexity of many enzyme systems; from the chemistry of the substrate/capture/visualising system adopted; from phenomena such as adsorption of reaction intermediates and products onto non-specific sites, giving rise to false positive results; and inhibition of and loss of enzymes and products from the tissue section. The question of the proportionally of staining intensity to amount of enzyme will also be discussed. Finally some account will be given of the need to use control procedures. Readers who have noted Chapter 11 in the Table of Contents may be puzzled by the existance of this discussion. Its justification is that the evaluation of enzyme histochemical staining presents a number of peculiar problems, best discussed adjacent to the general account of the methods.

Hopefully all the histochemical methods can be rigorously related to well established biochemical findings. Sometimes however the necessary biochemical background is lacking, as it is with peptidases. For these enzymes biochemical classification is usually attempted, not very satisfactorily, on the basis of the chemical characteristics of their active sites and their points of action on the peptide. No satisfactory correlation of biochemically defined peptidases with histochemically demonstrable aryl peptide hydrolysing enzymes has been achieved. Another example is the case of «nothing dehydrogenase». Under neutral or alkaline conditions the reduction of NAD or NADP may be catalysed by some tissue species jokingly labelled nothing dehydrogenase. Even in the absence of a formal substrate, other than the NAD or NADP, tetrazolium salts may be reduced to formazans. It now appears that the enzyme reponsible is alcohol dehydrogenase, though no satisfactory mechanistic explantion is forthcoming.

Even if an enzyme system is well defined biochemically one may not know which piece of it is being visualised histochemically. This arises for instance when demonstrating certain oxidative enzymes. If tetrazolium salts are used as electron acceptors, giving rise to coloured formazans as final reaction products, it may be asked at which point (points?) along the electron transfer chain does the tetrazolium salt pick up the electron? Precisely *what* is being demonstrated? It appears that different tetrazolium salts may be reduced at different points, and most at more than one point, in the chain. There are also species differences. As a particular case consider the attempted demonstration of glucose-6-phosphate dehydrogenase using the sugar phosphate as substrate. Probably many workers have, in carrying this out, merely succeeded in demonstrating a mitochondrial NADPH tetrazolium reductase. These issues are well reviewed by PEARSE (1972, p. 910).

Another instance of biochemical uncertainty arises when two or more isoenzymes are present in the tissue. This is well illustrated by a study of a histochemical model system, in which fractions containing mitochondrial and soluble aspartate aminotransferase were separated from homogenised liver (PAPADIMITRIOU and VAN DUIJN, 1970). The fractions were embedded in polyacrylamide gel slices, fixed, stained by trapping the oxalate initial reaction product with lead ions which were subsequently visualised by conversion to lead sulphide. Various differences between mitochondrial and soluble isoenzyme preparations were noted, such as their response to fixatives. The mitochondrial isoenzyme for instance was less inactivated by aldehydes. The composition of the optimum incubation medium also differed. It is worth pointing out here that analogous problems can and do arise in straight-forward biochemical enzyme assays.

Other unexpected complications include the solvent being an enzyme substrate. When demonstrating steroidal dehydrogenases the common substrates are insoluble in water, necessitating the use of organic solvents. Isopropanol or propylene glycol are now known to be unsuitable as they can be oxidised by certain tissue dehydrogenases.

Biochemical complications apart *it is desirable that the chemistry of the processes is understood and controlled.* Perhaps the simplest divergence from this, semmingly innocent, requirement is the problem of reagent purity. In the past at least some naphthyl phosphate substrates prepared by standard procedures have contained as major contaminants isomers, homologues and even free naphthols (LOJDA et al., 1964). The isomers may be of varying reactivity, and the free naphthols may be adsorbed onto non-enzymatic sites in the tissue leading to false positive staining. Certain naphthyl substrates (e.g. benzoyl-2-naphthyl phosphate) even if originally uncontaminated by free naphthol become so, due to non-enzymatic solvolysis, on being dissolved. Similar ambiguities can arise from the free fatty acids contaminating the Tweens used as substrates in certain lipase demonstrations. Even when reagents are well defined the chemistry of the demonstration reaction may be uncertain. Methods demonstrating oxidases using the oxidative polymerisation of arylamines give rise to pigments of uncertain structure and size for instance. Again, the tetrazolium salts widely used as histochemical reagents sometimes give rise to formazan pigments of two distinct colours, even if the original salt was homogeneous. The chemistry of this is still disputed: a mixture of mono- and bis-formazans? Or of a formazan and a half formazan? Or of a bis-formazan and a free radicle? As a final example note that certain enzymic processes may be mimicked by non-enzymic reactions.

An example is the hydrolysis of ATP to ADP plus inorganic phosphate. Lead ions, present as trapping agents in the Gomori-type demonstration systems for ATPase, can also catalyse this hydrolysis.

A major cause of *positive artifacts is from the uptake of enzyme, or initial reaction product, or visualising agent, or final reaction product at non-enzymic sites* on the tissue. In the early days of enzyme histochemistry the possibility was raised of enzyme diffusing from tissue into incubation medium and then back again to some new tissue site. Pertinent literature has been reviewed by NACHLAS, YOUNG and SELIGMAN (1957) who concluded that the phenomenon was rare. However binding of initial reaction products at sites distant from those of the generating enzyme has been clearly demonstrated, especially when using naphthyl substrates. Naphthoic initial reaction products, particularly when hydrophobic due say to benzoylation or halogenation, may be lost into the incubation medium. Such naphthols are readily taken into hydrophobic tissue sites such as elastic fibres, and lipid rich entities such as mitochondria, myelin sheaths and fat droplets; and also into collagen fibres and keratin. A leading reference is a paper by AHLQVIST (1963). Though use of naphthyl substrates derived from the poorly soluble Napthol AS series reduces such effects they can still occur, especially if the procedure requires solubilisation of the substrate with organic solvents such as acetone or formamides. The predominant sources of initial reaction product-tissue affinity are probably van der Waals attractions and, when aqueous incubation media are used, hydrophobic bonding. This topic is further discussed in Section 11.14. When the trapping agent used is a metal ion this too is readily taken into the tissues due to non-enzymic processes. For instance, in certain methods demonstrating carbonic anhydrase the intermediate product of carbonic acid is trapped using Co^{++} ions. However under these conditions Co^{++} is also deposited in the tissues as a result of such non-enzymic processes as pinocytic uptake of some colloidal Co^{++} species, and the coulombically-favoured binding of Co^{++} ions onto nuclei (CHURG, 1973). As shown in a critical study of the localisation of the ATPase activity in striated muscle, using a lead precipitation method, such artifacts may even occur at the putative enzymic sites themselves (GILLIS and PAGE, 1967). In addition to the trapping agent, several types of visualising reagents have been found to react with the tissue as well as with the initial reaction product, and so colour non-enzymic sites. Diazonium salts, as used in azo-dye procedures, may give rise to a general background staining, fortunately not necessarily of the same colour as the final reaction product (FRIED and KNOLLER, 1965). Or, as is the case with certain diazonium salts, the reagent may have a particular affinity for lysosomes. A leading reference here is a paper by BEADLE (1974). In these cases the general staining appears to be due to reaction of the diazonium salt with tissue groups such as tyrosyl, with the mechanism of lysosomal staining being less clear. Another visualising agent, the naphthoic acid hydrazide used to detect aldehydes in methods demonstrating monoamine oxidase, similarly generates weak general background staining plus localised stronger staining of dense, hydrophobic sites such as elastic fibres and myelin sheaths (BALLANTYNE, 1969; KOELLE and VALK, 1954). It has been suggested that the mode of this colouration is by reaction with tissue aldehydes, though it seems probable that uptake into non-polar structures will be aided by hydrophobic effects. Retention will be aided by the dense nature of these substrates. Yet another group of visualising reagents giving positive artifacts are the tetrazolium salts. Again hydrophobic effects and van der Waals forces appear important sources of salt-tissue affinity, with the larger tetrazolium salts being most strongly adsorbed onto non-polar sites such as elastic fibres and mitochondrial membranes. The uptake of these cationic reagents also resembles that of basic dyes, with binding onto proteins most pronounced at higher pHs whereas binding onto phospholipids still occurs under weakly acid conditions. In keeping with this basic dyeing pattern, chrome mordanting reduces tissue basophilia and also lowers binding of tetrazolium salts (PEARSE and HESS, 1961). A final factor in binding is the direct reduction to formazans by tissue groups such as thiols (BALLANTYNE, 1969; JAUBERT et al., 1979). Final reaction products may also be lost from the enzymic site, to be adsorbed elsewhere. An example arises in the demonstration of leucine aminopeptidase using a leucylnaphthylamide as substrate. Here the final reaction product is a copper co-ordination complex of an azo-dye, which in fresh preparations is well localised. However on storing mounted preparations nuclear staining becomes noticeable, with the cationic complex acting as a basic dye (NACHLAS et al., 1957). Another probable example is the uptake of uncharged azo-dye or formazan final reaction products onto the surface of lipid droplets and lipofuchsin granules (see respectively NOVIKOFF, 1963; KUGLER and WROBEL, 1978).

A major group of *negative artifacts,* namely those *arising from inactivation of the enzyme* will now be discussed. Inactivation due to fixative agents has already received extensive consideration in Section 7.1. Losses of activity due to other pre-incubation processing steps can also occur due to dehydration or denaturation or to mechanical damage to tissues organelles. Consequently high temperatures, non-aqueous media and low humidities are avoided whenever possible.

However it is the inhibitory effects of the various components of the incubation media which have received most attention, perhaps because nearly all types of component seem able to inhibit. The substrate itself can cause loss of enzyme activity, for instance the hydrogen peroxide present in benzidine media used for demonstrating peroxidase. Substrate impurities can also inhibit. For example the ADP present in most samples of ATP, especially after storage, can act as a competitive inhibitor of myosin triphosphatase (MEIJER, 1970). Capture and visualisation reagents of various types can also reduce enzyme activity. Lead salts are a case in point, and their influence apparently varies from the trivial to the substantial (compare MEIJER, 1970 with MOSES et al., 1966; also KLOHS et al., 1978). Similarly the ferric ferrocyanide used as a visualising agent in many indigogenic procedures, and also many of the diazonium salts used in azo-dye methods, will reduce enzymic activity though this is probably important only at sites of low enzyme concentrations (LOJDA et al., 1964). Impurities in such reagents may also be inhibitory, such as the zinc ions often present in commerical samples of stabilised diazonium salts. Even the nature of the buffer salts or the presence of polymeric materials used to prevent losses of enzyme from the tissue may be significant. Changing from tris to borate or veronal buffer systems markedly reduced staining intensities in a model system study of the histochemical demonstration of alkaline phosphatase (VAN DUIJN et al., 1967) and analogous effects have been found in work on tissue sections. For a description of the influence of polyvinyl alcohol on enzyme activity, see the work of Henderson and colleagues (HENDERSON et al., 1978).

A question often asked by practical men, histopathologists and others, is *does the intensity of staining give a reliable indication of the amount of enzyme present?* A related query asks if it is possible to accurately determine biochemical properties of an enzyme, such as its pH optima and Michaelis constant, using a histochemical system? In this case the hope is that application of enzyme histochemistry will allow information to be gained from individual cells, or even single organelles. Before considering questions such as these we must check out possible positive and negative aritifacts of the type already discussed. Then, assuming a system in which such problems are avoided or allowed for, we are obliged to at least acknowledge the possibility that, since biochemical procedures may disrupt the *in vivo* enzymatic milieu, biochemical and histochemical measurements may still differ. Finally we should remind ourselves that even if an enzyme's characteristics *were* the same during biochemical and histochemical assay, the amount of final reaction product at a given site might, due to histochemical complications of the kind already discussed, not be linearly related to the amount of enzyme. Conversely a linear relation could exist and be exploited even when the enzyme did differ in character during the two varieties of assay. Thus a spectrum of possibilities exists, from cases where histochemical determination of characteristics such as the Michaelis constant is possible, through examples in which, regardless of such niceties, the relative amounts of enzyme present at two tissue sites may be assessed, to the more gloomy circumstances where one is only able to say «Xase was here».

It is not surprising in view of these ifs and buts that few studies allowing comparison of biochemically and histochemically derived data have been carried out. Points from such studies are displayed in Table 41. In considering these and similar investigations one should remember that for a given enzyme it may not be possible to use the same incubation medium in a satisfactory histochemical assay system as in the standard biochemical system. The Table does show however that at least certain enzymes under

Table 41. Studies of enzyme properties in histochemical systems

The enzyme and its source	Methods and conditions of the histochemical study	Information obtained histochemically, compared to that found by standard biochemical procedures	Reference
Acid phosphatase of cells washed from the peritoneal cavity of rat	Microphotometry, after an azo dye method using a Naphthol AS phosphate as substrate and hexazotised pararosanilin as visualising agent; on a formalin-fixed monolayer of cells	Lineweaver-Burk plots gave K_m values 1.5 and 0.08 mM, comparable to the values reported for acid phosphatase isoenzymes using biochemical assays with another substrate	Bacsy and Rappay (1971)
Acid phosphatase of human neutrophil leucocytes	Microfluorimetry, using naphthol AS-BI phosphate as substrate; on cold-acetone fixed cell smears	Lineweaver-Burk plots gave a K_m value of 0.76 mM, rather higher than the 0.31 mM obtained biochemically using the same substrate	Prenna et al. (1977)
Alkaline phosphatase of rat kidney	Microfluorimetry, using l-naphthyl phosphate as substrate; on unfixed cryostat sections	Lineweaver-Burk plots gave a K_m of 0.8 mM, very similar to that obtained biochemically using the same substrate	Rost et al. (1973)
Alkaline phosphatase of duodenum and kidney of rat	Interferometry, after using β-glycerophosphate as substrate and calcium ions as capture reagent; on dewaxed sections from acetone or ethanol fixed, or on freeze dried unfixed, paraffin embedded material	Lineweaver-Burk plots yielded K_m values varying with the tissue over the range 1–2 mM, and so broadly comparable to the biochemical values; as were the pH optima	Casselman (1959)
Glucose-6-phosphate dehydrogenase of mouse islets	Microfluorimetry, after incubation in glucose-6-phosphate or NADP; on unfixed cryostat sections	Lineweaver-Burk plots yielded K_ms of 0.2 mM (with G-6-P as substrate) and 2.0 mM (NADP as substrate), rather higher than the biochemical determinations. The pH optima corresponded to the biochemical findings	Taljedal (1970)

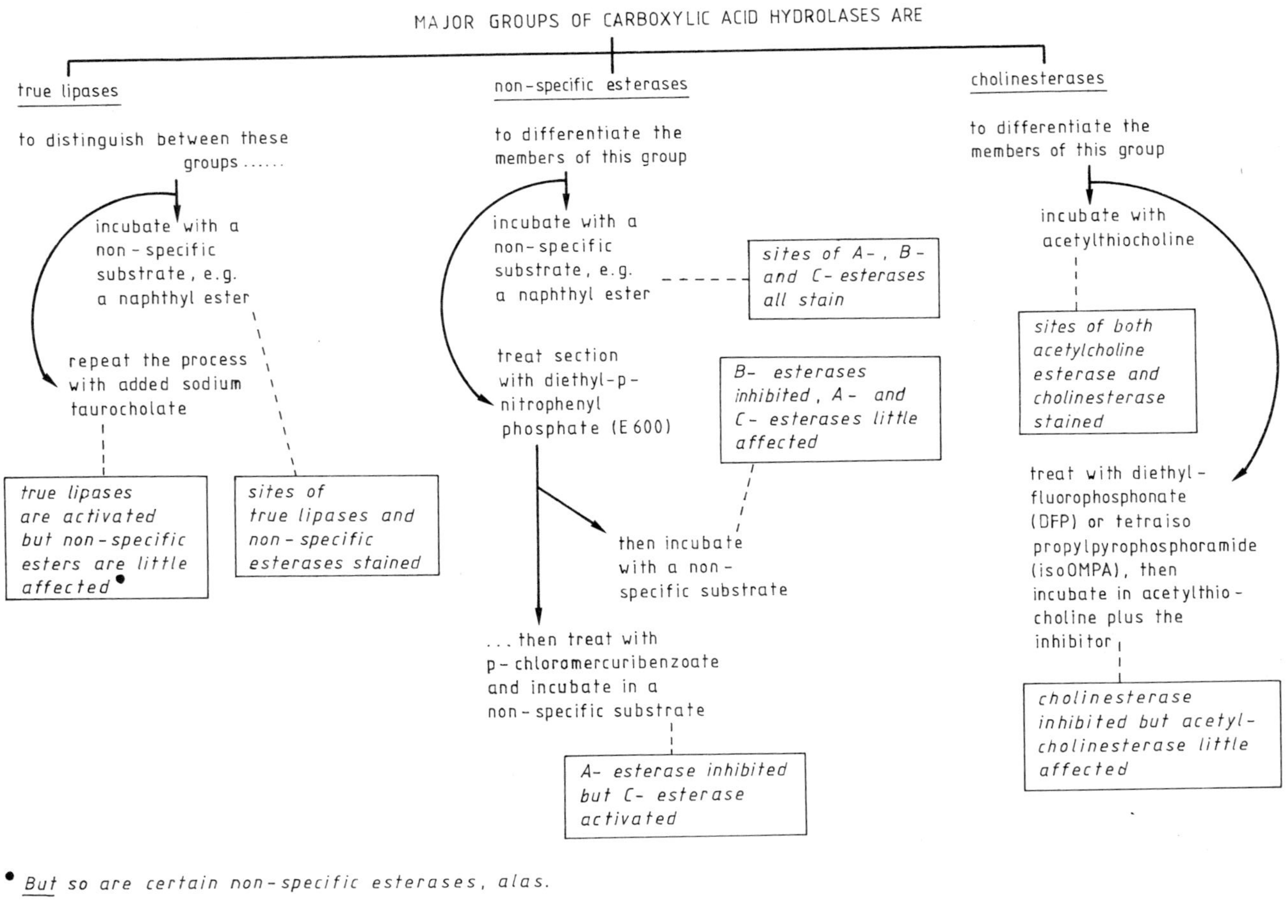

Fig. 53. Use of activators and inhibitors to increase selectivity in esterase histochemistry. There are very marked species differences in most such activator/inhibitor behaviour; the relevant biochemical findings must be followed with care

certain conditions may be studied histochemically and yield results directly comparable to those obtained via traditional biochemical procedures. To view this limited achievement in perspective remember that the biochemical studies themselves are subject to intrinsic artifactual distortions, perhaps not emphasised in elementary texts. Processes such as tissue homogenisation and subsequent fractionation and analysis can modify enzyme activities and distributions, or be difficult to standardise, or both.

Because of these interpretational problems a variety of *control procedures including the use of enzyme inhibitors and activators* are widely used. The most basic check is to see if staining is enzymatic, using interventions calculated to inactivate *any* enzyme. Thus the section may be heated, or treated with a vigorous protein denaturant such as glutaraldehyde. Further checks on enzymatic involvement in staining include the omission, where appropriate, of the substrate, or the capture reagent, or the visualisation agent, or of combinations of these from the staining procedure. The next level of control is to check that the enzyme giving rise to the staining is in fact the enzyme which one believes one is demonstrating. For instance, numerous esterases exist (such as «true» lipases, non-specific esterases, cholinesterases) capable of catalysing the solvolysis of a variety of esters. To distinguish between these, various methods using inhibitors and activators have been developed as outlined in Fig. 53; as well as methods based on the use of (hopefully) specific substrates, and combination methods (see Pearse, 1972, p. 795 for a review of this topic). Numerous other examples of the use of specific inhibitors and/or activators are provided by phosphatase histochemistry. For example myosin adenosine triphosphatase has been distinguished from actomyosic adenosine triphosphatase in myofibrils by utilising the activating effect of calcium ions, and the inhibitory effects of magnesium ions and of ADP, on the former enzyme

Table 42. Differentiation between isoenzymes using activators and inhibitors

Isoenzymes of ...	Biochemical nature of isoenzymes	Mode of selective demonstration	Special problems	References
Alkaline phosphatase	R (Regan or placental), and non-R (non-Regan or non-placental types)	R-type is selectively inhibited by L-phenylalanine; and non-R-type by L-homoarginine		Sasaki and Fishman (1974)
α-Glucan phosphorylase	A-type is twice the size of, and more active than, the B-type	Total phosphorylase is demonstrated by incubating in the presence of ATP and cAMP, under which conditions all enzyme is present as A-type. The A-type is active in the absence of adenylic acid, whilst the B-type has an absolute requirement for this activator	The A- and B-types readily interconvert; to prevent this it is necessary to inhibit the relevant enzyme with fluoride and EDTA	topic reviewed by Pearse (1972, p. 829)
Lactate dehydrogenase	Five types, one has 6 H (heart type) subunits, one 6 M (muscle type) subunits, and the other 3 are mixtures	H-type subunits are selectively inhibited by pyruvate, and M-type by urea		McMillan (1967)

(MEIJER, 1970). Again, alkaline phosphatase has been distinguished from such specific phosphatases as ATPase and glucose-6-phosphatase by the use of inhibitors such as Levimisole and R8231 (BORGERS, 1974 has some dramatic micrographs). Omission of soluble co-factors from the incubation medium can, when relevant, constitute a further check on staining specificity. An example is the omission of NAD from media made up to demonstrate lactate dehydrogenase (MCMILLAN, 1967).

A further refinement of inhibitor plus activator techniques, is their use to distinguish between iso-enzymes, Table 42 gives information on several such procedures.

Consider finally the efficacy required of activators and inhibitors for their successful histochemical use. The inhibitors used to distinguish between similar enzymes are not commonly specific in their action, only selective. Addition of excess inhibitor will some-what decrease the activity of the enzyme under study as well as grossly decreasing the activity of the «unwanted» enzyme. Adding too little inhibitor will allow some unwanted enzymic activity. Hopefully there will be a safe and useful range of inhibitor concentration, but histochemical usage must rest on a base of quantitative biochemistry plus good fortune. A review has been made of this problem, in the context of cholinesterase histochemistry, by PEARSE (1972, p. 768). The overarching question here is: what are the necessary and sufficient criteria for judging the validity of enzyme histochemical methods? In this context a careful perusal of a recent critical essay by STOWARD (1980) is recommended.

7.8 Peculiarities of electron-microscopic enzyme histochemistry

The discussion of enzyme histochemistry presented hitherto in this section has drawn on studies using both light and electron microscopy. There are however features of special importance involved in staining for the electron microscope. Several detailed reviews of this area are available, for example HAYAT (Volume 1 of a series was published in 1973, and is currently still in progress), PEARSE (1972, Chapter 32), and LEWIS and KNIGHT (1977). Here we will merely sketch those factors of peculiar concern to the electron microscopist.

Firstly *a basic requirement for the final reaction product is* not that it should be coloured, though this is desirable since parallel optical study is then facilitated, but *that it is electron dense.* A substantial electron scattering character may be variously achieved. For instance some methods designed for light microscopy in which the final reaction product is an organic pigment deposited as particles of high physical density have been used. Though some of the original results of this strategy are nowadays considered unsatisfactory, successful methods using this approach have been developed. Examples include an azo-dye method to demonstrate β-glucuronidase (HAYASHI et al., 1968) and a procedure for DOPA oxidase in which the final reaction product is a polymeric DOPA melanin (NOVIKOFF et al., 1968). Quite another approach is to stain with materials of low intrinsic electron scattering power but of characteristic morphology. The action of phosphorylase has been demonstrated using electron microscopy by observing the formation of the characteristically shaped polyglucose reaction product (SASAKI and TAKEUCHI, 1970). This is an interesting parallel to the use of virus particles as antibody labels for electron microscopy, see Section 8.4.

However a reaction product of high electron density is usually sought. A common way of achieving this is to ensure that the final reaction product contains atoms of high atomic number, with consequent high electron-scattering power. Again some procedures have been adapted, with only technical modifications, from pre-existing methodologies. Examples include many methods for hydrolases in which the reaction product is an insoluble lead salt, and also the thiocholine-copper ferricyanide procedures for demonstrating choline esterase in which the final reaction product is cupric ferrocyanide. Other metal ions present as insoluble-salt final reaction products include barium, cobalt, gold, silver and thallium. The non-enzymic uptake of metal ions, discussed previously as a source of positive artifacts in enzyme histochemical procedures involving metals ions, will be more troublesome and critical in the electron microscopic circumstances, as illustrated in Fig. 54. Much less common is the use of reagents which are themselves electron dense, due to their being organometallic compounds or metal co-ordination complexes. Examples include a tetrazolium salt osmate for the demostration of dehydrogenase (TSOU et al., 1968) and a diazotised acetoxymercurianiline and amino organolead compounds in azo-dye methods for demonstrating, respectively, β-glucuronidase and acid phosphatase (SMITH and FISHMAN, 1969; LIVINGSTON et al., 1970).

A number of examples of the other major strategy, namely choosing reagents such that the penultimate reaction product is reactive with a heavy metal ion or compound, are listed in Table 43. Several interesting points arise from the Table, for instance that

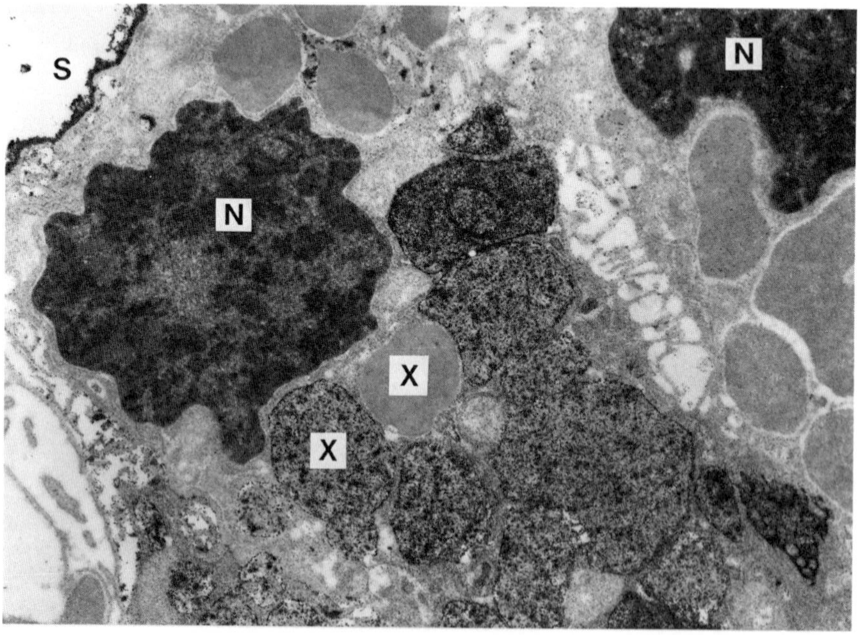

Fig. 54. An illustration of various staining artifacts arising in electron microscopical enzyme-histochemistry. The preparation of rabbit parotid acinar cells was stained by a Gomori-style lead method, to demonstrate Acid Phosphatase. On biochemical grounds this enzyme was expected to be localised in the zymogen granules (x). However, as can be seen, there was extensive artifactual staining of nuclei (N) and of cell surface (S). The failure of many mature zymogen granules to stain may also be artifactual. ×14500

Table 43. Some examples of the formation of electron-dense final reaction products via metallophilic penultimate reaction products

The nature of the penultimate reaction product	and how it was arrived at	References
Azo pigments which can react with osmium, or chelate Cu^{++} ions which may then be bridged to osmium via thiocarbohydrazide ligands	azo-coupling naphthylamines (themselves derived from the action of tissue peptidases on naphthylamide peptides) with diazonium salts	SELIGMAN et al. (1970), RUTENBERG et al. (1969)
An osmiophilic azo-indoxyl pigment	azo-coupling an indoxyl derivative (itself derived from the action of tissue cholinesterases on indoxyl acetate) with an osmiophilic diazonium salt	KAWASHIMA and MURATA (1969)
Osmiophilic diazoethers	reacting diazonium salts with thiols, themselves produced by the action of tissue cholinesterase and non-specific esterases on benzanilides	BERGMAN et al. (1967)
Osmiophilic formazan pigments	produced from osmiophilic tetrazolium salts, either directly by the action of tissue dehydrogenases, or indirectly by aldehyde groups which were themselves generated by the action of tissue monoamine oxidase on tryptamine	HAYDON et al. (1968), BOADLE and BLOOM (1969)
Various polymers (e.g. indoanilines, azines and so on), which are reactive towards osmium or Cu^{++} ions	generated by the action of tissue oxidases (such as peroxidase or cytochrome oxidase) on arylamines, or on mixtures of arylamines with phenols or naphthols	for example KERPEL-FRONIUS and HAJOS (1967)
Various azo dyes containing the lead phthalocyanin grouping	generated by the capture of naphthols (themselves derived from naphthyl substrates by the action of tissue acid phosphatases, β-glucuronidases and non-specific esterases) by diazotised amino lead phthalocyanin	LIVINGSTON et al. (1969)

the group reactive to metal may be part of the substrate, or the visualising/trapping reagent, or may be generated by the visualising/trapping process. The enthusiasm for osmium as the archetypal electron-dense metal is also intriguing. In part this probably reflects the efforts and preferences of SELIGMAN, who is more than once disguised behind the *et alia* of Table 43. Perhaps also the osmium enthusiasm reflects the advantages to the innovator of using reagents which are familiar to potential users. Finally note the tactic of enhancing the staining intensity by attaching further metal atoms to those of the initial stain, which is not restricted to the examples in Table 43. The bridging of osmium to a metal atom via thiocarbohydrazide ligands, as in the Table, is the most common amplification procedure, with the initial stain being osmium, copper or mercury. An earlier process, still sometimes used, involves converting cupric ferrocyanide precipitates to cupric sulphide, and then reacting this with osmium or silver.

The problem of resolution discussed more generally in Section 7.6 *is more acute when dealing with ultrastructural localizations.* To improve resolution one must minimise the mobility of the reaction product, and the size of the particles of stain. For instance the formation of reaction products which are both water insoluble *and* lipophobic is even more important here than when using light microscopy, as is discussed by SELIGMAN and his collaborators (1970). Two-step methods in which the ions constituting a crystalline precipitate of initial reaction product are altered by an ion-exchange process (lead for calcium, in certain acid phosphatase methods; or sulphide for ferrocyanide, in some cholinesterase procedures) fail to give good ultrastructural localisations. Most progress in this field has been made on an empirical suck-it-and-see basis. An example is the substitution of aurous gold for lead as capture reagent in the thiocholine method for demonstrating cholinesterase. For no obvious reason this resulted in a marked diminution of the size of the reaction product particles (KOELLE and GROMADZKI, 1966). As a final magical manoeuvre, consider the use of prolonged formalin fixation to reduce the «relatively large size» of crystals of cupric ferrocyanide occurring as final reaction product in a demonstration of cholinesterase (ERÄNKÖ et al., 1967). The dominant influences of such practical details certainly serve to underline the difficulty of achieving rationally devised electron microscopic, enzyme histochemical procedures, applicable to a wide range of tissues.

Finally consider a potpouri of problems, coming under none of the headings used so far. First a reiteration, that the need to preserve both morphology *and* enzymic activity constitutes a cruel dilemma. If we fix the tissues there will often be loss of enzymic activity, and if we do not fix there may be no ultrastructural detail left to study, for discussion see Section 7.2. Secondly consider the use of plastic embedding media. Unfortunately for the flexibility of experimental design it is often not possible to demonstrate enzyme in plastic embedded tissues. There are exceptions, for instance β-glucoronidase has been demonstrated in Epon embedded tissue after first etching the plastic with cold ethanol (HAYASHI et al.,1968). This is paralleled by certain procedures in labelled antibody staining, as discussed in Section 8.8. A general account of staining plastic embedded tissues is given in Chapter 10. Note finally that if enzyme staining is carried out prior to embedding, and if the stain consists of lead or barium sulphates, it may be extracted during subsequent staining of the ultrathin sections with lead citrate or uranyl acetate (KALIMO et al., 1968).

8 Immuno-chemical Staining Using Labelled Antibody and Related Methods

8.1 Some strategies of staining

A number of biochemical systems involve specific interactions of certain regions on protein molecules with particular chemical groupings on a variety of other molecules. The interactions of most interest to us here are those between immunoglobulins and haptens on proteins and other polymers, namely antibody-antigen binding. Other such interactions are enzyme-substrate, complement-immunoglobulin, lectin-saccharide, and RNA-DNA complexes.

If one or more of the compounds applied to the tissue is «labelled», such specific interactions can be used for staining tissue antigens or antibodies, enzymes or substrates, and so on. That is if the compound carries a marker, the label, which may directly or indirectly be seen down the microscope.

A detailed discussion of such labels is given in Section 8.4, but a few preliminaries will be useful. Firstly the label may itself be the stain. This is so when the label is a fluorescent grouping and can be seen visually, or when it contains high atomic number atoms which can be seen directly in the electron microscope. Alternatively the lable may be capable of generating the stain, that is light absorbing or emitting or electron scattering products. In this latter case the methods are indirect since they require a second visualising step, which step is non-immunological in character.

Set out in Figs. 55 and 56 are some of the staining strategies used. The methods illustrated include the detection of tissue antigens using antibodies as reagents as well as the detection of tissue antibodies, and also methods detecting saccharides and enzymes and enzyme substrates by analogous procedures. Direct methods such as those shown in Fig. 55, are straight forward to use but depend on the user possessing the necessary specifically labelled reagents. If several tissue substrates are studied this will require an equal number of labelled reagents. This practical limitation also applies to certain of the indirect methods used to detect antibodies in the tissue.

Other indirect methods however, as for instance one of those illustrated in Fig. 56, require more stages but are less dependent on the user possessing a wide range of specifically labelled reagents. In the case in Fig. 56 a single labelled anti-immunoglobulin antibody will enable many different tissue antigens to be demonstrated. A further advantage of indirect as compared to direct methods is that the amount of label bound

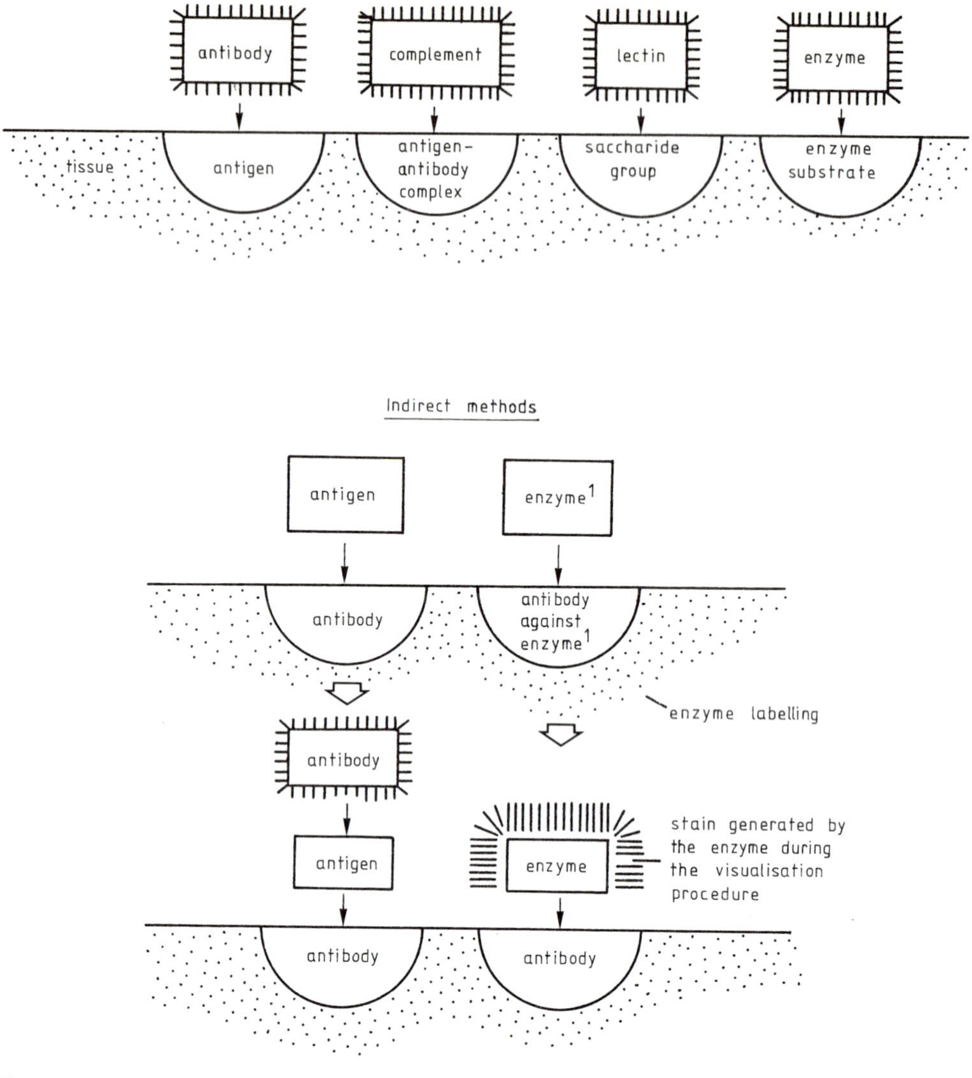

Fig. 55. Staining methods using labelled antibodies and analogous reagents in which only a single immunochemical interaction occurs. The applied reagents are indicated by rectangular boxes, and labelled reagents shown as «hairy» boxes

per tissue site is greater, since each bound antibody will bind several labelled antibodies. This advantage is multiplied when a multi-layer is applied. Enzyme labelling is also applicable to such multi-layer systems, though for diagrammatic simplicity this was not shown in the figure. A typical varient exploits the multivalent character of the lectin Concanavalin A (ConA). Tissue is first incubated in ConA, which binds to mannopyranosyl and glucopyranosyl residues. The complex is then visualised by treating with fluorescent-labelled serum glycoprotein (Thoss and Roth, 1976).

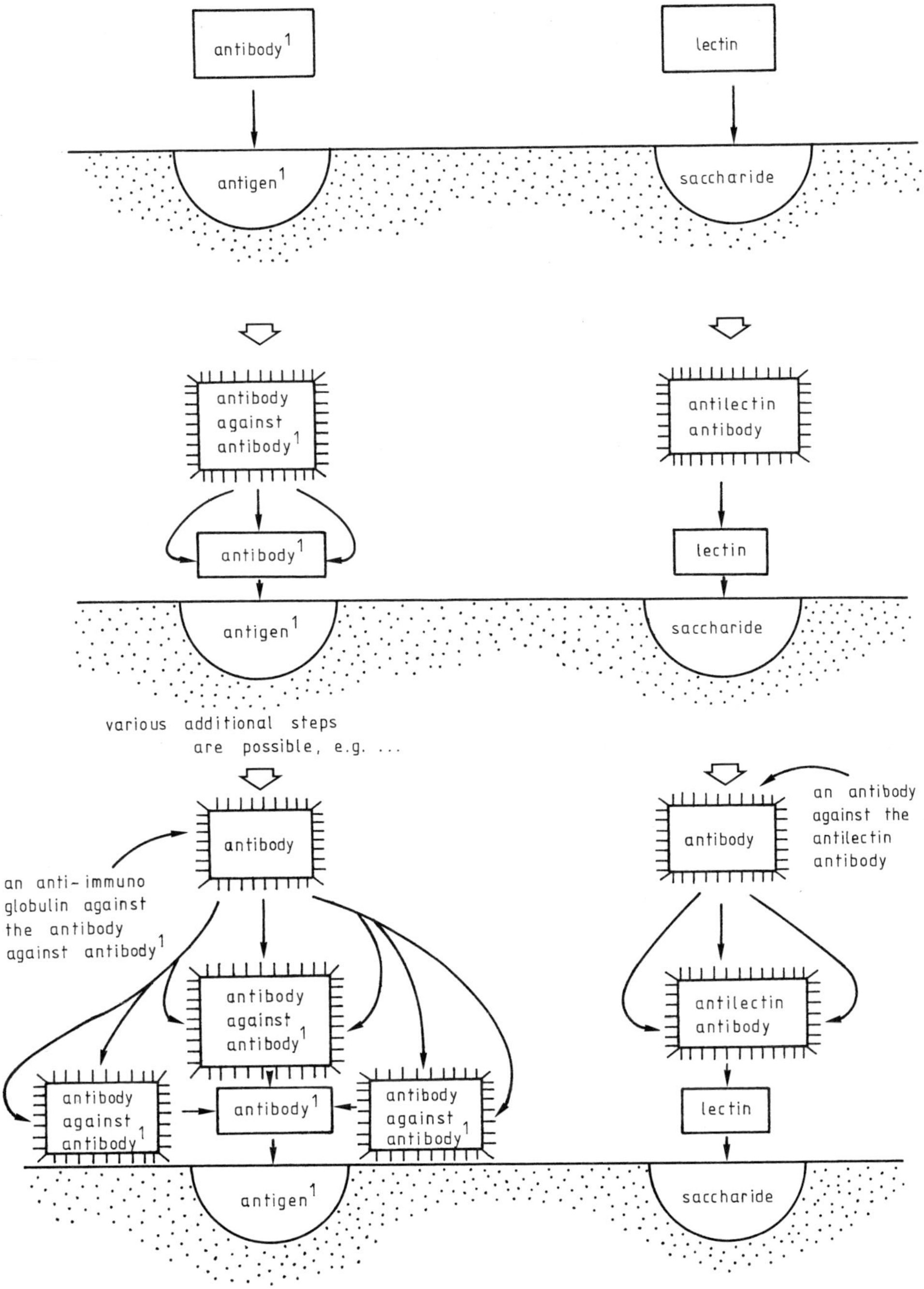

Fig. 56. Staining methods using labelled antibodies and analogous reagents in which two or more immunochemical interactions occur. Reagents applied are indicated as rectangular boxes, and labelled reagents shown as «hairy» boxes

8.2 A sketch of the chemistry of antibodies and antigens, and of their interactions

The antibodies, or immunoglobulins, are large proteins ranging from 150 to 900 thousand daltons, which possess a complex molecular form, see Supplement 12. Not all the antibody molecule is involved in binding antigen. Consequently though the Fc portion of the heavy chains play important biological roles after antigen is bound *in vivo*, when immunoglobulins are used as stains the Fc portion is often, as it were, spare. So this portion of the molecule can be linked to a label or be involved as antigen to another antibody, without directly interfering with antigen binding.

The chemical natures of antigens are more varied than are antibodies, and so are less clearly defined. To understand antigens and antigenicity it is useful to first introduce the idea of a hapten. A hapten is a small, chemically well defined structure which will combine with an antibody once formed, but cannot itself stimulate antibody formation. For example compounds such as aminobenzoic or sulphonic acids have been used as experimental haptens. Comparing haptens with antigens, note that the latter can combine with and induce the formation of antibody. However if haptens are coupled to polymers they will then also induce specific antibody formation.

Studies with haptens have shown that it is their overall configuration which is overridingly important in defining the antibody-hapten specificity and affinity. Antibody and hapten must have complimentary steric features. In fact each antiserum raised by immunisation against a given antigen, however simple its structure, contains a variety of antibodies with a range of binding affinities. These probably correspond to a range of degrees of complimentarity of hapten with different antibody binding sites. The actual physico-chemical factors involved in producing antibody-hapten affinities will be familiar from Chapter 4, namely Coulombic interactions, hydrogen bonding, hydrophobic effects, and van der Waals forces.

Analogous specific protein-protein interactions can occur with other classes of molecule. For instance between certain of the Complement proteins (large molecules, up to 400,000 daltons molecular weight) and the Fc regions of certain IgG and IgM subclasses. This occurs after the Fc conformation has been altered by formation of an antibody-antigen complex. Such interactions have been exploited histochemically, as have Protein A–Fc interactions (Dubois-Darcq et al., 1977).

The lectins, a structurally diverse group of proteins, apparently bind in a similar fashion to saccharide residues. The histochemically widely used metalloprotein Concanavalin A for instance binds with high selectivity and affinity to mannopyranose and glucopyranose residues present in sugars, polysaccharides, glyco-proteins and mucosubstances. Another histochemically applied lectin is *Ricinus communis* agglutinin which binds to galactosyl residues (Yamada and Shimizu, 1977). Another protein-saccharide interaction exploited as the basis of a stain involves the glycoprotein Avidin, which binds to sialic acid residues (Bayer et al., 1976).

Formation of other types of complexes have occasionally been adapted to histochemical purposes. Enzyme-substrate complexes for instance, as between chitinase and chitin (Benjaminson, 1969), and cellulase and cellulose (Seibert et al., 1978). Enzyme-inhibitor interactions too have been used, for example those between DNase I and Actin (Wang and Goldberg, 1978), and between Trypsin and soya bean Trypsin inhibitor (Avrameas, 1972). Even specific protein-lipid interactions can sometimes be made use of histochemically, as are those between Myeloma Proteins and phosphorylcholine (HOOGE-PETERS et al., 1979).

Another specific interaction occurs between complementary nucleic acid sequences. This too has been exploited cytochemically. Specific DNA sequences have been localised microscopically by cytochemical hybridisation with fluorochrome-labelled RNA (BAUMAN and VAN DUIJN, 1980).

8.3 Preparation of antibodies and analogous reagents

In outline, preparation of antibodies is straightforward. Immunise an animal with an antigen to raise an antisera containing the required antibody, then if necessary separate the specific immunoglobulins from the plasma proteins. In practice many empirical procedures are involved, and practical immunological experience is invaluable. Some common antibodies, including many required for clinical diagnosis, are however available in labelled form as articles of commerce. Consequently only a brief account of some problems arising and methods adopted will be given here. The reader hoping to engage in this branch of biological staining will be obliged to proceed from here to more detailed accounts, perhaps via some of the many reviews in the immunological and histochemical literatures, for instance AVRAMEAS (1972), PEARSE (1968, p. 198), STERNBERGER (1972) and WILLIAMS (1977).

Even obtaining pure antigens can be troublesome. This is especially so when tissue antigens are required, when procedures such as tissue fractionation via homogenisation and ultracentrifugation may prove necessary. It is convenient to start from a hyperimmune serum of high antibody content, and various techniques favour this. With non-living antigens it is commonplace to administer adjuvants during immunisation. When the antigen is of low molecular weight, as are the polypeptide hormones, adsorption of antigen onto a particulate material such as polymethylmethacrylate beads increases the response.

The next stage in the preparation of antibody, namely obtaining immunoglobulin of a suitable degree of purity, is again achievable in a wide variety of ways. At its simplest this will entail selective precipitation of the immunoglobulin component of the plasma proteins by the addition of appropriate amounts of sodium or ammonium sulphates, followed by dialysis to remove the salt. This naturally will not produce an homogeneous antibody preparation, and if such is wanted a separation process based on some form of antibody-antigen interaction is required.

For instance some antibodies can be selectively precipitated from an antiserum by titration with antigen to the equivalence point. Sometimes the complex may be converted to the soluble form by treatment with excess antigen and this used directly; as with the horse raddish peroxidase-antihorseraddish peroxidase complex, useful in certain indirect staining methods. More generally, antigen in an insoluble form can be used to selectively adsorb antibody from an antiserum. This antiserum can then be replaced by a solvent into which the antibody can be released by suitable alterations in pH and ionic strength. Insoluble, or at least solid, antigen preparations can be obtained in various ways. Proteins for instance can sometimes be treated with glutaraldehyde yielding insoluble but still antigenic derivatives. Alternatively proteins may be crosslinked with glutaraldehyde to an insoluble matrix such as polyacrylamide; or reacted with a reactive derivative of cellulose, for instance a diazonium salt. More complicated antigenic materials, such as a basement membrane preparation, have been immobilised in an analogous fashion by mechanical inclusion in a polymeric matrix of acrylamide (Martinez-Hernandez et al., 1975). These solid antigenic preparations can be used batchwise or can be packed into a tube and the antisera and other solutions passed through the solid material sequentially. Currently the former procedure is generally known as immunoadsorption whilst the latter is often termed affinity chromatography.

Immunoglobulins so obtained will usually be in fairly dilute solution. Solid proteins may be obtained by using salt precipitation or freeze drying. Alternatively the solution may be used directly, or after a concentration step. This latter can be achieved by dialysis against a concentrated polymer solution, say of polyvinylpyrrolidone; or by treatment with dehydrated beads of hydrophilic though insoluble polymer.

Whatever procedures are adopted it will often be necessary at this stage to assess the purity of the antibody. Standard immunological techniques may be used, for instance immunoelectrophoresis or gel diffusion as tests for homogeneity; and titration with antigen as a quantitative measure of antibody content. A final problem is that of storage. Fortunately most antibodies can be kept in a refrigerator without deterioration for several months.

A few analogous reagents are also used. Thus the Fab fragments of immunoglobulins have found some application, with their users suggesting that the smaller size of Fab relative to the original antibody is an aid to reagent penetration. The fragments are prepared by subjecting the relevant antibody to enzymatic hydrolysis with papain, and separating Fab from undigested immunoglobulin by gel-filtration. The lectins are another widely used reagent type. The widely used Concanavalin A is now an item of commerce, and is prepared from the jack bean. This is also the case for chitinase which, after labelling with Fluorescein or Rhodamine isothiocyanates or with Ferritin, has been used to stain its substrate, chitin (Benjaminson, 1969).

8.4 Nature and preparation of the label

Labels may be conveniently divided into two types. Some labels are themselves stains which strongly absorb or emit light, or scatter electrons, or possess a distinctive shape. Other labels generate stain at some subsequent stage in the staining procedure, after the antibody-antigen interaction step.

Lables which are stains may be sub-divided into several distinct varieties. The original successful immunohistochemical labels, still used for light microscopic studies, were simple fluorescent compounds which carried groupings reactive towards protein amino groups. Those in widespread use are the isothiocyanate derivatives of Fluorescein and of Rhodamine B. In spite of vigorous efforts no significantly improved general alternatives are available. Alternatives having some specific advantage in particular instances do exist. For instance the successful use of a reactive furanone derivative which yields fluorescent products with proteins has been reported (Weigele et al., 1973). This compound has the advantage that after immunoglobulin-label coupling any unreacted reagent may be destroyed by mild hydrolysis, so chromatographic purification of labelled protein is not required. In studies using the electron microscope the metalloprotein ferritin plays an analogous role to these simple fluorescent labels, since its core of hydrated iron is visualised as a black dot. Other electron-dense labels include colloidal gold, low molecular weight organometallic compounds such as a ferrocenyl derivative of phenylisothiocyanate, and simple protein-reactive inorganic species containing high atomic number atoms such as lead and uranyl ions and iodine molecules. Of such only uranyl ions are persistently recommended, notably by STERNBERGER (for instance, 1972). At least one attempt has been made to produce a single lable which was both electron dense and fluorescent. Unfortunately the reagent, diazotised tetraacetoxymecuriarsanilic acid, proved of low stability in the electron beam (PEPE and FINCK, 1961).

Success has also been claimed for dual labelling of a single antibody, with ferritin and fluorescein (HSU et al., 1963) and with colloidal gold and Rhodamine (HORISBERGER and VONLANTHEN, 1979). Instances exist of labelling by shape, e. g. with a label comprising virus particles of characteristic morphology (HAMMERLING et al., 1968), or of whole red blood cells (ALROY et al., 1978).

Most labels which generate a stain after the labelling stage are enzymes. The stain is typically produced by visualising the enzyme label using the methodology of enzyme histochemistry, as discussed in Section 8.6. Several enzymes have been used as labels, including acid and alkaline phosphatases, glucose oxidase and tryrosinase. The most commonly used is however horseraddish peroxidase. Chemically modified enzymes have also been proposed, used and recommended for various purposes. An example is the use of a cytochrome fragment. This is a relatively small molecule and this small size was said to be penetration of antibody into the tissue (KRAEHENBUL et al., 1974). The field of enzyme labelling remains in a state of technical ferment. Discussion of the various compromises involved, such as balancing label stability (peroxidase is very stable) against size (but the cytochrome fragment is smaller) and resolution of stain (with acid phosphatase said to be unusually good) is given in the many reviews available (such as AVRAMEAS, 1972; STERNBERGER, 1972). There have been a few non-enzymic stain-generating labels. One involved labelling immunoglobulin with dextran. The labelled antibody was reacted with antigen in the tissue, then the materal post-fixed in osmium tetroxide-lead citrate. The dextran label chelated lead ions, resulting in a localised black blob when the tissue was viewed in the electron microscope (KENT and WILSON, 1975).

With regard to the preparation of labels the situation is quite unlike that of obtaining specific antibodies. All the popular labels from Fluorescein isothiocyanate to ferritin to horseraddish peroxidase are currently items of commerce. Less commonly used labels, such as cytochrome fragments or ferrocenyl derivatives, are less likely to be produced commercially, a chicken and egg situation.

8.5 Labelling methods: preparation of labelled antibodies and related materials

Here is considered the process of labelling, discussing both direct labelling strategies and the use of conjugating agents; and also procedures used to minimise loss of active sites on the proteins, and to control the size of the products. Purification of labelled products is discussed, with emphasis on obtaining optimally labelled material. Finally the labelling of substances other than immunoglobulins is considered.

Labelling may be direct when the label itself carries a protein reactive group. Common examples are the isothiocyanate groups of Fluorescein and Rhodamine isothiocyanates, which react with protein amino groups. Periodate oxidised dextran, and oxidised horseraddish peroxidase (a glycoprotein), and the 2-methoxy-2,4-diphenyl-3(2H)-furanone reagent also react with protein amino groups. Other labels attach to other protein side chains. For instance mercuriacetoxy groups react with thiols, iodine molecules with tyrosyl residues, and uranyl cations with a variety of ligands. In some instances the label may bind by an antibody-antigen process, as when using horseraddish peroxidase as an enzyme label. Here the enzyme may be bound to a sandwich filling which is itself an anti-peroxidase antibody.

Other labels are not themselves reactive with immunoglobulins, so here *use may be made of bifunctional conjugating agents.* These are compounds reactive towards both label and immunoglobulin, and normally both these latter will be proteins. The conjugating compounds used have been varied. Many carry electrophilic reactive groupings and react with amino groups. Examples include dicyclohexyl and l-ethyl-3(3-dimethylpropyl) carbodiimides, cyanuric chloride, difluoro-m,m'-dinitrophenyl sulphone, glutaraldehyde, N-hydroxysuccinamide ester of p-formyl benzoic acid, toluene diisocyanate and m-xylylene diisocyanate. Also used are diazonium salts reactive with residues such as tyrosyl, e. g. diazotised o-dianisidine and benzidine. A marginally distinct process is to convert the label to an immunoglobulin-reactive compound (Nakane and Kawaoi, 1974).

We can now ask several general questions concerning labelling. Firstly, and this is applicable to direct labelling and to conjugation, *how do we minimise the loss of activity of antibody or of enzyme label, or of both, due to the denaturation caused by the reagents?* Many of the conjugating agents are also used as fixatives, indicating that they are good protein denaturants. Coping with this is partly a matter of choosing the reagent appropriately. For example the conjugating agent toluene diisocyanate causes denaturation even at low concentrations (Williams and Gregory, 1966). Glutaraldehyde is also particularly effective as a denaturant. Here the only advice is to be cautious, don't assume other worker's recipes will work on your problem. A more positive strategy, recommended by Sternberger (1972) specifically for use when uranium-labelling immunoglobulins, is to apply the label to an antibody-antigen complex. The degree of inactivation has been little studied in any rigorous way however. Many workers appear to regard it as minor or even trivial, as typified by the rather cavalier comments of Miller (1972). What hard results there are indicate that such optimism is, at the very least, not *always* justified; and that there are marked differences between one conjugating agent and another in this respect (see Boorsma 1976; Clyne et al., 1973; Williams and Gregory, 1966).

A second general question is *how to expose the antibody to a satisfactorily high concentration of label or conjugating agent?* When all components are water soluble (as are antibodies, ferritin and enzyme labels, uranyl ions, and conjugating agents such as glutaraldehyde or bis-diazonium benzidine) no problem exists. When the label or conjugating agent is non-ionic, and so poorly soluble in water (as is the case with Fluorescein isothiocyanate and toluene diisocyanate for instance), technical difficulties do arise. The simple approach is to saturate the aqueous solution with reagent, though the presence of particles of organic solids sometimes proves strongly denaturating. A cautious varient places such a saturated solution inside a dialysis bag which is itself immersed in the aqueous protein solution. Alternatively the label can be added to the protein in a water miscible organic solvent, such as acetone or dioxane. Such solvents themselves are protein denaturants however. This ploy can be made more sophisticated with conjugating agents, since if the conjugation is carried out in two steps the presence of an organic solvent may be required only for the first. Once attached to the label-protein or antibody-protein the conjugating reagent is then effectively solubilised (KRAEHENBUL et al., 1974). Another approach used with Fluorescein isothiocyanate is to adsorb the non-polar reagent onto an innocuous solid medium such as paper or cellite, which is then immersed in the aqueous protein solution.

A closely related question is *how to apply the label or conjugating agent under effective reaction conditions if these conditions tend to cause denaturation of the antibody or protein label?* Perhaps the only useful suggestion here is to make careful choices, whenever possible, between the various labels and conjugating agents. Thus horseraddish peroxidase has the advantage of not being readily denatured, even under the alkaline conditions needed for the efficient reaction with the conjugating agent difluoro-m,m'-dinitrophenyl sulphone. This is not so for acid phosphatase, consequently when using this enzyme as a label it is necessary to make use of a conjugating agent reactive under neutral conditions, such as l-ethyl-3(3-dimethylpropyl)carbodiimide. It will be no surprise to note that little quantitative study has been made of the reactivities of conjugating agents. The available information disconcertingly indicates that even standard conjugating agents such as toluene diisocyanate or water soluble carbodiimides may sometimes be quite ineffective (WILLIAMS and GREGORY, 1966).

Attention can now be directed towards two *questions peculiar to techniques using conjugating agents.* Firstly, how do we minimise self-coupling or, in the extreme, polymerisation of label-proteins and immunoglobulins? Secondly, how do we ensure that free reactive groups derived from the conjugating agent are absent from the immunoglobulin preparation when reaction with the tissue occurs?

Though *self-coupling* certainly occurs, it does so in an extremely varied fashion, with the outcomes dependent on both the specific protein and the conjugating agent involved. One study used glutaraldehyde as conjugating agent, in a single-step procedure with peroxidase-label plus the immunoglobulin plus glutaraldehyde all in a single solution. Though products with various antibody: label ratios occurred there was no self coupling of the peroxidase. A two-step conjugation procedure was also studied. In this instance the peroxidase label was first exposed at pH 7.0 to an excess of glutaraldehyde. Peroxidase does not self-couple under these reaction conditions. Then after removal of conjugating agent the peroxidase, now carrying reactive aldehyde groups, was reacted with an alkaline solution of the antibody. Under these conditions conjugation yielded an homogeneous product in which the antibody-label ratio was 1:1 (AVRAMEAS, 1972). Another study compared the conjugation of horseraddish peroxidase to various antibodies by three coupling agents: glutaraldehyde, toluene diisocyanate and dicyclohexyl carbodiimide (CLYNE, 1973). Again there was little self-coupling of peroxidase, but the antibodies were polymerised. After using glutaraldehyde the resulting conjugate contained trimeric immunoglobulins; after toluene diisocyanate a mixture of monomer and dimer; and after the carbo-diimide an undefined polymer. So to minimise self-coupling one merely makes what choices are open. Peroxidase does not readily self-couple, but perhaps you do not wish to use peroxidase. Two-step methods are preferable, but sometimes less practical, and so on. Then, appreciating the complexity of possible outcomes, be sceptical as to the nature of your product.

More detail can be usefully provided of a few general tactics. For instance markedly different results may be obtained at different absolute and/or relative concentrations of label, conjugating agent and immunoglobulin.

Consequently investigation of a variety of concentrations is useful though time consuming. With any conjugating agent the use of two-step procedures, perhaps with different conditions of pH or solvent in each step, increases the degree of control over the reaction. The use of conjugating agents carrying two different reactive groups will still further increase control, and decrease self-coupling. Examples of such agents include toluene diisocyanate whose two isocyanate groups are of different reactivities, being at different positions on the aromatic ring; and the N-hydroxysuccinamide ester of p-formylbenzoic acid, which carries two chemically different electrophilic groupings. Consider as a final tactic the use of selective blocking agents. For instance the coupling of horseraddish peroxidase to an antibody by reaction of the periodate-oxidised moiety of the peroxidase with the amino groups of the immunoglobulins (NAKANE and KAWAOI, 1974). In this instance self-coupling of the peroxidase was prevented by prior treatment with dinitrofluorobenzene, which converts amino groups to unreactive derivatives.

When using bifunctional coupling reagents it is possible that the labelled antibody will also carry *residual protein-reactive groups from non-bridging conjugating agent residues.* Such groups could react with tissue proteins and so are a potential cause of non-specific staining. These groups can sometimes be removed by conversion into non-reactive derivatives. Example include treatment of glutaraldehyde coupled materials with a water-soluble reducing agent such as sodium borohydride, and the reaction of product conjugated with a bis-diazonium salt with a naphthol.

Methods for the *purification of the labelled antibody,* once a difficult technical area, are now well defined. A typical procedure when labelling with a fluorescent marker is as follows:

1. If necessary remove low molecular weight impurities such as unreacted fluorescent label, or its hydrolysis products, by gel-filtration chromatography.
2. Apply the high molecular weight material to a cellulose anion exchange column, in a low ionic strength buffer.
3. Elute with buffers of increasing ionic strength. Unlabelled immunoglobulin is eluted first, then «correctly labelled» material with an average of 1.5 labels per immunoglobulin molecule, with heavily labelled («overcoupled») products being eluted only if high ionic strength buffers are used.
4. The correctly labelled eluate is concentrated, for instance by dialysing against a strong polyvinylpyrrolidone solution or by salt precipitation.

With other labels, other purification methods are used. When using ferritin as the label an initial centrifugation step may be used to separate the ferritin-containing species from non-ferrugenous proteins. As a general technique gel-filtration chromatography may be used to separate labelled from unlabelled and overlabelled conjugates when the label is a protein, whether ferritin or an enzyme.

This discussion should be concluded with a cautionary note: several of the conjugating agents mentioned are extremely unpleasant and toxic substances. Thus benzidine and its analogues are carcinogens, toluene diisocyanate is a potent skin sensitizer, and glutaraldehyde reacts vigorously with wet body surfaces.

8.6 Nature of visualising procedures

The usual stain-generating labels are enzymes. Perhaps the least complicated visualising procedure for an enzyme is the production of an electron dense stain. The label used most commonly in such circumstances is horseraddish peroxidase. This enzyme is demonstrated by incubating in a hydrogen peroxide-diaminobenzidine solution («Karnovsky staining»). The action of peroxidase on this system generates a coloured and osmiophilic polymer which is subsequently reacted with osmium tetroxide yielding an electron dense final reaction product. A few other enzyme labels have been used in electron microscopy, for example glucose oxidase which also uses the peroxide-benzidine system; and acid phosphatase which generates an electron dense lead salt precipitate from a Gomori-type incubation medium (Nakane and Pierce, 1967). All the complexities of enzyme histochemistry, discussed at length in Chapter 7 and in particular Section 7.8, must be considered when assessing such methods.

To describe the production of stains for light microscopy is rather more awkward. The range of enzyme labels which have been successfully used is larger, as consequently are the number of visualising methods. A review by Avrameas (1969) for example notes the visualisation of peroxidase by the hydrogen peroxide-diaminobenzidine system; alkaline and acid phosphatases by azo-dye methods; glucose oxidase using a tetrazolium salt procedure; and tyrosinase by the use of a dihydroxyphenylalanine procedure. There is also the possibility of staining several different antigens on a single tissue section in different colours. An elegant example of this is provided in an immunohistochemical study of the distribution of the pituitary hormones (Nakane, 1968). Anti-growth hormone, anti-luteinizing hormone and anti-thyroid stimulating hormone antibodies were prepared. All three were labelled with horseraddish peroxidase. A section of pituitary was first exposed to a solution of anti-GH IgG and the peroxidase label visualised by Karnovsky staining to yield a brown polymer. The antibody was eluted from the section by several washings with acid buffer, and the tissue exposed to labelled anti-TSH IgG. This time the peroxidase label was visualised by incubating with l-naphthol to produce a colourless polymer, which was stained red with the dye Pyronin Y. This second antibody was in its turn eluted from the tissue with acid buffer, and the section exposed to labelled anti-LH IgG. The peroxidase was demonstrated on this final occasion by incubating in 4-chloro-l-napthol, which yielded a blue polymeric oxidation product. Alternatively enzyme labelling can be combined with fluorescence (Lechago et al., 1979).

Enzyme labels such as acid phosphatase and horseraddish peroxidase may be visualised by the same procedure for both light and electron microscopy. This good fortune allows comparisons to be made between serial sections, or indeed on the same material if a re-embedding procedure is adopted after viewing a thin section optically.

As a tailpiece consider visualisation, or at least the intensification, of non-enzyme labels. For example the contrast of ferritin particles can be intensified by exposure of ferritin stained sections to a bismuth subnitrate solution (Ainworth and Karnovsky, 1972). Analogously the staining of uranium-labelled antibodies can be made more intense by reacting the uranyl cations with thiosemicarbazide, and then reacting the sulphur containing derivative with osmium tetroxide (Sternberger, 1972).

8.7 Specificity of the staining

Antibody-antigen interactions are very specific processes, as described earlier in Section 8.1. Non-specific staining will normally derive either from the use of heterogeneous immune sera or from non-immunological binding of stain. The former effect take us back to purification of immunoglobulins discussed in Chapter 8.3, leaving non-immunological staining to be considered here.

Most studies of non-immunological staining by labelled-antibody preparations consider immunoglobulins labelled with Fluorescein isothiocyanate, and such work has been reviewed by PEARSE (1968, p. 185). This account will be largely of the same area, though it will be extended where possible to consider more recently studied topics or additional factors.

The simplest cause of non-specific staining when using fluorescent reactive dyes as labels is that residual *uncoupled label may remain in the antibody preparation* and react with the tissue section. After gel-filtration was introduced for purifying the immune serum however this particular artifact was no longer a serious hazard. An analogous problem can occur when using ferritin or enzyme-labels if label or the antibody-label conjugate carries free reactive groups derived from the conjugating agent (e.g. PALLADINI and SOLHEID, 1976). This may be dealt with by chromatographic purification or chemical inactivation, see Section 8.5. Note that enzyme-labels, even when lacking such reactive groups, may interact non-specifically with the tissue. This is presumably due to the same mechanisms as discussed below for immunoglobulins. An additional possibility in this case is the formation of enzyme-substrate complexes. An example of such non-specific staining occurs when acid phosphatase is used as a label. It was found that even pure enzyme bound selectivity to cell nuclei (NAKANE and PIERCE, 1967). The occurrence of this can be investigated by incubating tissues with unconjugated enzyme and then staining for the enzyme.

More generally *immunglobulins can have a non-immunological affinity for certain tissue proteins,* such as those of myelin sheaths. Furthermore the labelling process can markedly increase such affinities. The reaction of immunoglobulin with the isothiocyanate groups of the Fluorescein and Rhodamine labels for instance takes place via the amino groups. Hence such labelled proteins carry a more negative overall charge than their unlabelled precursors. Consequently labelled antibodies tend to act in a manner analogous to acid (anionic) dyes. So non-specific staining is less marked if the pH of the antibody solution is high, or if the tissue has been fixed in formalin; both of which procedures reduce the number of tissue cations. Pre-staining treatments of tissue with an oxidant and/or albumin have also been used to reduce non-specific staining, and may be understood on the same basis (e.g. ZEHR, 1978). Conversely non-immunological staining is more marked in regions of the tissue rich in basic proteins (or even in the antigen! see BIGBEE et al., 1977), or if the tissue is exposed to buffer solutions which selectively extract anionic tissue components. Since such non-specific staining increases with the magnitude of the overall negative charge on the labelled protein, that is as the number of label molecules per immunoglobulin increases, the least desirable material is in fact removed during the standard ion-exchange purification of the anti-serum, see Section 8.3. A further consequence of attaching labels such as the fluorescent dyes is that the proteins acquire a more polarisable and hydrophobic character, and so become more liable to bind by van der Waals and hydrophobic bonding. Note finally that any process increasing uptake of immunoglobulin, such as pinocytosis if the label is applied to live cells, or pre-staining protease treatments, could enhance non-specific as well as specific staining (WILLIAMS, 1977; MEPHAN et al., 1979).

At this point one thinks of the parallels between such labels and the conjugating agents. Substances such as difluoro-m, m′-dinitrophenyl sulphone, glutaraldehyde and the toluene and xylidine diisocyantates also yield derivatives which are hydrophobic in character. However no studies seem to have been carried out on such effects.

There are quite different non-immune mechanisms for the binding of immunoglobulins, e.g. mediation by complement fractions present in the sera (BUFFA et al., 1979).

There are certain *problems of specificity arising only with use of enzyme labels,* such as staining of endogenous enzyme (either due to lack of substrate specificity or to the presence of label-type enzyme in the tissue), and also of non-specific uptake of enzyme. Substrate specificity can be investigated by omission of ingredients from the incubating medium, such as the hydrogen peroxide from the Karnovsky staining method. The problem of endogenous enzyme has been tackled using inhibitors. When using horseraddish peroxidase as label, a problem arises if the tissue of interest is rich in peroxisomes. Such endogenous peroxidase has been inhibited by treatment with methanolic sodium nitroferricyanide prior to exposure to the antibody solution (STRAUS, 1972). The enzyme label may itself be inhibited, for example phenylhydrazine can be used to inhibit horseraddish peroxidase used as a label, so allowing a subsequent double staining with a different peroxidase-labelled immunoglobulin (STRAUS, 1973), without the need to elute the first antibody. Yet another artifact due to enzyme labelling is diffusion of the final reaction product. Thus the diaminobenzidine oxidation product appears to be able to diffuse from the site of its formation and bind to certain adjacent structures such as ribosomes, chromatin and mitochondrial membranes (NOVIKOFF et al., 1972).

Mention has already been made of certain control procedures and precautions. It is useful to conclude this discussion by listing additional generally applicable controls and precautions which may, and in critical studies should, be used. Firstly staining should only occur in tissues containing the antigen, and there should be no staining of tissues of similar type but antigen free. Contrariwise staining should not occur with labelled sera free of specific antibody, or with a labelled specific serum after exposure to its specific antigen. Finally, staining should be inhibited by pretreatment of the tissue section with unlabelled specific serum, the blocking test.

8.8 Some factors limiting sensitivity

Consider the usage, or rather the usages, of the term sensitivity. A precise meaning arises when a single immunoglobulin molecule can be unambiguously detected, as with labelling by ferritin or sometimes with a dextran-lead chelate or with peroxidase. In this instance an increase in sensitivity means an increase in the proportion of antigenic sites which bind antibody and so stain. When individual antibodies cannot be separately identified, as with fluorescent- and many enzyme-labels, this definition becomes invalid and the concept of sensitivity blurs into that of staining intensity. Another precise, but quite different, usage of the term sensitivity has arisen when considering the resolution of staining. Here an increase in sensitivity means an increased ability to relate the position of the stain to that of the antigenic site. Yet another usage involves the signalto-noise concept. In this case the most sensitive staining is that with the optimal ratio of specific to non-specific staining. The following discussion inevitably reflecting this confused nomenclature, may be summarised as follows. Perhaps the staining reagents cannot penetrate the material being stained? Even if they can do so, perhaps insufficient labelled antibody is bound to the antigen? Finally, even if strong staining of antigen does occur, could this be obscured by non-specific staining?

Most studies on the penetration of staining reagents have centred on the nature and state of the tissue. When considering the penetration of intact tissues rather than that of individual cells or organelles certain findings do seem generally agreed. Even quite thick cryostat sections are usually found to be penetrated by staining reagents, whilst staining of blocks of similar tissues is usually restricted to a surface layer only a few cells deep (see LEDUC et al., 1968; MILLER, 1972; SPALL and AINSWORTH, 1978). Reagents can also penetrate certain embedded tissues, for example when the embedding medium used was polyethylene glycol (COLEMAN and PRATT, 1974). Again however penetration was superficial and could be further limited due to formation of an impervious surface crust of stain.

In spite of these problems several water-miscible embedding media have been investigated including a cross-linked hydroxypropylmethacrylate, glutaraldehyde cross-linked serum albumin, glycolmethacylate, an undefined «polyampholyte», and polyethylene glycol. Published micrographs vary from black dots in a grey fog to those showing excellent morphological detail. The usual non-polar embedding media used in electron microscopy do not allow ready penetration of polar reagents (see Section 10.4), so limiting staining to a surface. Nevertheless some workers have reported success after staining sections embedded in such materials, usually after removal (etching) of at least some of the plastic. Such methods include etching araldite with hydrogen peroxide, and methacrylates with organic solvents or a water-ethanol-benzene mixture. Section 8.9 provides an account of the usefulness of such media in preserving ultrastructure.

There is some dispute regarding penetration into cells or organelles. However many workers now consider that penetration of large molecules such as immunoglobulins will not take place unless gross membrane damage, and hence some distortion of ultrastructural morphology, has occurred. So the disruptive effects of freezing and thawing in the cryostat methods should perhaps be regarded as beneficial. Indeed freezing and thawing, treatment with digitonin, non-ionic detergent, or even enzymic digestion have been used to facilitate reagent entry (e.g. BOHN, 1978; LAURILA et al., 1978). For a brief review of early work see SINGER and MCLEAN (1963). Treatments such as use of protease can however induce non-immunological staining (MEPHAM et al., 1979). If this viewpoint is valid, ready penetration of staining reagents should occur into cell compartments directly accessible to the cell surface region, as rough endoplasmic reticulum often is. Isolated organelles such as mitochondria and nuclei should be less readily penetrated. This does in fact seem to be so.

The influence of the size of reagent proteins on their penetration into cells has also been hotly disputed. Some workers consider that antibody fragments and low molecular weight labels, which give large reductions of molecular weights compared to those of whole immunoglobulins and of large protein labels, give superior results. These they attribute directly to the reduced molecular weights (AVRAMEAS, 1972; GONATAS et al., 1974). This conclusion is contested by others who point out that even large reductions in molecular weights correspond to only small diminutions in molecular cross sectional areas. Such critics suggest that most proteins probably penetrate through gross tissue holes, natural and artifactual (see a letter by SELIGMAN, 1971, but also the adjacent reply by KUHLMANN and AVRAMEAS).

Even when reagents do penetrate the tissues, the amount of staining resulting may be insufficient for satisfactory study. Causes include insufficient binding of antibody to antigen; insufficient labelling of antibody; and with enzyme-labels insufficient stain produced by the label molecules. One factor influencing the number of antibody molecules binding per antigen molecule is the antibody-antigen affinity, discussed in Section 8.2. This affinity is sometimes lowered by fixation of the tissue, as discussed in Section 8.9; or by the process of labelling, as discussed in Section 8.5. Another relevant factor is the number of haptenic sites on each antigen molecule. One advantage of the indirect staining methods illustrated in Fig. 56 is that each immunoglobulin acting as antigen has several sites possessing antigenic activity, and consequently each molecule of the sandwich filling can bind several labelled antibodies. The number of labels per antibody is a further significant variable. When using Fluorescein isothiocyanate a spectrum of labelled immunoglobulins are produced, as described in Section 8.5. It is those immunoglobulins with one or two labels per molecule which are required. Less labelling gives rise to weak staining, more labelling gives rise to increasingly non-specific staining, as discussed in Section 8.7. With other labels, such as the uranyl cations, increasing the number of label per molecule eventually causes denaturation and precipitation of the antibody.

Problems can still arise if sufficient staining does occur, since this may not be restricted or specific. Non-specific staining of non-antigenic materials has already been discussed, in Section 8.7, but analogous phenomena may usefully be described here. One persistant problem, amongst many, confronting workers hoping to carry out immunostaining directly on ultrathin plastic sections is non-specific staining of the embedding medium. Among the various ameliorating measures attempted have been the selective electrophoretic removal of ferritin (MOLENAAR et al., 1966), and the blocking of the binding sites of the plastic with unlabelled serum protein prior to staining with an indirect peroxidase method (STERNBERGER, 1972). A distinct problem arises from the varying sizes of the units of the stains. The stain particles may be beyond the practical resolution of the viewing system used, as with fluorescent labels studied visually or with the final reaction products of some enzyme labels. This is not always so. Ferritin molecules for instance are seen in the electron microscope as black dots some 100 AU (10 nm) in diameter, and lead-chelated dextran labels occur as even larger black blobs. Of course the precipitates resulting from enzyme labels are of indeterminate size, though it has been suggested that some are less indeterminate than others (NAKANE and PIERCE, 1967). Interpretation of resulting images in terms of the presumed sites of antigenicity presents the same, though in the present case often unacknowledged, problems as does the interpretation of electron autoradiographs.

Before leaving the subject of sensitivity there are several additional topics to be considered, not fitting into the frame of discussion used so far. For instance is sensitivity influenced by the type of label used? It is often said (for example AVRAMEAS, 1972) that with the light microscope enzyme-labelling provides much the same sensitivity as does fluorescence-labelling. For electron microscopic immunostaining the issues are more complex. Factors such as the almost unique unambiguity of presence of the ferritin label must be balanced against the greater intensity but also greater geometrical ambiguity of an enzyme-labels' final reaction product. No satisfactorily incisive discussion of these issues seems available. Comparisons have in addition been made of immunocytochemical staining, usually involving enzyme-labelling, with on-section radioimmunoassay methods. The immunocytochemical methods often have the greater sensitivity (for example MORIARTY et al., 1972). Attempts have also been made to quantitate immunostaining. Using the light microscope, immunostaining with fluorescent-labelled antibodies has been carried out in parallel with cytochemical and biochemical assays (WACHSMUTH and WOODHAMS, 1974). Enzyme-labelling involving direct blob counting using the electron microscope, or allied with spectrophotometry, is also thought promising for instance by AVRAMEAS (1972) and STERNBERGER (1972). The trial use of a series of progressively diluted, labelled immunoglobulin preparations is now a widely accepted technique. In this way workers hope to find an antibody concentration giving minimal non-specific staining, but retaining strong specific staining. How to objectively distinguish non-specific from specific staining is a thought I leave for the skeptics among the readers.

8.9 The conflicting problems of preserving antigen and antigenicity (or antibody and immunogenicity) and cell morphology

Here are outlined procedures adopted in attempts to reconcile these conflicting objectives. It is of course essential to retain the antigen or antibody in the tissues, preferably at the *in vivo* site. Since some antigens are of quite low molecular weight and so diffuse easily, and since immersion of frozen sections in aqueous solutions causes gross loss of substance (see Section 3.2), fixation is usually carried out prior to staining. However this risks altering the substrate and eliminating or reducing its antigenicity or immunogenicity. With electron microscopic immunocytochemical staining, where the preservation of ultrastructural integrity is the object of the exercise, this conflict is exacerbated. The more so since the useful fixatives are now fewer in number. A related problem is the use, or otherwise, of embedding media. Most workers find that good preservation of ultrastructure demands sections cut from embedded tissue. The embedding process, and indeed the embedding medium, can however extract or alter the antigen or antibody.

Generally these problems are least acute when preparing materials for study by light microscopy when only general tissue preservation is required. Useful results can often be obtained from frozen sections which are fixed in ethanol or methanol and then stained. Some antigens, for instance those of certain viruses, are altered by these alcohols, and here acetone has been found a suitable alternative. More chemically vigorous fixatives have sometimes been used, as in the study of certain polypeptide hormones. The retention of these small molecules is aided by formalin fixation, which in this instance does not destroy their antigenicity. In some other studies Bouins fluid has been used with success in this way, and even tissues stored in ethanol for decades (CELIO, 1979). Often of course the more reactive the fixative the better the morphological preservation and the more destructive of antigenicity, a double bind (e.g. HOFSTÄDTER et al., 1977; WATKINS, 1978). Different fixation regimes can however give rise to different distributions of retained antigenic materials. An example of this is provided by a study of mouse submandibular peptidoesterases (EKFORS and HOPSU-HAVU, 1971). In this case staining occurring after acetone fixation was intense and localised, whilst after fixation in ethanol or formalin the staining was evenly distributed and less intense: but which was the artifact? In another system intracellular «diffusion» was seen after fixation in stirred ethanol-acetic acid, but was minimised by avoiding stirring and by the use of paraldehyde (GUILLOUZO et al., 1978). Another problem is that cryostat sections do tend to be rather thick. For more detailed studies it is sometimes possible, without gross loss of antigenicity, to embed freeze dried tissues in paraffin wax. This enables thinner sections to be cut for subsequent staining. Vapour phase fixation has been investigated for such procedures and recently comparative studies of various agents have been made (PEARSE and POLAK, 1975; HOFSTÄDTER et al., 1977). A quite different approach was adopted to allow the demonstration of the extremely soluble cyclic adenosine monophosphate. Cryostat sections were treated directly with antibody, then the tissue, antigen-antibody complex included, was fixed (WEDNER et al., 1972). Sometimes the loss of antigenicity consequent upon fixation may be reversed, e.g. by treatment with trypsin or pepsin (BROZMAN, 1978). Other promising, and more intelligible, possibilities are: to apply a protective chemical mask prior to fixation, removing it later (TAKAMIYA et al., 1978); or to use an analogue to substrate protection (KURZON and STERNBERGER, 1978).

Tactics adopted to cope with the more severe problem of preparing tissue for study by electron microscopy fall into three main groups. Nearly all methods though involve an initial fixation step, since otherwise little ultrastructural preservation would be expected. Amongst the few exceptions is the study of materials such as erythrocytes, which cells may be successfully stained while still viable and then processed for electron microscopy (VOAK and WILLIAMS, 1971). Useful fixatives are formalin, sometimes used in a mixture with picric acid, glutaraldehyde, and mixtures of these two aldehydes. The effects of these agents on antigenicity seem unpredictable. For instance the antigenic activity of phytochrome was found to be grossly reduced by glutaraldehyde but not by formaldehyde (COLEMAN and PRATT, 1974). In many other studies however glutaraldehyde seems advantageous, as emphasized in a review of this topic by MILLER (1972).

After fixation paths diverge. Some use a special microtome so that ultrathin frozen sections may be cut. Such sections may be stained directly, minimising loss of antigenicity. This approach has been investigated with persistance and enthusiasm by a number of groups. The degree of ultrastructural preservation seems poor, and so the amount of information to be extracted from the final image limited. A less jaundiced view will however be obtained from a protagonist of ultracryomicrotomy, Singer for instance (SINGER et al., 1974).

Another minority approach has been to fix, embed, cut thin sections, and stain the embedded tissue. With such techniques ultrastructural integrity will hopefully be preserved. Loss of antigenicity (polymeric embedding media are formed from *reactive* monomers) and poor reagent penetration may be serious problems. As discussed in Section 8.8 there are several variations on this theme. For instance the embedding medium may be hydrophobic, and sections may then be etched before staining; or the medium may be hydrophilic; or a compromise may be made when a thick section of tissue embedded in a polar medium is stained, and then re-embedded in a non-polar resin and thin sections cut for microscopic study. Perhaps the best comment on the current state of this promising approach is to note that though new embedding media and procedures continue to be investigated no consensus seems in sight as to preferred procedures and materials.

Most electron microscopic immunocytochemical studies are still carried out by staining prior to embedding. The major period of fixation is often placed after the staining process, balancing the loss of antigen and structural detail on the one hand and of antigenicity on the other. Tissues may be prepared for staining either by chopping them up with a razor blade, the traditional method used for electron microscopic studies; or by slicing them up on a cryostat. The former technique gives the better ultrastructural preservation. The latter facilitates penetration of reagents and also allows different parts of the tissue sample to be more easily related.

9 Some Miscellaneous Biological Staining Methods

9.1 An introduction

Chapter 9 could have been given a more explicit though long-winded title such as «A hotch-potch of biological staining methods which have intrigued me, and which serve very well to illustrate the general applicability of the concepts of rate, affinity and reactivity: which methods do not however fit readily into any other chapter of this book».

The methods discussed here fall into three categories. Firstly procedures using staining reagents which are not dyes, but whose uptake into the tissues involve much the same physico-chemical processes as do the uptake of dyestuffs. Secondly methods in which soluble dyestuffs are converted into insoluble derivatives in the tissues, «trapped» in fact. Finally vital staining, in which reagent uptake is achieved or aided or influenced by some «vital» process or processes of living cells and tissues.

9.2 Dyeing with reagents that are not dyes

Dyestuffs, as considered at length in Chapter 4, are coloured organic compounds containing extensive systems of delocalised electrons. Such compounds are able to move from their solutions onto substrates, tissue sections and so on, immersed in those solutions. The driving force, the affinity, for such a process includes increases in the entropy of the solvent, hydrophobic bonding, and dye-substrate attractions such as coulombic or van der Waals forces. Note that dyestuffs do not, at least in the first instance, form covalent bonds with the substrates. In the present discussion we are concerned with staining by reagents which are probably taken onto the substrate for similar reasons, but are not coloured, or if so are not organic, or if so do not contain large delocalised electronic systems. Example to be discussed are colloidal iron, fluorescent brightening agents and tannic acid. Some brief notes on their chemical structures are for convenience collected together in Table 44.

Though little rigorous work appears to have been carried out in this area, some sources of reagent-tissue affinity can be suggested. For example since the fluorescent brightening agents and many tannic acid components contain numerous aromatic rings and are applied from aqueous solutions it is probable that in these instances van der Waals attractions and hydrophobic bonding (Supplements 4 and 7) are important. Van der Waals attractions also seem the basis of the affinity of binding of iodine/iodide to

Table 44. Characteristics of some reagents used in the dyeing with non-dyes methods discussed in Section 9–2

Reagent	. . . and its chemical nature	Leading references
Labelled Aprotinin	Aprotinin (or «basic pancreatic inhibiter», «Trasylol®») is a polypeptide containing 58 aminoacid residues, basic ones being in slight excess. Labelling was carried out with fluorescein or Rhodamine B isothiocyanates	KIERNAN and STODDART (1973), CHAUVET et al. (1966)
Labelled basic polysaccharides	*Aspergillus* polysaccharide, and deacetylated chitin: both fluorescein labelled	EVANS and KENT (1962), KENT and EVANS (1962)
Colloidal iron	Particles of ferric hydroxide, usually with an overall positive charge, though anionic colloids can also be prepared. Particle size varies with age and mode of preparation, but diameters are often in the range 30–90 Å	GASIC et al. (1968), GEYER and STIBENZ (1974)
Fluorescent brightening agents	Colourless though fluorescent organic compounds usually with molecular weights of 500 or above, containing numerous aromatic rings, sometimes sulphonated	SARKAR (1971)
Iodine/iodide	Molecular iodine (I_2) is coloured whilst anionic iodide (I^-) is not. The triiodide anion (I_3^-) which exists in equilibrium with mixtures of iodine and iodide is also coloured. All three species are highly polarisable	Standard inorganic chemistry texts
Tannic acid	The term loosely applied to the leather tanning extracts of such materials as oak galls. Such extracts are very complex in content with most of the compounds being phenolic, though carboxylic and saccharide groups are often present too. Molecular weights run up to 1000 or higher	Many reviews available with that of WHITE (1958), though dated, being particularly interesting to histochemists since adjacent reviews discuss topics relevant to fixation and to mordanting

polysaccharide polymers such as glycogen and starch, MORAWETZ (1965) offers an account of the mode of action of this beautiful stain. For ionic reagents, coulombic attraction and replusions are probably important. For example the usual, large particled cationic form of colloidal iron, the cationic polypeptide Aprotinin, and also various cationic polysaccharides, preferentially stain tissue anions such as those in mucins, mast cell granules and cartilage matrix. This staining pattern is similar to that obtained with a large basic dye. Again paralleling basic dyeing, increasing the number of tissue anions by sulphation, or decreasing the number of tissue cations by nitrous acid deamination, often increases the intensity of colloidal iron staining (for example GASIC et al., 1968; THOMPSON, 1966). Similarly, decreasing the number of tissue anions by methylation, or increasing the salt content of the staining solution, decreases staining intensities both with colloidal iron and labelled-Aprotinin (as previously, and also KIERNAN and STODDARDT, 1973). Analogous effects, but this time paralleling acid dyeing, were seen with an anionic colloidal iron preparation (GASIC et al., 1968). Some workers have argued that tissue anions actually bind colloidal iron by chelation of the Fe^{+++} ions rather than merely coulombically. There is some support for this ligand exchange concept to be found in the colloidal chemistry literature, for instance a study of YATES and Healy (1975). The same study however also indicates that such anion-colloidal iron interactions do not always involve coordinative links and can occur merely due to coulombic attractions. Whatever the contribution of coordination to stain-tissue affinities the relative charges of the colloidal iron and the tissue sites would still be expected to largely control the *selectivity* of staining.

As with dyestuffs, selectivity of staining with such substances is controlled both by the affinities of the sort already mentioned acting selectively, and also by selective rates of uptake. For instance it seems plausible that the selective uptake of the large cationic particles of colloidal iron and of the large, weakly basic, polypeptide Aprotinin is rate controlled. Thus only the fast staining basophilic sites are stained with either reagent. It is interesting that while this mechanism has been argued for colloidal iron (Goldstein, 1962b) the previously suggested mechanism for Aprotinin postulated a specific polypeptide-polysaccharide affinity (Kiernan and Stoddard, 1973). Rate control is an even more plausible suggestion for the staining of anionic mucosubstances by fluorescein-labelled, cationic polysaccharides (Evans and Kent, 1962; Kent and Evans, 1962). These authors do in fact note that in solution these compounds bind non-specifically to many macromolecules, but with tissue sections staining is «similar to results with Alcian Blue».

Various visualisation procedures are often required in these methods. Examples are the chelation by the phenolic groups of tannic acid of Fe^{+++} ions; of conversion of colloidal iron into the pigment Prussian Blue; and the illumination of tissue stained with fluorescent brightening agents or by labelled-Aprotinin or polysaccharide with ultra violet radiation. Except for the latter procedure any such visualisation procedure may itself play a role in determining the selectivity of the final staining, just as is the case in enzyme histochemistry.

9.3 A trapping method: the Gram Reaction

Trapping implies a procedure in which a dye is selectively retained in a particular structure by conversion of the dye into an immobile form. Many trapping agents are mentioned in the biological staining literature. The metallic mordants discussed in Section 4.4 may be considered in this way, as can the conversion of basic dyes on stained tissue into insoluble picrates, phosphomolybdates and phosphotungstates. In the present discussion we will concern ourselves only with a single method, hoary with tradition and still widely used, the Gram stain. This procedure has many variants but the essential practical steps are:

1. Stain a smear or section with the primary dye, usually Crystal Violet.
2. Treat the stained preparation with the trapping agent or mordant, usually an aqueous potassium iodide-iodine sulution.
3. Differentiate, using one of several possible organic solvents each with its own optimum period of action.
4. Counterstain with one of many possible dyes, but typically with Safronin.

What does this sequence stain? It depends mostly on step 3. When a vigorous differentiator such as acetone or 95% ethanol is used the only intensely staining structures occurring in the tissues which are usually studied will be the Gram positive bacteria and fungi, though tissue constitutents such as hair shafts and keratin also stain. After a milder differentiator such as aniline, additional structures retain stain, including fibrin, some mucins and, sporadically, polysaccharides such as glycogen and starch.

Before offering a theoretical model for this process some intellectual and bibliographic acknowledgements and clarifications are called for. Gram staining was devised in the nineteenth century and after only a few decades there were already numerous and conflicting explanations being offered. There is no attempt in the present account to offer a historical view nor even in general to attribute sources of information. Several reviews (BISWAS et al., 1970; PEARSE, 1968, p. 280; THOMPSON, 1966, p. 1002) are good starting points for those wishing to repair these omissions. I do however wish to emphasize the outstanding contributions of BARTHOLOMEW and his fellow workers, both for wide ranging and sceptical experimentation and also for critical theory building. A summary of much of this work is available (BARTHOLOMEW, 1962).

There follows a sketch of a general model, which will be presented quite simply and didactically. The crucial question when attempting to understand the Gram stain is as follows. In what way do Gram positive and Gram negative structures differ? As quantitative studies indicate that the uptake of primary dye and also of iodine is much the same in Gram positive and Gram negative bacteria, this question may be rephrased to: why do Gram negative structures lose stain more easily than Gram positive ones? The answer appears to be that Gram positive structures are those which lose stain slowly, and the process is rate controlled. Consider staining a mixture of Gram negative and positive bacteria with any one of a wide range of dyes, including Crystal Violet. If such a preparation is differentiated without prior treatment with an iodine-iodide reagent then dye is lost fastest from the Gram negative bacteria. Similar observations were made with alcoholic solutions of iodine or of picric acid, with these substances entering and leaving the Gram negative cells faster than the positive cells.

Such a rate-control model implies that any structure which stains in the initial step and is slow to destain during differentiation will be Gram positive. The reasons for the slow rate of destaining may well vary from one situation to another. So the differential rates of destaining of Gram positive and negative bacteria and fungi could indeed lie in some properties of the cell walls. Conversions of Gram positive (i. e. slower to destain) bacteria to Gram negative (i. e. faster to destain) by such varied treatments as mechanical crushing or treatment with proteases or surfactants or organic solvents may all be plausibly viewed as increasing the porosity of the cells, though perhaps by quite different processes. To cite another example, the action of RNAase (specificity?) on Gram positive bacteria converts them to Gram negative forms, and simultaneously increases their measured permeability to iodine. In keeping with this, amongst the most strongly Gram positive structures in tissue sections are keratinous entities (including hair shafts), elastic fibres, and RNA-rich basophilic cytoplasms. As indicated in Section 4.13 these are typical examples of structures slow to stain (and so to destain?) with dyestuffs. Also consistent with rate control are certain observations on dehydration/hydration. When staining bacterial preparations the amount of dehydration prior to differentiation and after treatment with iodine is a significant practical variable. If the preparation is well dried the rate of loss of stain is sharply diminished. However if the preparation is merely blotted, or if a dried preparation is briefly rehydrated, then the usual period of differentiation may be used. Such effects could be due to hydration increasing the swelling and hence the permeability of the cells. Consistent with this rate-control model, it has been reported that nitrous acid or chloramine T deamination treatments greatly reduce Gram positive staining, but do not eliminate basophilia (JOHNSON et al., 1957). Such blockades are extremely aggressive, and would be expected to increase the permeability of the tissues, see Section 11.3.

The question now arises as to why it is necessary to treat with iodine? The probable answer lies in the relative ease with which dye may be removed from Gram positive and negative structures. At least for bacteria this difference is so small that control (usually achieved by varying the time of differentiation) presents marked technical problems. However conversion of the primary dyes into poorly soluble derivatives (such as the iodine-iodide, or the picric acid complexes) decreases the rate of loss of stain, so allowing much greater control. In keeping with this suggestion, a range of trapping agents may be used in place of the iodine-iodide or the picric acid. It is noteworthy that almost the only common feature of such substitutes, and none incidentally are as effective as iodine or picric acid, is that they give poorly soluble complexes with the primary dyes. It may be further noted that if the complex produced is excessively insoluble, as with permanganate, the differential extraction of stain again becomes difficult to achieve. Analogous effects are seen with the various differentiating agents studied. Loss of stain into differentiator can be too rapid as with methanol, or too slow as with butanol. In these cases the periods of differentiation required are respectively too short to control, and too long for one's patience.

It is also necessary to consider the counterstaining procedure (step 4) in some detail. This process, during which Gram negative structures are stained, may result in the removal of further primary dye from the preparation. So the step may constitute a further period of differentiation, again emphasising the need to convert the primary dye into a poorly soluble derivative. A number of dyes studied as alternative primary stains proved satisfactory until the application of the counterstain, when they were masked or extracted. Perhaps certain dye-trapping agent complexes are more soluble in, or more easily dissociated by, the aqueous counterstain than by the organic differentiator? It is worth noting that if the extent of step 4 is sufficiently lengthened then even the Crystal Violet-iodine complex can be extracted by the counterstain. The nature of the dye in the counterstain solution does not seem to be very important since a variety of dyes, both acid and basic, have been successfully used.

Now consider step 1 of the Gram staining procedure, the uptake of the primary dye. Though the dye must have affinity for some substrate in the cells or tissues, the nature of the substrate does not seem of overwhelming importance. A variety of basic *and* acid dyes have been used, with some success, as primary dyes in the Gram staining of bacteria. The acid dyes were probably staining protein, and as the basic dyes were applied from neutral or weakly alkaline solutions they too would probably be staining protein in addition to nucleic acid-rich sites. This suggestion is consistent with a variety of observations. For instance if the pH of the primary dye staining-solution is reduced, some Gram positive bacteria show a stippled appearance. This could be due to protein-rich structures no longer staining, leaving only the nuclei-acid-containing areas still basophilic. A gross reduction in pH converts Gram positive bacteria to Gram negative, perhaps due to protonation of the phosphates of nucleic acids, and perhaps also to general increases in permeability due to enhanced swelling of protein at low pHs. The reverse also holds and a large rise in pH results in Gram negative bacteria apparently becoming Gram positive, perhaps due to more protein binding sites being generated. Histochemical blocking processes are also intelligible in this light, with Gram positive bacteria being converted by methylation to Gram negative when the primary stain was the basic dye Crystal Violet, but remaining Gram positive when the stain used was the acid dye Acid Fuchsin. Note also that the uptake of the primary stain is quantitatively much the same for Gram positive and negative bacteria.

Finally consider the significance of Gram staining in the light of the previous discussion. It seems that for a structure to be Gram positive it must have an affinity for the primary stain, though this is very non-specific since many Gram negative structures also fulfil this criterion. Also Gram positive structures are for some reason, or more probably reasons, slow in releasing the stain during differentiation. Though this latter effect is more selective, correlation with characteristics of biological interest such as pathogenicity will be essentially coincidental.

9.4 Vital staining procedures

These involve application of a reagent to living cells or organisms so that some or all of the still living structures become stained. Consequently the distribution of stain reflects in part the physiological and metabolic activities of tissues and cells and may be used as probes of such. Such a description avoids the question of whether observations are made on the still living cells (an attractive prospect) or merely on processed material derived from vitally stained tissue. This ambiguity reflects current nomenclature, where both possibilities on occasion have been described as vital staining.

The modes of application of vital stains to cells are various. Aquatic animals, varying in size from protozoans to tadpoles, and also isolated cells of metazoans such as the cells of the blood, may be merely suspended in a suitable solution of the stain. With larger creatures, stain may be injected into cavities or circulating fluids, or administered dissolved or mixed with water or food. Alternatively live fragments may be detached from an organism and placed in the stain. This latter is termed supravital or survival or, following BAKER, supervital staining. More recently ionic reagents (metal ions and anionic reactive dyes) have been introduced into single neurones by inserting a hollow reagent-filled microelectrode and passing current until the required amount of stain ions has entered the cell.

What type of staining reagents are used? A wide variety, as the only rigid requirement is that the reagent is not too toxic. The most commonly used are dyestuffs, usually basic (Janus Green, Methylene Blue, Neutral Red) but sometimes acid (Eosin Y, various bis- and tris-azo dyes), or more recently reactive dyes (various Remazols and Procions, especially Procion Yellow). Some of the dyes are used as fluorochromes (Acridine Orange, Coriphosphine), and non-dye fluorochromes such as tetracycline and sulphonated fluorescent britening agents have also found application. Colloidal iron as well as cobalt and silver ions have also been used as vital stains, though these require some visualisation procedure. Finally note that enzyme histochemistry has been carried out vitally with bacteria, with the enzyme substrate incorporated in the growth medium. Of the large organic reagents applied directly to cells or tissues it is usually the cationic compounds which stain cell interiors, whilst staining by the anionic compounds is often limited to cell surfaces or to extracellular material such as bone matrix and connective tissue fibres. A possible explanation is that, as well documented by pharmacologists (e.g. ALBERT, 1979), ionic species pass through cell membranes less readily than do non-ionic and hence more hydrophobic compounds. If this seems no explanation consider the following. The anionic compounds are usually sulphonated and remain ionic throughout the pH range of cell viability. However the pKa's of the basic dyes used often fall into the physiological range and hence solutions of such dyes will contain a proportion of their molecules in uncharged form. Consistent with this is the observation that intracellular staining by acid dyes is commonly of dye particles ingested by phagocytosis, which did not therefore have to penetrate the plasmalemma. One of the few acid dyes to stain other intracellular structures is Eosin whose pKa is, unusually for an acid dye, near the physiological range.

Many structures have, over the last 100 years or so, been stained vitally prior to microscopy. This account considers only the more topical aspects of the process. Readers with more general interests may find BAKER's review (1958, chapters 16 and, in part, 15) useful, and other general sources include LILLIE (1965), PEARSE (1972) and THOMPSON (1966) whose indices should be consulted for «supravital» and «vital» staining. Several areas of current interest may be identified.

There is a substantial diagnostic role for dyes and staining reagents. Commonly however this does not really concern the present book; for example dyes may be used to assess the viability of bone fragments after fractures (MILLIGAN, 1979). Even when staining of a more orthodox kind is taking place, microscopic study may not be used. Consider for instance staining the microorganisms of dental plaque (GALLAGHER, 1977). A diagnostic field in which conventional vital staining has been applied is the detection of a variety of bacteria and protozoa, e. g. microfilariae trypanosomes and plasmodia in cultured blood (THOMPSON, 1966, p. 1054).

Vital staining is also applied to mark the sites of physiologically interesting processes. Most dramatic perhaps is the assessment of viability as such. For instance WIDHOLM (1972) treated cells from culture with a mixture of Fluorescein diacetate, which is colourless and non-ionic, and an ionic dye such as Evans Blue or Safronin. The dead cells were more permeable and hence stained rapidly with the ionic dyes. The live cells excluded these charged species but were penetrated by the non-ionic compound. After uptake tissue esterases converted this to the readily detected fluorescent dye Fluorescein. Single dye methods for viability assessment of cell cultures have also been developed, exploiting the fact that the fluorescence metachromasia of Acridine Orange varies with the viability of the cells (SCHWARTZ et al., 1974). The Acridine Orange system has also been proposed as a way of detecting cell damage prior to death and is allegedly more sensitive than phase contrast microscopy for this purpose (SÖDERSTRÖM et al., 1977). This poses the problem of when is a damaged cell dead. Other more specific uses have been to mark sites of bone mineralisation (MALTHA et al., 1977) and of phagocytosis (STOREY, 1968). There is no clear boundary between vital staining and the use of dyes and microscopy by physiologists (e. g. BROWN et al., 1977).

Cell types or organelles as well as non-cellular structures are also demonstrable by vital staining. The ramifications of neurones for instance have long been traced in this way, orginally with leuco-Methylene Blue which was reconverted to the coloured form intracellularly. Currently ionic dyes such as Procion Yellow or metal ions such as Co^{+++} are introduced into selected neurones with the aid of a microelectrode, following which the stain diffuses throughout the cell. Perusal of a symposium volume edited by KATER and NICHOLSON (1973) provides numerous examples of such work. Non-neurological examples include the use of Procion Navy Blue M3RS to stain collagen and elastic fibres (HENRY, 1968); of 4, 5, 6, 7-Tetrachlorofluorescein to stain the plasmalemma of cells of onion bulb scales (TAMPION et al., 1973); of saccharated iron oxide to stain fungal hyphae (CIBIS and YAMASHITA, 1959); selective staining of olfactory receptor cells with Trypan Blue (HOLL, 1965); and detection of phagocytes in lymphocyte preparations (SHAALA, 1979).

Finally some cautionary comments. The usual requirement for a vital stain, namely low toxicity, is more subtle than it first appears. Consider the final effect of simple toxicity: the cells die. The staining of such dead cells will often differ from that of the live ones. Think of mammalian white blood cells being stained with basic dyes. In live cells organelles such as mitochondria and lysosomes are stained, but nuclei are not. In dead cells nuclei stain intensely. Hence an independent measure of viability has often been made during vital staining process, such as cell mobility or motility, or cell division, or in a more recent study of crab axons (KOMISSARCHIK and LEVIN, 1964) the presence of an action potential. Use of such criteria will not necessarily indicate localised necrotic changes however. There is a fundamental dilemma: staining reagents must be highly interactive molecules, which not surprisingly commonly produce very marked physiological and metabolic changes in the cells in contact with them. This point is extensively illustrated in a review by BARBOSSA and PETERS (1971). A further problem is that vital staining may be masked by or confused with non-vital staining (DEBARI and NEEDLE, 1978).

10 Peculiarities of Staining Plastic Sections

What are «plastic sections»? In this discussion attention will be directed towards sections about 1 µm thickness. These «thick» sections may be stained to aid orientation when using ultrathin sections in related electron microscopical work; or for histological identification and cytochemical studies using the light microscope. Some of the information discussed below does however originate from electron microscopic studies. Both the general theoretical model and the practical implications described and discussed in Sections 10.5 and 10.6 apply to the staining of both ultrathin and thick plastic sections.

In this Chapter «plastic» almost always means a resin suitable for sectioning on an ultramicrotome, usually an epoxy or methacrylate or polyester polymer. Embedding media used primarily for light microscopy, the polyethylene glycols for instance, are not within this brief. Note further that the term plastic section often merely means «sections cut from a tissue block embedded in plastic» since the staining may be carried out on tissue freed, wholely or in part, from the polymer. Enzyme histochemistry and labelled antibody staining on plastic sections are discussed elsewhere; in Section 7.8, and in Sections 8.8 and 8.9 respectively.

10.1 The state of the art of staining plastic sections

The first question to be considered is why is there a continuing spate of staining methods, many involving only very minor modifications of previous procedures? The answer is probably that a method which works in one laboratory often fails in another, or even in the same place if attempted a couple of years later. Arising from this we can ask why are there no general theoretical models to direct empirical enthusiasms? Perhaps because little experimental effort appears to have been directed towards understanding the processes of staining. This may be dramatically appreciated by inspecting reviews concerned with staining plastic sections, for example that of Hayat (1975). Such accounts are predominantly summaries of practical recipes. Finally one may enquire why, in spite of the lack of theoretical models, no rule-of-thumb concensus on how to stain plastic sections has emerged? To which question, in the light of the discussion below, I have no answer.

Consequently Section 10.1 will outline the state of the art, noting the range of variation of the dyebath composition, temperatures, resins, etch methods and so on.

Later, in Section 10.5, a theoretical model of plastic section staining will be outlined, and in Section 10.6 an attempt will be made to offer some general, practical strategies for successful staining procedures. For readers desiring a wider background several bibliographies are available (for instance DAVIE, 1970 and HAYAT, 1975). Journals such as the Journal of Microscopy and Stain Technology also provide a continuing trickle of publications on this topic.

Though the tactics used to assist the staining of plastic sections are very varied and have been largely developed on an *ad hoc* basis some general threads can be followed. For instance it has been suggested, though see Section 10.2, that fixation in osmium tetroxide inhibits staining. So chemical pre-staining treatments to remove the osmium, for instance with aqueous solutions of hydrogen peroxide or potassium permanganate, have been used.

It has also been proposed, as discussed in Section 10.4, that the use of hydrophobic resins (such as Araldite or Epon or butyl methacrylate) as embedding media impedes access of polar stains to the embedded tissue.

Practical responses to this include use of polar resins such as glycol methacrylate or the polyester Vestopal W. Alternatively the non-polar resins have been treated with organic solvents such as acetone, dimethylformamide or xylene; or aggressive reagents such as halides (bromine or iodine) or oxidants (hydrogen peroxide, peracetic, performic or periodic acids and potassium permanganate); or strong bases such as sodium ethoxide or methoxide; or plastiziers such as aniline or phenol. Such treatments may soften or remove the embedding medium completely or in part, see Section 10.3.

Many treatments have been used in efforts to facilitate stain penetration into tissue embedded in non-polar resins, Fig. 61 gives some indication of their frequency of use. One of the most usual is an elevated staining temperature. The range 60–80 C is common, though higher temperatures have been used, for instance 100 C in an aqueous solution containing a quantity of polyethylene glycol (ALSOP, 1974). The applications of haematoxylins (Ehrlich's, Harris', iron and so on) to un-etched sections of non-polar resins always appear to involve heated staining solutions. Often, and apparently either as an alternative to heating or for use with thermally labile substances such as Schiff's reagent or Alcian Blue, prolonged staining times are adopted. Several hours is not uncommon and several days has been reported (KOSAKAI, 1970; SCHWALBACH et al., 1963). Another option is to use organic rather than aqueous solvents for the staining reagents. In the light of the use of solvent etching, see above and Section 10.3, it is surprising that this approach has not been applied more generally. After all, a solvent such as dimethylformamide readily dissolves ionic and hydrophobic reagents and, to some extent, the resin. However ethanol is usually the chosen solvent, though dimethylsulphoxide has also been used for an aldehyde reagent (COTTELL and CAMPBELL LIVINGSTONE, 1976). Manipulation of the ionic strength and pH of the dyebaths so as to give maximum dye-tissue affinity is also often adopted. Hence when attempting a general staining of tissue, essentially when staining protein, anionic (acid) dyes are often applied from extremely acidic dyebaths, e. g. pH 1.3 with Acid Fuchsin and Aniline Blue, pH 1.4 with Picric Acid and pH 2.8 for Light Green. Contrariwise cationic (basic) dyes are commonly applied from extremely alkaline solutions, e. g. pH 11.0 for Azur A and Toluidine Blue, and pH 12.5 for Methylene Blue. The addition of calcium or magnesium salts to enhance the dye-tissue affinity of anionic fluorescein derivatives Eosin and Phloxine, a technique known from paraffin section technology, has also found application in staining plastic sections.

The considerable exploration of the use of different stains may be given special mention. In a sample of papers read to provide background information for the present discussion the use of some 58 different stains was reported. Half were cationic dyes, and a few of these were metal-complex (mordant) dyes. Rather fewer anionic, plus a couple of non-ionic, dyes were also used. The sample included less than a dozen non-dye stains.

This topic provides a bridge to a discussion of successes and failures in staining plastic sections. Significantly only half of these 58 stains were used on more than a single occasion. Certain stains were however used on more than half a dozen occasions. These were the cationic dyes Basic Fuchsin, Methylene Blue and Toluidine Blue; the anionic dye Acid Fuchsin; and some reactive staining procedures, namely the Feulgen nucleal, the periodic acid-Schiff and various silver methods. Other multiple-use stains included the cationic dyes Alcian Blue and Safronin, and the anionic dyes Eosin and Fast Green. These few stains may justifiably be regarded as successes.

The *failures however do have a great deal to teach us.* Firstly note that, compared to the paraffin section or cryostat technologies, only a narrow range of staining techniques have been successfully applied to plastic sections. Even these techniques are dramatically capricious. The remainder of Chapter 10 offers various viewpoints on this issue. Secondly consider the numerous technical snags of the published methods. For instance high temperatures, extremes of pH, use of organic solvents, and long staining times can, singly or in combination, encourage section loss or wrinkling, and stain precipitation or decomposition. Some workers have reported that chemical etching of the plastic sections seriously alters tissue morphology. Others report that staining is sometimes strongly influenced by such difficult to identify factors as the degree of tissue dehydration, the particular buffer used during fixation, or whether sections were stained floating or mounted on glass slides. Certain of these effects are intelligible or can be rationalised in terms of the model presented later, but the benchworker approaching the primary literature is presented with a mass of apparently arbitrary options. To add to the complication, sometimes even the embedding media stain. For instance Epon is coloured by anionic dyes if the basic hardener DMP 30 has been used, or after staining with high affinity dyes such as Luxol Fast Blue. Others have found Epon to be stained by the cationic dye Basic Fuchsin, but point out that the plastic may be decolourised by differentiation with aqueous ethanol. The frequently noted basophilia of glycol methacrylate appears to be due to the presence of acrylic acid in some batches of monomer. Finally note that Vestopal W is coloured by both anionic and cationic dyes but, in spite of this, staining schedules allowing the successful use of dyes have been devised.

10.2 What is the effect of the fixative?

Are the difficulties experienced in staining plastic sections due, in part or in whole, to the fixation of tissue by osmium tetroxide or glutaraldehyde? The results discussed are again in large part derived from work on thick plastic sections.

Can tissues fixed in glutaraldehyde and osmium tetroxide be stained at all? One simple approach is to study tissues fixed with these reagents and embedded in paraffin. As this embedding medium does not react covalently with the tissues the effects of fixation may be seen more clearly. Glutaraldehyde has in fact been widely used in this way by histopathologists and others. The acidophilia-basophilia staining balance is strongly influenced, with tissues being markedly basophilic as described in Section 3.7. However this fixative does allow staining with acid or basic dyes, and with reactive stains such as the Feulgen nucleal and the periodic acid-Schiff procedures. Few workers use osmium tetroxide as a fixative for light microscopy as it is so expensive and at any rate it blackens tissues, necessitating oxidative bleaching prior to staining. The available evidence however shows that staining can be carried out after such fixation.

The relevance of this information could be questioned, as osmium tetroxide or glutaraldehyde fixation *followed by* embedding in a polymerising medium may be responsible for inhibition of staining. To investigate this possibility we will consider several situations, in each of which the fixative was glutaraldehyde or osmium tetroxide, or both. The situations are staining tissues embedded in hydrophobic resins; staining tissues embedded in hydrophilic resins, through which ionic reagents may be expected to penetrate readily; and staining sections embedded in hydrophobic resins from which all plastic has apparently been removed. In each case the complication of monomer-tissue chemical interactions is present, see Section 10.3.

Studies comparing different fixatives whilst using a single hydrophobic resin have produced few surprises. Staining procedures which are successful after one fixative will usually be successful after another, even when the two fixatives are as different as formalin and osmium tetroxide. Differences of degree are seen (e. g. de BUIJN and MCGEE-RUSSEL, 1965; SHIRES et al., 1969; SNODGRADD et al., 1972) and may usually be attributed to basophilia-enhancement by glutaraldehyde and by osmium tetroxide.

Tissues may be embedded in hydrophilic resins, such as glycol methacrylate or the hydroxylated polyester Vestopal W. Staining with acid, basic or metal-complex dyes and also with reactive stains such as the periodic acid-Schiff and Feulgen nucleal procedures is then possible after fixation in acrolein, or glutaraldehyde or osmium tetroxide for instance (FEDER and O'BRIEN, 1968; SCHWALBACH et al., 1963).

Finally, what of sections from which the plastic has been removed? Methacrylate sections may be completely freed of resin by the action of organic solvents such as acetone or xylene. Stains of the paraffin section technology are then generally applicable (for example MUNGER, 1961; RUNGE et al., 1958). To free epoxy sections of plastic needs more vigorous reagents such as iodine in acetone, or sodium methoxide in methanol-benzene, as described in Section 10.3. Once freed of resin however staining is comparable to that achieved with resin-free methacrylate sections (e. g. HOFF and RAYBURN, 1974; LANE and EUROPA, 1965; SNODGRADD et al., 1972).

So though fixative effects are seen, the generalised difficulties experienced when staining tissues embedded in hydrophobic plastics is not due to peculiarities of osmium tetroxide or glutaraldehyde. To conclude this account note that various types of fixative-related phenomena discussed in Chapters 3 and 4 are also seen when staining plastic sections. For instance the enhancement of certain stains by certain fixatives, such as silver stains after an oxidising fixative (GOLDBLATT and TRUMP 1965), and phenylene diamine staining after osmium tetroxide (LEDINGHAM and SIMPSON, 1970). Such enhancement may be misleading, as with the increased intensity of periodic acid-Schiff staining following glutaraldehyde fixation especially if the cause is not understood (CARDNO and STEINER, 1965).

10.3 Some properties and chemistry of plastic embedding media, and of the related monomers

The monomeric forms of the polymeric embedding media discussed above are of three chemical types. Those monomers with commerical names such as Araldite or Epon are epoxides. The reactive epoxide groupings derive from glycerol residues but the non-reactive portion varies, typically being aromatic in Araldites and aliphatic in Epons. Another major class of media are the methacrylates, whose monomers are esters of methacrylic acid. The esterifying alcohol is usually methanol, n-butanol or ethylene glycol. Finally there are polyester resins such as Vestopal, formed by the reaction of malic acid with glycerol or other polyhydric alcohols.

There are two variables which probably influence stain penetration but are not related to the nature of the reactive group of the monomer. These factors are the numbers of cross-links, and the number of hydrophilic groupings. There is no ready way of anticipating the degree of cross-linkage. Thus Vestopal W, unlike Alradite, Epon or the methacrylates, is regarded as intrinsically cross-linked. In practice however the degree of cross-linkage of the resin, whatever the nature of the monomers, is controlled by the amount of low molecular weight cross-linking agent (hardener) added to the polymerisation mixture. The hydrophilic nature of the resin is however more strongly dependent on the nature of the monomer. Monomers such as those of methyl and butyl methacrylates and of Araldite have no polar substituents. Such monomers fail to penetrate incompletely dehydrated tissues, and produce resins quite impermeable to ionic or polar substances. On the other hand hydrophilic monomers such as that of glycol methacrylate will infiltrate wet tissues, and give rise to resin a much more permeable to polar reagents. Yet other are betwixt and between. Vestopal W contains some free hydroxyl groups, and some fifty per cent by weight of Epon 812 may be water soluble.

Resin properties are dependent on the precise ratio and nature of monomer, hardener, plasticiser and accelerator, and also on temperature and the batch or source of the monomer. Since a vast range of permutations of these variables are in common use no further account of this side of resin chemistry will be provided here. Those seeking further information have available an impressively detailed review by GLAUERT (1974, Chapter 5).

The reactivity of the monomers with the infiltrated tissues must however be noted. Unfortunately little relevant work appears to have been published by those directly concerned with biological staining.

Reviews discussing the reactions of acrylic ester (LUSKIN and MYERS, 1964) and epoxide (STONE and STRATTA, 1967) monomers with organic functional groups are available. Biochemical work includes studies of reactions of various epoxides with several proteins and with DNA (FRAENKEL-CONRAT, 1944; STACEY et al., 1957–58). Carboxylates and phosphates were found to be esterified, amines and thiols converted to N- and C-alkylated derivatives, and phenols to ethers. As the iso-electric points of the proteins were usually raised by up to 3 pH units after such treatments, it seems that esterification of carboxylates was the dominant reaction. One can also expect that formation of polyester polymers from monomers containing hydroxylic and acid groups will reduce the number of free acidic groupings and hydroxyls in tissues.

Finally we will consider the efficacy of various etching solutions for removing embedding media from plastic sections. Note remove rather than, say, dissolve. Gross loss of material from 1 μm plastic sections could well occur without complete dissolution.

Though various etching solutions are reported to aid staining there are few explicit accounts as to what such reagents actually do to the plastic. A major exception is the concensus that non-crosslinked methacrylate resins are totally soluble in many organic solvents. It has also been noted that after the action of such solvents on Araldite or Epon sections «... the resin is not removed but appears to undergo a softening and swelling...». Explicit evidence is hard to find. An exception is the use of scanning electron microscopy of the surfaces of plastic sections to assess gross loss of Epon after exposure to etching solutions (ERLANDSEN et al., 1979). This was coupled with labelling the embedding medium with $Na^{125}I$. Loss of ^{125}I into the etch solvents was then used to provide numerical indication of loss of embedding medium. This work indicated that only the alkaline alkoxide treatments removed plastic *en masse*. Organic solvents alone did remove some plastic. The aqueous media studied, even if containing hydrogen peroxide or sodium hydroxide, had only trivial effects.

It is interesting that two simple and rapid means of investigation have apparently not been exploited. It has been reported that hydrophobic dyes may be used to stain hydrophobic embedding media (BRONNER, 1975; DE MARTINO et al., 1968; HOROBIN and TOMLINSON, 1976; MCGEE-RUSSEL and SMALE, 1963). Such staining could be used to check for gross resin-loss after etching. In addition resin blocks could be weighed, exposed to etching solutions, and re-weighed to assess loss into etching media. Mr. John Proctor and I therefore applied these procedures to study various etching processes. Some of our results are shown in Table 45. Of the etching methods we investigated only halogens in acetone, and alkoxides in ethanol or in methanol-benzene appeared to remove all the plastic from epoxy sections. The latter observation agrees with the findings of ERLANDSEN et al., (1979). In keeping both with the general impression that etch methods increase permeability of sections to stains, and with

Table 45. The effects of etching solutions on epoxy resins (note 1)

Nature of the etch solution (prepared according to ...)	Effects of etch solutions on thick plastic sections (note 2)	Weight loss (or other relevant consequences) of resin exposed to etch solutions (note 3)
Aqueous, acidified hydrogen peroxide (POOL, 1973)	No gross effects	12%
Iodine in acetone (YENSEN, 1968)	Total removal of resin from around tissues	Resin fragmented: fragments were heavily iodinated, with a weight gain
Acetone	No gross effects	0.5%
Sodium methoxide in methanol-benzene (MAYOR et al., 1961)	Total removal of resin from around tissues	73% (perhaps an underestimate, due to alkali contamination)
Methanol-benzene	No gross effects	1%

Note 1: This table comprises previously unpublished results of PROCTER and HOROBIN concerning the epoxy resin Araldite CY212

Note 2: The presence of plastic was demonstrated by staining with Sudan Black B

Note 3: Araldite was weighed, immersed in etch solutions, washed, dried under vacuum and re-weighed

ERLANDSEN et al., most etch methods, including organic solvents alone, did remove some plastic.

Even if effective, etching is not without its own hazards. One investigation showed that although performic acid etching did «soften» the plastic and enhance staining, the membranes of the endoplasmic reticulum and Golgi apparatus were lost, and other structures (e. g. eosinophil granules) were grossly distorted (HEATH, 1977).

10.4 Penetration of reagents into plastic sections: some observations and a rationale

This discussion makes use of information from studies on both thick and ultrathin plastic sections. Most concern is focussed on hydrophobic resins, unetched sections, and ionic or at least polar staining reagents applied in polar solvents.

This discussion will hinge around a series of propositions, as follows. Firstly that the penetration of reagents into plastic sections is spatially an extremely selective process, reagents do not enter all parts of a section. Secondly that polar reagents only penetrate through certain tissues and not through non-polar embedding medium. Thirdly that within the tissues, those structures readily penetrated by polar reagents are relatively or absolutely free from plastic, i. e. infiltration by the plastic is incomplete. Finally that poorly infiltrated structures are usually those which are hard to dehydrate or are physically dense, or both.

Differential penetration of stains into plastic sections is dramatically demonstrated when a thick, stained section is re-embedded and re-sectioned at right angles to the original plane of section. Direct observation in the microscope then displays the stain penetration patterns (HOROBIN and HAGUE, 1976; LOCKE and KRISHMAN, 1971; PETERS, A. et al., 1971). These preparations show that the ionic stains phosphotungstic acid or uranyl acetate, when applied from aqueous or polar solvents to tissues embedded in hydrophobic resins, penetrate most readily into structures such as chromatin, collagen, elastic fibres and glycogen. Most other structures of nucleus and cytoplasm are not deeply penetrated.

Further information on reagent penetration comes from observing the efficacy of certain enzymic and solvent extractions of embedded tissues, used as control procedures (compare Section 11.2). Consider staining a section directly, with a serial section stained after being extracted. The differences of staining intensity between the two sections gives some indication of which structures are accessible to extractants. Consider extraction of Araldite or Epon embedded, glutaraldehyde-osmium tetroxide fixed tissues with hot perchloric acid (DOUGLAS, 1970). Such treatment applied to plastic-free sections resulted in rapid removal of RNA from ribosomes and nucleoli, followed by loss of DNA from chromatin (PEARSE, 1968, p. 277). However after extraction of plastic embedded sections, staining of chromosomes was indeed prevented, but no change was seen in the staining of the nucleoli. Analogous anomalies, oddities that is compared to results seen in tissues not embedded in plastic, are also found with enzymic extractions. Exposure of Araldite or Epon embedded, glutaraldehyde-osmium tetroxide fixed tissues to non-specific proteases such as chymotrypsin, pepsin or pronase results in specific patterns of extraction (DOUGLAS et al., 1970; ANDERSON and ANDRÉ, 1968).

Next consider the gross and clear-cut example of selective reagent penetration provided by staining an embedded suspension of tobacco mosaic virus particles (SHALLA et al., 1964). The cylindrical virus particles were fixed in osmium tetroxide-dichromate, and embedded in an epoxy resin (Araldite or Epon) or in methyl/butyl methacrylate. Sections were stained by flotation on an aqueous solution of strontium permanganate. It was observed that only virus particles cut through by the surface of the section were stained. Further when a tissue-free plastic section was placed between the stain solution and the virus-containing section no staining occurred. In this dramatic instance it is clear that the ionic staining reagent could only penetrate the embedded material and could not pass through the hydrophobic plastic.

So we have established that when using polar staining reagents and embedding in non-polar plastics the reagents penetrate certain portions of sections very much more easily than others. Furthermore in at least one instance a polar reagent was quite unable to pass through a non-polar plastic.

In the light of these findings we can now consider *evidence concerning the distribution of plastic within a piece of embedded tissue.* One way of observing the distribution of non-polar embedding medium is to stain with non-ionic dyes. Suitable dyes are those used in biological staining for colouring lipids, such as Oil Red O and Sudan Black B; or those used industrially to colour hydrophobic varnishes and resins, for example the metal-containing NeoZapons. Table 46 displays information obtained from such staining studies. Overall the results show very marked differential infiltration by these non-polar embedding media into a variety of normally processed tissues. In such studies the possibility exists that complete infiltration did occur but that, for some reason, staining is incomplete. So an independent indicator of the presence or absence of plastic is desirable. One approach is to use the electron microscope to observe the surfaces of metal-shadowed, thin plastic sections. In these preparations structures such

Table 46. Observations on staining of plastic sections with non-polar dyes

Embedding medium	Dye used	Staining observed	References
Epon	Sudan Black B	Lipid droplets were intensely stained. The embedding medium itself was strongly stained. Cytoplasms, especially if rich in mitochondria, were partially stained. Chromosomes and nucleoli were completely unstained	McGEE-RUSSELL and SMALE (1963)
Methyl methacrylate	Sudan Black B	Lipid droplets and the embedding medium were strongly stained. Sperm heads and basement membranes were only partially stained	McGEE-RUSSELL and SMALE (1963)
Araldite	Sudan Black B	Lipid drops and the embedding medium were strongly stained. Starch granules and cellulose cell walls were completely unstained	BRONNER (1975)
Araldite	Oil Red O, or Sudan Black B, or NeoZapon Red G	The embedding medium was stained. Structures which were unstained or showed diminished staining were collagen and elastic fibres, glycogen deposits, red blood cells, zymogen granules and some basement membranes	HOROBIN and TOMLINSON (1976)

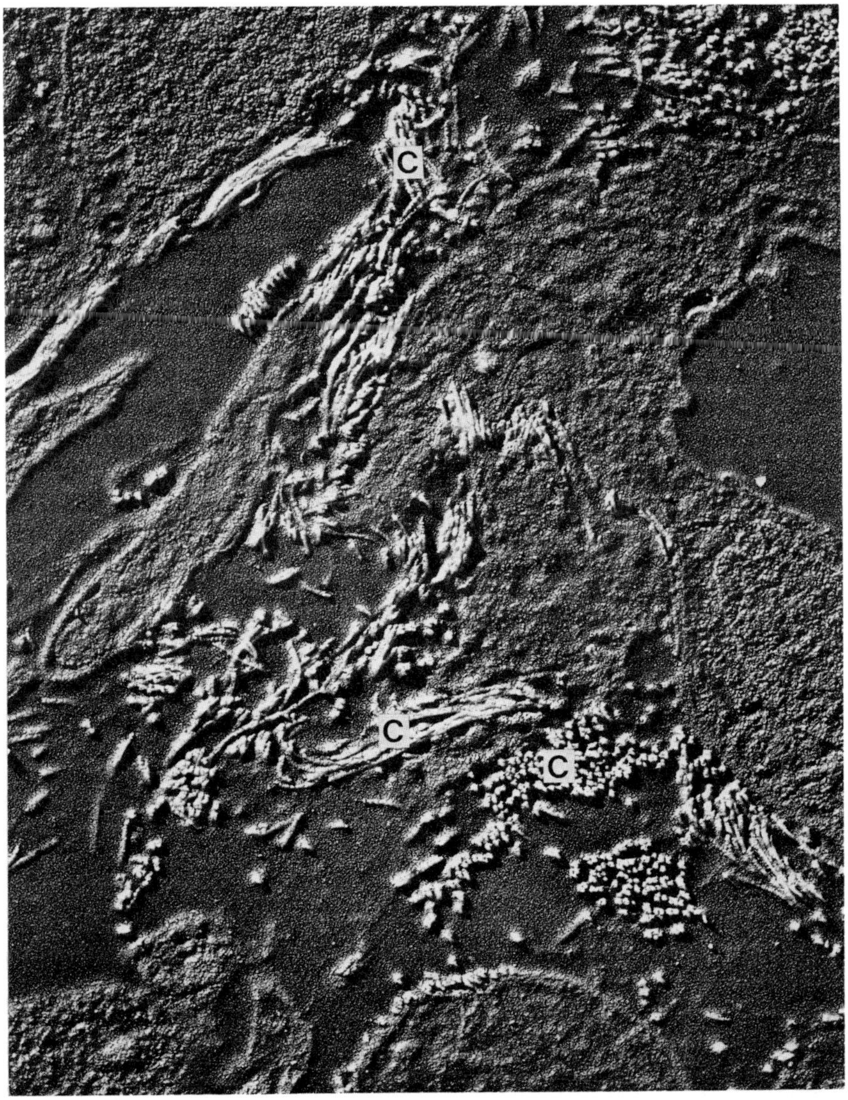

Fig. 57. An illustration of the gross protrusion of certain structures, for instance collagen fibrils (C), from the surface of thin sections following sectioning. The tissue shown is the alveolar wall of cat lung, fixed in glutaraldehyde-osmium tetroxide and embedded in Araldite. ×12500

as collagen fibrils, elastic fibres and red blood cells are seen to protrude well beyond the surface of the plastic. So differential swelling of those structures shown by the staining procedure to be poorly infiltrated by the plastic occurs, after the dehydrated section falls into the water-filled trough of the microtome. An example of this is shown in Fig. 57 and other examples may be seen in Horobin and Tomlinson (1976) and Williams (1969, Figs. 5 and 6). Williams also discusses, and dismisses, possible causes of such swellings other than differential dehydration of tissue elements. Note that those structures which are plastic-free and those structures which are easily-penetrated by polar reagent structures form overlapping categories.

Now consider *factors differentially inhibiting complete infiltration of the tissue by the embedding media.* The plastic-free structures mentioned in the discussion thus far are listed in Table 47. Consider the infiltration process, in the course of which a viscous, hydrophobic liquid diffuses through the dehydrated tissues. The rate of infiltration will be expected to fall as both physical density and the hydration of the tissues rise. This would simply account for the significance of one of the above categories, namely density. In this context note that McGee-Russel and Smale pointed out in 1963 that when staining Araldite sections with Sudan Black B certain areas of tissue are colourless, and that these are «the densest elements of the tissue... from which, presumably, the plastic is largely excluded». The significance of the periodic acid-Schiff positivity probably derives from the fact that such staining indicates the presence of saccharide residues. These are strongly hydrated, and resistant to complete dehydration. Penetration of hydrophobic monomer will be slower into such hydrophilic structures.

Table 47. Some characteristics of structures poorly infiltrated by hydrophobic embedding media

Structure	Common property for each group of structures
Chromosomes, elastic fibres, mitochondria, nucleoli, red blood cells, sperm heads and secretion granules	High physical density
Basement membranes, cellulose cell walls, collagen and elastic fibres, glycogen deposits and some secretion granules	Stain with the Periodic acid-Schiff method

Having provided some basis for the view that staining non-polar plastic sections is strongly influenced by differential infiltration, *several previously puzzling observations can be rationalised.* For instance it has been suggested that poor dehydration can give poor staining (Ansell and Marshall, 1973). The published micrographs however show that staining was poor because the background staining was too intense. That is, there was more staining than was usual or desirable. Perhaps because insufficient dehydration, and so wetter tissues, keeps the plastic monomer out of more tissue elements than usual? Other intriguing phenomena are the increased staining intensities seen when sections are not allowed to dry out before staining (Shalla et al., 1965), or at least are stained by flotation of the section rather than by first mounting the section on a glass slide (Moe et al., 1962). In this latter case use of flotation dramatically reduced the staining time from 4 days to 15 minutes. These effects could be due to surface-sealing of those plastic-free portions of the section into which the polar stains enter, since mounting a section on a glass slide entails heating to a fairly high temperature for some time. Such a process may well cause surface sealing, compare surface sealing of paraffin sections described by Lillie and Fulmer (1976, p. 106).

10.5 A general approach to staining plastic sections, with some applications and complications

An overview of plastic section staining, derived from the information and arguments presented in Sections 10.1 to 10.4 is given in Fig. 58. In the discussion and examples

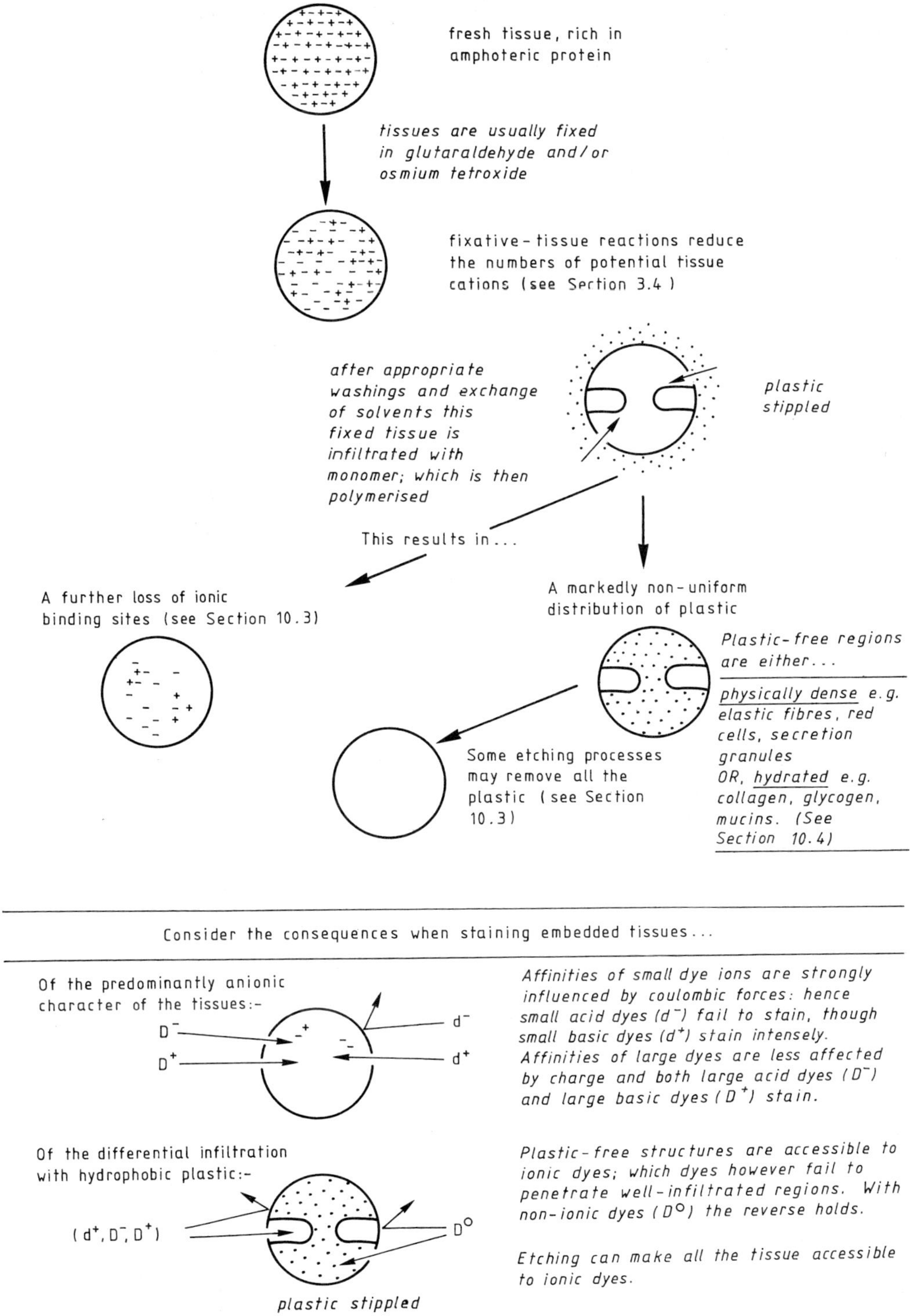

Fig. 58. An outline of the factors influencing the staining of plastic sections

following, relatively little attention will be given to phenomena already discussed in Chapter 4.

The marked *effects of glutaraldehyde and osmium tetroxide in reducing the acidophilia of tissues* is clear from a consideration of which anionic (acid) dyes and dyebath conditions may be successfully used. Small and/or hydrophilic anionic dyes such as Acid Fuchsin and Orange G will be highly influenced by coulombic effects and either fail to stain tissues so fixed, or stain well only from unusually acid dyebaths. A low pH will maximise the coulombic contribution to dye-tissue affinity by ensuring that tissue proteins carry on overal positive charge, c.f. Sections 4.4 and 4.9. The low pH will also increase tissue hydration, possibly aiding stain penetration, c.f. Section 9.3. It is not however the penetrability of these anionic dyes which is the critical factor. Large anionic dyes such as Chlorazol Black E and Phloxine stain the plastic-free tissue elements readily enough. Such dyes however are only weakly hydrophilic and possess large conjugated bond numbers hence dye-tissue affinities will be less dependent on coulombic effects, c.f. Sections 4.4 and 11.14.

As a final dye-tissue affinity topic consider the staining pattern seen after treating tissue embedded in hydrophobic resins with alkaline solutions of Toluidine Blue. This seems to illustrate salting-off, a phenomenon discussed earlier in Sections 4.4 and 4.9. Applications of the dye from a highly alkaline solution ensures high dye-tissue affinity, since most protein carboxyl groups will be in their anionic form and amino groups in their uncharged form. However, such a high pH involves a solution of high ionic strength, apparently inhibiting staining of the salt-sensitive collagen fibres. With other stains, involving less salty solutions and often using dyes less sensitive to salt, the plastic-free collagen fibres are in fact coloured.

Table 48. Usage of various types of embedding media and staining regimes in the case of the most widely used stains: based on a sample literature survey by the present author

Staining reagent or method	Number of different procedures noted using the reagent...			
	with intact hydrophobic resins		using etched hydrophobic resins	using hydrophilic resins
	used directly	used with the penetration aids described in Section 10.1		
Acid Fuchsin	–	–	3	3
Alcian Blue	–	–	3	2
Basic Fuchsin	2	5	4	1
Eosin	–	1	2	1
Fast Green	1	–	3	1
Feulgen nucleal procedure	1	–	–	6
Haematoxylins (Harris', Ehrlich's etc.)	1	–	10	2
Methylene Blue	1	3	2	1
Periodic acid-Schiff procedure	–	2	6	3
Safronin	1	–	3	2
Various silver stains	1	3	3	2
Toluidine Blue	1	3	2	3

Now consider embedding media-dependent exclusion of reagents from tissues. Some embedding media are more permeable than others. Compare hydrophobic plastics (such as the epoxy resins Araldite and Epon, and also the butyl and methyl methacrylates) with the resins which are to some degree hydrophilic (such as glycol methacrylate, the hydroxylated polyester Vestopal W, and the hydroxylated epoxy resin Durcupan). The advantages of polar resins for easing the penetration of polar stains was clearly seen displayed in the literature surveyed while preparing this discussion, a typical sample of which is summarised in Table 48. Few successful methods use intact sections of hydrophobic resins. On the other hand, if the resin is removed by etching, or if a hydrophilic resin is used, then the tricks to aid penetration are no longer required. Incidentally hydrolytic enzymes also penetrate hydrophilic resins more readily than into hydrophobic, as shown by LEDUC and her collaborators (LEDUC et al., 1966). The hydrophobic resins themselves are not however identical. Tissue embedded in hydrophobic methacrylates often stains faster or more intensely than if embedded in hydrophobic epoxy resins. This effect has been reported with stains as diverse as strontium permanganate (SHALLA et al., 1964), alkaline Fast Green (DOGE, 1964), and Malachite Green-Toluidine Blue mixtures (GRIMLEY, 1964). Increased penetrability could in certain circumstances decrease staining, for instance by allowing over-exposure to some reagent. A possible example occurs in an application of the Feulgen nucleal procedure (DODGE, 1964), when staining only occurred with tissue embedded in Araldite and not with methacrylate embedded tissue. Perhaps the methacrylate allowed greater penetration of reagents and so an excess of acid hydrolysis occurred, lowering the staining intensity? Speculation apart, it is certainly true that variables such as the amount of hardener or the precise source of the monomer can have a greater influence on staining reagent penetrability than does the chemical type of the plastic.

Table 49. Some characteristics of intensely stained structures

Structure	Common property for each group of structures
Chromosomes, elastic fibres, mitochondria, myelin sheaths, nucleoli, red blood cells, secretion granules (e. g. of the juxtaglomerular apparatus, and of the endocrine and exocrine pancreas)	High physical density
Basement membranes, collagen and elastic fibres, glycogen, goblet cell mucin, mast cell granules, microvilli	Stain with the periodic acid-Schiff method

The key concept, that differential infiltration of plastic leads to dramatic differences in subsequent penetrability of reagents, is a powerful tool when interpreting the staining of thick plastic sections. First consider *a simple case, embedding in hydrophobic plastic and staining with a single polar reagent.* In the animal tissues often studied much the same limited range of tissue components is intensely coloured whatever the staining reagent. Such structures may be placed into one of two categories, as shown in Table 49. Firstly the plastic-free entities already discussed; the plastic-containing structure chromatin will be considered later. Further emphasising the view that staining is largely controlled by reagent access rather than by specific dye-tissue affinities, note that such diverse materials as elastic fibres, red blood cells, and secretion granules are stained by a whole spectrum of reagents, so:

Labelled antibodies,
Large anionic dyes such as Chlorazol Black E and Phloxine B,
Small anionic dyes such as Acid fuchsin and Ponceau 2R,
Large cationic dyes such as Alcian Blue and Gomori's Aldehyde Fuchsin,
Small cationic dyes such as Basic Fuchsin and Toluidine Blue,
Metal complex dyes such as chromium-haematoxylin and phosphotungstic acid-haematoxylin,
Metal co-ordination compounds such as cobalt thiocyanate,
Silver stains.

Now consider *embedding in hydrophobic resins and staining with a polychrome system*. Some polychrome methods involve a mixture of dyes, others a sequence of dyes or reagents. In either case, when applied to those tissues considered above no additional intensely stained entities are visualised. The same strongly stained structures appear but are now multicoloured, further evidence for penetrability-control. In such polychrome staining systems the selectivities observed, once the overriding effect of differential plastic distribution has been noted, may be rationalised using the principles outlined in Sections 4.10 to 4.19, but see especially 4.18. For convenience however the question will be discussed in a later paragraph in this Chapter.

Returning to the central theme of differences between resins, the question of what happens after *embedding in hydrophilic resins* can now be considered. Does differential penetrability have any role in this instance? There is unfortunately no evidence available on the distribution of such resins within tissues. However comparisons of the staining patterns of tissue embedded in hydrophobic and hydrophilic resins do suggest, though no more, a similarity. For instance metal impregnation of tissues embedded in the hydrophilic Vestopal W produces intense staining of collagen fibres, mitochondria, nucleoli and red blood cells (MOE et al., 1962; TZITSIKAS et al., 1962). These are structures which are also stained intensely after silver impregnation of tissues embedded in hydrophobic plastics.

To further emphasise the importance of penetrability, consider staining tissues embedded in hydrophobic plastic, noting *the effects of removing the plastic by etching*. Systematic comparisons of etched and unetched sections has rarely been carried out deliberately. Though there are surprisingly few comparisons to be made from the plethora of published techniques, examples are provided by silver staining methods (see ADAMO and DAIGNEAULT, 1972; DOUGLAS et al., 1970; GOLDBLATT and TRUMP, 1965; MOVAT, 1961; ROSENQUIST et al., 1971; SEVIER and MUNGER, 1968). Mitochondria, nucleoli and red blood cells were intensely stained in both intact and etched sections. Additional intensely stained structures included basement membrane, elastic fibres and myelin sheaths (all probably poorly infiltrated) in un-etched sections. Etching however resulted in the intense staining of nerve processes and not of myelin, also of Nissl substance and nuclei which were hardly stained at all in un-etched preparations. A very similar case arises if nervous tissue is embedded in paraffin or celloidin (a hydrophobic medium), and stained with iron haematoxylin (METZ 1976). After dewaxing the paraffin sections (equivalent to etching plastic sections), staining followed by differentiation resulted in strong and selective colouration of myelin sheaths, probably due to rate control, c.f. Rate Model 2, Section 4.18. However when celloidin embedded tissue (equivalent to plastic sections) was stained progressively, the myelin sheaths were selectively coloured without requiring differentiation, as were red blood cells and nucleoli. Perhaps due to these dense structures excluding the hydrophobic celloidin during embedding?

Two possible strategies for *staining tissues embedded in hydrophobic plastics with non-polar reagents* will now be considered. Firstly reagent penetration could be aided by using a stain with some limited solubility in the plastic, or secondly a fully plastic-soluble but tissue-reactive reagent could be used. Probable examples of the former strategy include methods using such standard cationic dyes as Basic Fuchsin and Toluidine Blue. In such methods a final rinse or differentiation step is commonplace. Significantly it is occasionally noted that this is required to remove stain from the plastic (CHANDRA and SKELTON, 1964). Fascinatingly, another possible example is the traditional lead citrate reagent used to stain ultrathin sections for electron microscopy. Some evidence that this stain can penetrate hydrophobic plastic, including the observation that a lead containing species may be extracted from lead citrate staining solution by a hydrophobic organic solvent, has been noted by HOROBIN and TOMLINSON (1976). In keeping with this, some preliminary experiments carried out using the thiol-reactive, Araldite-soluble reagent Mercury Orange, show that tissue staining is possible though technically unsatisfactory (MEEK and HOROBIN, unpublished results).

Two topics remain to be discussed, namely staining of apparently infiltrated structures, and polychrome staining of plastic sections.

Firstly, *staining of apparently plastic-infiltrated structures by polar reagents.* What is the evidence that materials such as chromatin are in fact infiltrated with plastic? Nuclei are not usually left unstained when hydrophobic plastic embedding media are coloured with hydrophobic dye, and no deep penetration of nuclei by ionic stains is seen in sectioned-section preparations. Moreover no gross swelling of nuclei occurs after hydration of an ultrathin section of embedded material. However, nuclei are clearly stained with cationic dyes and uranyl cations, and also by the Feulgen-nucleal procedure. A possible explanation is that the chromatin beads are not well infiltrated, containing as they do DNA which is both a polyhydroxy and a polyionic polymer. Since the degree of connection between the beads of chromatin is probably rather low (for a review see BRADBURY, 1976), it may be that only surface chromatin is accessible by polar stains, the remainder being encapsulated within plastic. Alternatively the chromatin may be well infiltrated but the affinity of the exposed chromatin for stain is so high that sufficient stain binds, to produce visible colouration.

A somewhat less speculative topic is *the polychrome staining of plastic sections by acid and basic dyes.* Only the more commonplace hydrophobic plastics will be discussed, methods using plastic-etching processes will be excluded from consideration and arguments will be based only upon intensely staining structures. A wide variety of staining sequences have been used including:

1. D_1^+, followed by D_2^+,
2. a mixture of (D_1^+ plus D_2^+), followed by D_3^+,
3. D_1^+, followed by D_2^+, followed in turn by D_3^-,
4. D_1^+, followed by a mixture of (D_2^- plus D_3^-).

(a cationic dye is represented by D^+ and an anionic dye by D^-).

The results of the application of such varied staining systems to animal tissues show, after allowing for affinity effects (acidophilia and so on), a remarkable simplicity. Thus the structures staining intensely with the first dye, or one of the first dyes if a mixture is used in the first staining step, are consistently among the physically dense-plastic free entities discussed previously in Section 10.4. Examples are mitochondria, myelin sheaths, nucleoli, red blood cells, zymogen and many other granules. On the other hand, those structures staining intensely with the final dye, or one dye of the final dye mixture, are consistently from amongst entities described above as periodic acid-Schiff positive, plastic-free. Examples are cartilage matrix, collagen and elastic fibres, goblet cell mucin and glycogen. A rate-control model of staining can readily rationalise these observations. Thus both types of plastic free structures stain more readily than those materials which are plastic filled. However, there will be great differences between the two classes of plastic free materials. Those structures which exclude plastic due to their high physical density will also tend to exclude dye ions. They will be slow to stain, but once stained will be slow to destain. Hydrated plastic-excluding structures will tend however to swell in aqueous dyebaths, and hence allow ready entry and loss of dye ions. Consequently they will be fast staining and destaining, and will readily exchange the first dye for the second in sequence staining procedures.

Although this rate-control model rationalises so many methods it is still useful, in view of many statements in the literature, to offer some additional argument on the improbability of affinity control of the staining patterns. Consider the final dye of a sequence. This dye may be anionic or cationic, large or small, without affecting the staining pattern. As a delightful example consider the use of sequence staining methods involving Basic Fuchsin and Methylene Blue (for example APARICO and MARSDEN, 1969; HUBER et al., 1968). When Basic Fuchsin is applied before Methylene Blue then myelin sheaths are red and collagen fibres blue. When the dyes are applied in the reverse order, myelin sheaths are blue and collagen red.

If a mixture of two or more dyes is used, as part of a sequence or as the total staining process, then both rate and affinity factors will be important. Often this leads to complex results. Consider two polychrome staining methods using mixtures of Basic Fuchsin and Methylene Blue. One results in red colouration of collagen fibres, mitochondria and red blood cells, with a purple colouration of elastic fibres and zymogen granules (SATO and SHAMOTO, 1973). The other procedure results in purple colouration of zymogen granules, with mitochondria, myelin sheaths and many other structures being coloured various shades of blue (ALSOP, 1974). These staining patterns are both unlike each other *and* unlike the rate controlled patterns described previously. However, some dye mixtures do give rise to interpretable results. A clear example of affinity control occurred when a mixture of Acid Fuchsin and Aniline Blue was used (REEMPTS and BORGERS, 1975). In this case the blue dye has the higher conjugated bond number and is less hydrophilic, and as expected from Sections 4.4 and 11.14, stained the elastic fibres. A further example is seen when using a mixture of Methyl Green and Crystal Violet (SIEVERS, 1971). Once again it is the more hydrophobic violet dye which stained the elastic fibres.

As the question of polychrome staining is of practical significance, a summary of the preceding paragraphs is presented diagrammatically in Fig. 59.

10.6 Some practical implications of the general model

Various methodological suggestions, derived from work previously discussed more exploratively, can now be offered. Some practical information concerning *staining of common tissue elements embedded in hydrophobic plastic* has already been summarised,

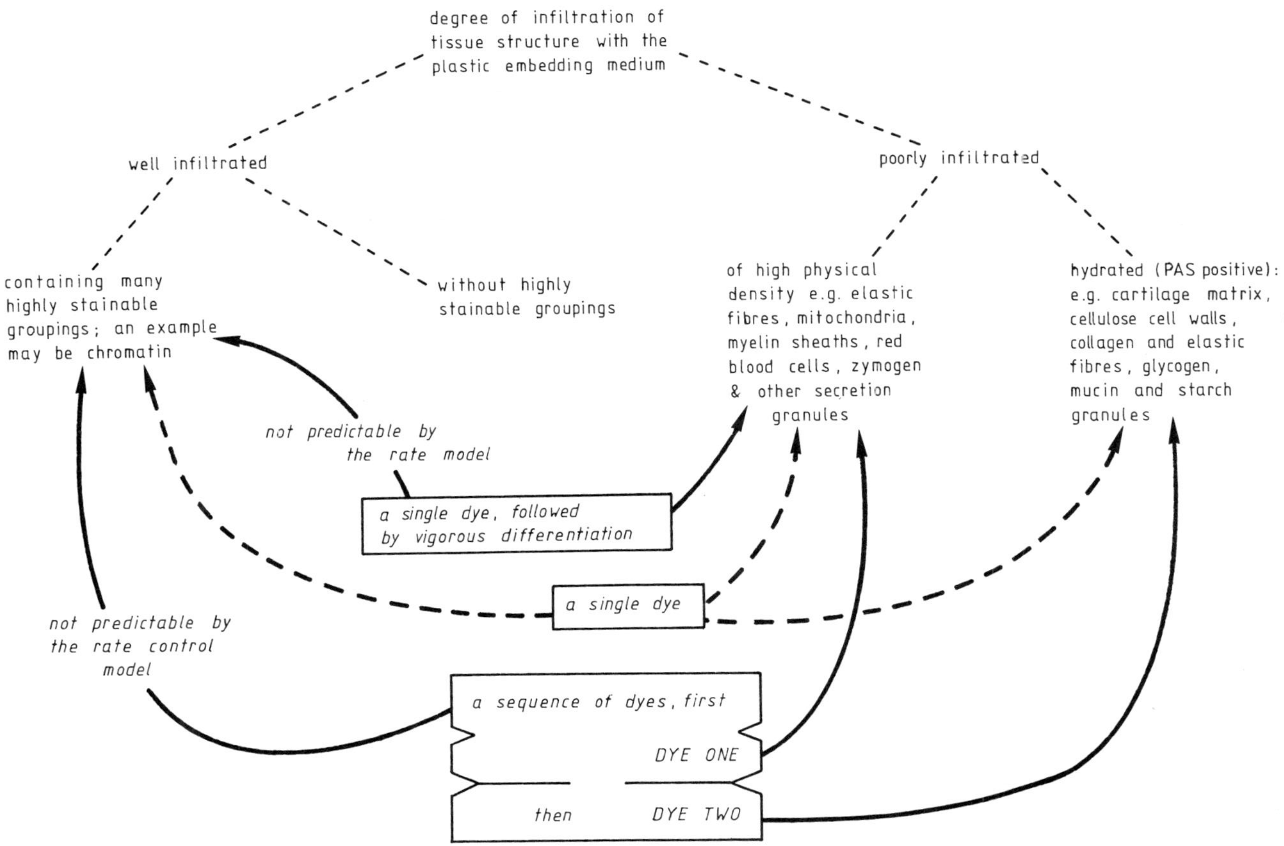

Fig. 59. The usual pattern of polychrome staining achieved when using ionic dyes on tissues embedded in hydrophobic plastics

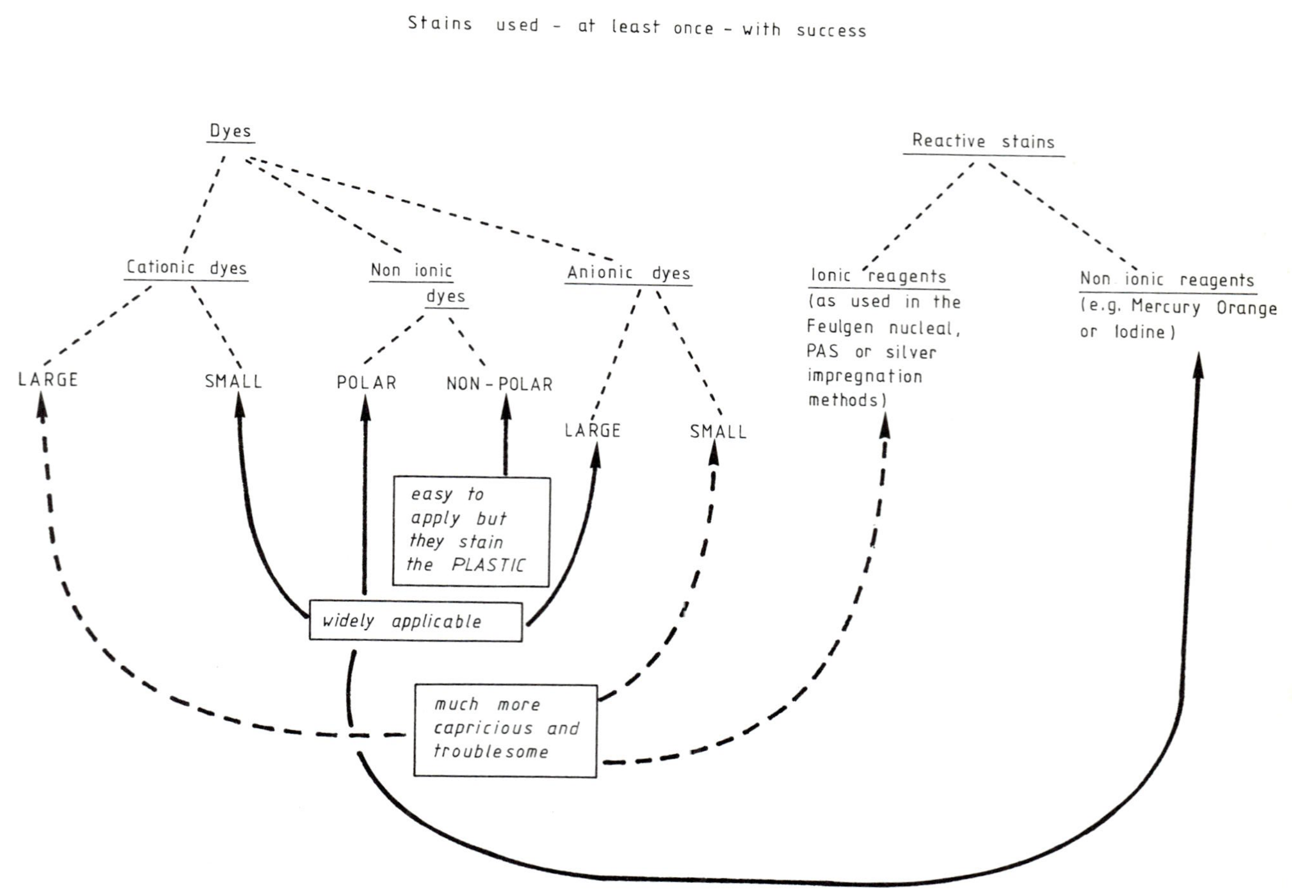

Fig. 60. Applicability of different types of stains to tissues embedded in unetched, hydrophobic plastic. *Note:* a large anionic dye would be Phloxine, a small one Orange G; a large cationic dye would be Alcian Blue, a small one Toluidine Blue; a polar non-ionic dye would be Alizarin Blue, a non-polar one Sudan Black B

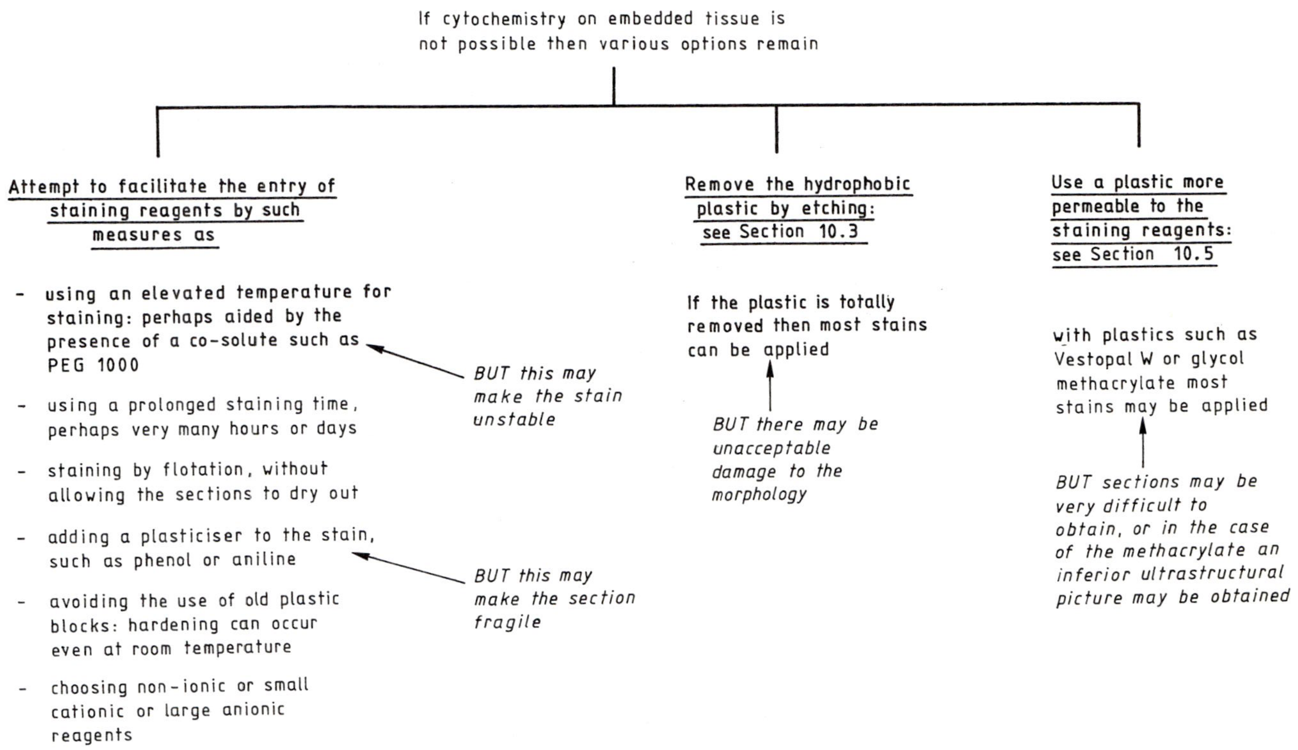

Fig. 61. Some varied procedures for facilitating staining of thick plastic sections

in Fig. 59. To change the perspective, we can now ask *which stains are likely to be most readily applicable to tissues embedded in hydrophobic plastics?* A summary answer is provided diagrammatically in Fig. 60. Attention has been focussed on tissue embedded in un-etched hydrophobic plastics, since it is embedding media such as Araldite and Epon which give optimal ultrastructural preservation, and suffer least damage in the electron beam of the microscope. What procedures may be adopted however if these first-choice media fail to allow the use of required cytochemical procedure? *Various more exotic possibilities* are displayed diagrammatically in Fig. 61.

11 The Interpretation and Analysis of Histochemical Observations

This discussion attempts an overview of the variety of ways in which the results of biological staining can be checked and interpreted. The discussion falls into two parts. In Sections 11.2 to 11.10 the assessment of staining or no-staining in particular instances is considered. The topics considered and their inter-relationships are set out in Fig. 62. The second part of the discussion, Sections 11.11 to 11.13, considers more generally some ideas and methods relevant to the overall aim of this discussion.

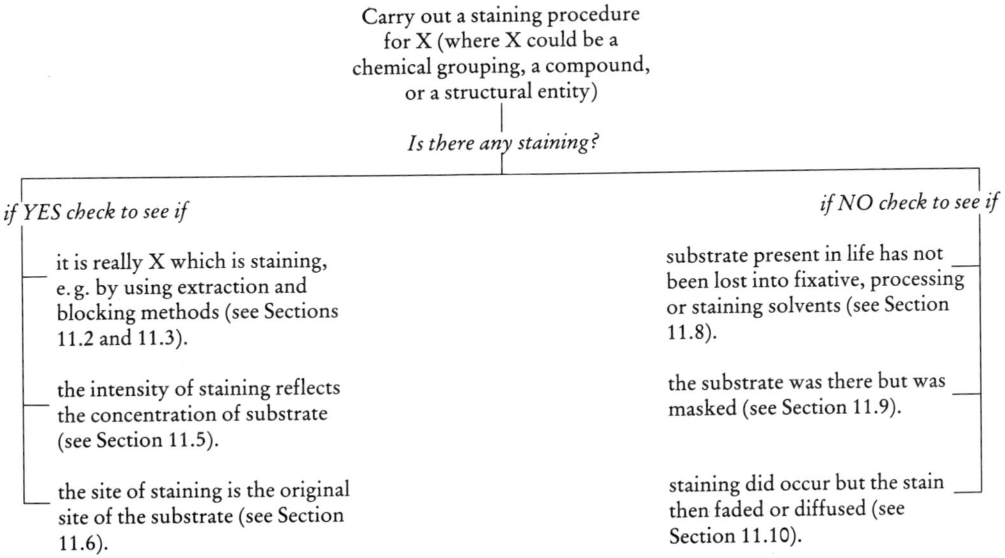

Fig. 62. An illustration of the relationship between various checks of significance

11.1 Checking the meaning of staining: an introduction

When a staining procedure intended to selectively colour substrate X results in staining occurring, various problems arise:

1. How to know that X is indeed the substrate responsible for staining («specificity»),

2. how to tell if there is any relation between staining intensity and amount of X present? («quantitation»),
3. how to discover the relation between distribution of stain and that of substrate X («localisation»).

Since these points are complex, some general introductory comments are offered. Practical approaches to the first problem have commonly been of two sorts: to modify the tissues prior to staining, or to alter the staining procedure. Thus chemical modifications may be made to X in order to produce a derivative incapable of interaction with the stain. This is usually termed blocking. There are various complications, e.g. the blocking strategy requires an accurate knowledge of the mode of interaction of X and the staining reagents, and of X and the blocking reagents. In addition procedures for both blocking and extraction of X should not cause significant changes in non-X tissue components, either chemical or physical. Both occur but are often overlooked, perhaps due to the lack of appreciation of rate factors in staining mechanisms. Other control procedures commonly involve omitting reagents or steps from staining procedures, then checking to see if staining still occurs. Such methods also require an understanding of the staining mechanism. The question of quantitation, though complex, does not appear to have such hidden assumptions. Localisation however has, since unless the mode of action of the stain is well understood there is always a tendency to circularity of argument. For instance reticulin fibres are those structures coloured by reticulin fibre stains, which stains are of course those reagents colouring reticulin . . . so why shouldn't different reticulin fibre stains be *expected* to show fibes of varying shape and size?

11.2 The use of selective extraction

This approach is based on the following assumption. When staining is due to interaction between substrate X and the staining reagent, then selective removal of X from the tissue prior to staining should prevent colouration. So the extraction process hinges around finding ways of removing X without causing other changes in the tissue. Such selective extractions are typically based either on extraction of X from the tissue into a solvent having low affinity for other tissue components, or on fission of X into fragments which are lost into the solvent, using enzymes or other reagents having no effects on non-X tissue components.

Some examples can be briefly described. *Solvent extraction* is widely used in removing lipids from tissues. The extraction of cryostat sections with anhydrous acetone to selectively remove hydrophobic lipids, leaving polar lipids and other polar materials in the tissue for instance (Elleder and Lojda, 1971). *Enzymic fission* of tissue components, after which the resulting fragments are lost into the aqueous solvent, has been widely used histochemically. For example to remove nucleic acids (DNA by DNase, RNA by RNase), glycogen (by amylase), hyaluronic or sialic acid-containing mucosubstances (with hyaluronidase or neuraminidases respectively), or proteins (with trypsin, pepsin or sometimes elastase or «collagenase»). An extensive discussion of enzymic extraction is available elsewhere (Pearse, 1972 Chapter 25). *Chemical fission* followed by extraction of the fragments into the solvent is also widely used. For instance there are a variety of chemical extraction procedures for nucleic acids. Many, as reviewed by Pearse (1968), use acid hydrolysis. Hot aqueous solutions of perchloric or trichloracetic acids extract both RNA and DNA, whereas a cold aqueous solution of perchloric acid extracts RNA preferentially. An even more selective extraction of RNA has been achieved using a saturated solution of barium hydroxide (BRUCHHAUS and GEYER, 1974), that is using alkaline hydrolysis. Such a division of extraction methods into three categories is sometimes a formality. Consider the selective extraction, using aqueous acid, of basic protein from myelin (ADAMS and BAYLISS, 1968 a), or of histones from chromosomes (e.g. COWDEN and CURTISS, 1970). Is this solvent extraction or is there selective hydrolysis?

Such questions do not exhaust the problems associated with extraction procedures. The remainder of this discussion considers two of the most common snags, along with briefer mention of several of the less common. First the marked influence of fixation on extraction will be discussed, then the various kinds of non-specific extraction and tissue modification. In all cases our interest will be centred on how staining is influenced and so how interpretation is made more complex.

The influence of fixation on extraction of substances from tissues has been widely commented on. However, the importance of such effects is still underestimated. Fixative influences have for example been reported with solvent extraction methods and with procedures using enzymic or chemical hydrolysis. Fixative effects can be found whether the material to be extracted is lipid, nucleic acid or protein. One group of effects arise from different fixatives differentially influencing the ease of extraction of a given substrate from a tissue by a given reagent. For instance after fixation in osmium tetroxide-containing solutions, elastic fibres are digested much less readily than when glutaraldehyde or formaldehyde alone are used (MUNZ and MEVES, 1974); and after fixation in fluids containing dichromates or mercury compounds it is difficult to remove nucleic acids from tissues either with nucleases or with aqueous acid (reviewed in PEARSE, 1968, 1972). The mechanisms of such effects, perhaps involving enzyme inhibition, or substrate denaturation, or trapping in other fixed tissue components such as protein, have been little studied. Another type of effect reported is a change in the specificity of the extractive agent after the tissue is fixed. An example occurs when pepsin and trypsin are used to digest collagen. After fixation, collagen is attacked by both enzymes whereas only pepsin is active against the native form (PEARSE, 1972, p. 1039).

Several kinds of *non-specific extraction* occur due to several causes. For instance extraction of additional substances closely related to the target material may occur. Consider for example the use of cold anhydrous acetone to selectively extract hydrophobic lipids. The presence of traces of water (such as might arise from atmospheric contamination of the acetone if stored loosely stoppered, or from condensation onto a cryostat section), or extraction for longer periods or at higher temperatures than usual, causes considerable amounts of polar lipids to be extracted (ELLEDER and LOJDA, 1971). Another problem is extraction of substrates unrelated to the target material. An example is found with the perchloric acid extraction of RNA. Though results reported by different workers do not always coincide (perhaps due to use of different experimental regimes?) it does appear that at least some DNA, glycogen, phospholipid and protein can also be extracted (ALDRIGE and WATSON, 1963; BLOCH and HEW, 1960; KASTEN, 1965). Complications in such instances include contamination of enzyme preparations by additional hydrolytic enzymes, amylase contaminated by proteases for example (KLESSEN, 1979); and solvent extraction in the presence of a hydrolytic enzyme. RNA, and sialic acid-containing mucosubstances, for instance may be soluble in aqueous buffer alone, even without the presence of the appropriate hydrolase.

In such examples the important event for staining is that after extraction not only the target but also certain non-target substances will have been lost, and so naturally will not give rise to colouration. However staining can also apparently be influenced in other rather more subtle ways. Consider the following observations. Mucins in sections of insect gut were stained with Alcian Blue after which the nuclear chromatin, as usual, was unstained. After treatment with β-glucuronidase or with hyaluronidase to extract the mucosubstances, the chromatin was stained intensely by Alcian Blue (SALTHOUSE, 1961). In another case the protein component of mucosubstances, normally unstained by the anionic dye Biebrich Scarlet, stained intensely after treatment with the protease papain followed by barium hydroxide (LEV and GERARD, 1966). A further example occurs when staining growing teeth or bone with Sudan Black B or acidified Methylene Blue. Using these dyes the calcification line can be stained only after extraction of the tissues with benzene, ethanol, or pyridine (IRVING, 1959). As an arbitrarily final example consider acetone or trypsin extraction of formalin-fixed myelin sheaths, or phospholipase A extraction of unfixed sheaths. This increases the intensity of subsequent staining with colloidal iron (ADAMS and BAYLISS, 1968b). As discussed later in Chapter 11.9 such effects are often considered under the heading masking. However another mechanism for these enhancement-of-staining phenomena can be suggested, namely the easier access of staining reagents due to generalised loss of tissue components. In support of this view note that such effects are often seen with large reagents which will usually penetrate tissues only slowly, such as Alcian Blue or colloidal iron; or with dense tissue components such as myelin. Furthermore stain enhancement is achieved with a strange variety of reagents: organic solvents, lipases, proteases. These do however have the common feature that they will be expected to remove something from the tissue.

A list of such problems could be extended, and extended . . . but as a finale consider just two. Firstly the possibility that large reagent molecules such as enzymes will not have ready access to all tissue compartments (PEARSE, 1972, p. 1033; GOLDSTEIN, 1962b), and secondly the further possibility that even if enzymes do penetrate, stable enzyme-substrate complexes may form (ALFERT et al., 1963).

11.3 The use of histochemical blocking

For convenience the layout of topics in this discussion parallels that of Chapter Section 11.2.

The use of blocking to check the selectivity of a staining method entails making the following assumption. When staining is considered due to a selective interaction between substrate X and the staining reagents, then chemical modification of X into a non-reactive derivative prior to staining should prevent colouration. Hence inhibition of staining after prior application of a blocking method is taken as evidence that staining was due to the presence of substrate X. The blockade process is thus dependent on finding methods for the quantitative and irreversible transformation of X.

Often several selective chemical inactivation (blocking) procedures have been developed for a given substrate. One reason for such multiplicity arises from the need for blocking methods to be directly related to the staining methods subsequently applied. As an illustration consider a nitrous acid treatment of tissue, during which primary amino groups are converted into hydroxyl groups. This reaction will be effective as a blockade both for reactive staining methods directed at the $-NH_2$ moiety, and for acid dyeing which depends at least in part on the presence of tissue cations such as $-NH_3^+$. However other chemical modifications of amino groups may produce less general results. Treatment with methyl iodide for example yields quaternary ammonium salts as derivatives ($-N(CH_3)_3^+$) and so will only constitute a blockade against amino-reactive stains, not against acid dyeing. Another cause of multiplicity of methods is the fact that each blockade procedure will have its own peculiar set of snags suiting it to some particular histochemical problem and making it unsuitable for another. Note that a procedure analogous to blocking, namely the selective inactivation of enzymes by inhibitors, has already been discussed in Chapter 7.

Fixation is a major complication of blocking procedures. For instance the fixative-dependance of the reaction times required for effective blockades. Thus nitrous acid deamination of alcohol-fixed blood smears was found to be complete in 2 minutes, but progressive exposure of smears to formaldehyde increased the deamination times needed. 24 hour formaldehyde treatment resulted in a requirement for a 12 hour deamination (LILLIE et al., 1970). Another example concerns the times required for boron trifluoride-methanol methylation to abolish all tissue basophilia. About 6 hours blocking was needed after fixation in Bouins fluid, which is acidified aqueous picric acid-formaldehyde; 10–12 hours after dichromate solutions such as Helly's or Zenker's, or after Carnoy's fluid; and about 24 hours after aqueous 4% w/v aqueous formaldehyde solution (COALSON and SCHWARTZ, 1970).

Several other snags are conveniently viewed as non-specific reactions. For instance *the chemical reactions involved in the blockade may be more complex or even different to the reactions usually assumed.* Several examples of this are shown in Table 50. The practical consequences of these chemical ambiguities are fortunately minor, since the method would not have been adopted if a blockade were not achieved. However potential errors of interpretations exist, since a blocked section may be stained for a nominally unblocked entity which may in fact no longer be there. Alternatively unblocking procedures may be adopted which are not relevant to the actual blockade reaction. One example arises from the observation that methylation following deamination produced an additive reduction in the staining by Gallocyanin-chrome alum, whilst deamination following methylation showed no effect beyond that of methylation alone (TERNER and CLARK 1960). This was attributed to the previously unsuspected alkylation of tissue amines by the hydrochloric acid-methanol mixture, which occurred in addition to the conversion of tissue anions to uncharged species. Another case was seen when the use of sulphide ions to convert tissue disulphides to sulphydryls was being studied. Sulphide was effective but did introduce artifactual positive staining by converting tissue carboxylates, via aldehydes, to sulphydryls (BARRNETT and SELIGMAN 1954). Exploratory and critical papers such as those cited are too often overlooked, leading to a certain false optimism on the part of some workers, and even by writers of some practical staining manuals.

An analogous effect is the *loss of tissue components by extraction into the blockade reagents.* For example various polysaccharides and mucosubstances are considered to be acid hydrolysed, and the fragments extracted, during blockades using hydrochloric acid-methanol (TERNER, 1964; SORVARI and STOWARD 1970a). Indeed an apparent blockade can in reality be an extraction, as when applying the aniline-acetic acid

Table 50. Some examples of histochemical blockades involving chemistry of unappreciated complexity

Objective of the blockade, and the chemical basis usually* cited	Reagents used in the blockade	Alternative or additional chemical bases of the blockades	References
Blockade of primary amino groups, achieved by their oxidative deamination to aldehydes	Aqueous neutral hypochlorite solutions	Both the amino blockade and the Schiff positive character of blocked tissue have been attributed to the formation of chloramines	BURSTONE (1955), STOWARD (1975)
Blockade of primary amino groups by their conversion into hydroxyl derivatives	Nitrous acid, often in acetic acid solution	It has been suggested that at least part of the blockade involves total loss of α-glutamyl residues, via lactone formation	BURNS and STOWARD (1971), GLAZER et al. (1976)
Blockade of tissue basophilia, achieved by conversion of anionic groups to uncharged esters: for instance the conversion of carboxylic acids to their methyl esters	Methanolic hydrochloric acid	At least part of the blockade appears due to the formation of uncharged lactones and, at least at high temperatures, to conversion of primary amines to cationic methyl ammonium salts	Compare SORVARI and STOWARD (1970b) with LILLIE and JIRGE (1975), and also TERNER and CLARK (1960)

* that is, as commonly described in practical manuals of histochemistry

aldehyde blockade to acid hydrolysed tissue sections. Subsequent staining of DNA with Schiffs reagent (that is, Feulgen staining) is prevented. This is however due to extraction of apurinic acid, and not to the blockade of its aldehyde groups (TANDLER 1974).

Before moving to more complex snags it is useful to consider some general questions. Firstly *how complete need a blockade of substrate X be?* Secondly even if all substrate X were blocked can it be assumed that the blockade is stable? And finally, even if a blockade is complete at one tissue site can it be assumed to be complete at all? The answer to the initial, key, question depends on the particular staining procedure used. Often a complete blockade is not necessary since even a reduction in staining intensity may be meaningful. The answer is also dependent on the mechanism of the staining process. Consider staining involving the primary amino groups of proteins. When staining is by low molecular weight acid (i. e. anionic) dyes, such as Eosin Y or picric acid, dye uptake will usually only occur when the protein carries an overall positive charge. That is when the number of ionised amino groups exceed the number of ionised carboxylate groups. Hence an effective blockade will only require the elimination of a minor proportion of the tissue amines. However with a reactive dye, such as a Procion or Remazol, intense staining of the tissue proteins is achieved when only a small proportion of the amino groups have been converted to coloured derivatives. In such cases a much more complete blockade will be needed before staining is markedly reduced. So it is not surprising that blocking reactive staining of tissues has been reported as difficult (WEINEL and GEYER 1967). The answer to the question concerning *reversability of blockades* is dependant on both the blocking and staining reactions. Consider for instance various methods applied to block native tissue-aldehydes prior to applying the periodic acid-Schiff stain. Here the various aldehyde derivatives formed during the blockade are exposed first to the periodic acid solution and then to the strongly acid Schiff reagent. Whilst those derivatives resulting from reaction with aniline-acetic acid and phenylhydrazine are stable at such pH's, those derived from dimedone, semi- and thiosemicarbazide, and sodium borohydride are acid hydrolysed back to aldehyde (BAYLISS and ADAMS, 1976; STAPLE, 1958). An analogous situation arises when the derivatives resulting from a blockade are in fact themselves reactive with a particular stain. An example is the reactivity of dihydroxydinaphthldisulphide with the mercuriphenyl derivative resulting from tissue blockade with phenylmercuric nitrate, as discussed by PEARSE (1968, p. 154).

Now for the final question, *is a blockade found to be effective at one tissue site likely to be effective at all sites?* Unfortunately the answer is no. Such unwanted variation is probably more common than the occasional mention (for instance STAPLE, 1969) implies. Interestingly even the snags may be tissue-dependent. For instance the amount of mucosubstance lost by extraction into a methanolic-methyl iodide blockade mixture has been noted as varying from one tissue to another (TERNER, 1964).

Returning now to more subtle snags. Consider the *effects on staining of non-specific modifications of tissues by the blockade processes.* A wide variety of blocking reagents, and indeed of the solvents used in blockade processes, can modify tissues. For instance sulphydryl blockades may well produce further denaturation of proteins. Also many blocking systems will extract tissue components, such as proteins and mucosubstances. Even when losses do not occur reagents such as formic and acetic acids or strong alkalies will often produce gross changes in the swelling of the tissues. Such alterations can affect staining rates, with swelling often increasing rates; and dye-tissue affinities, with denaturation often increasing affinities.

Effects like these could underlie the observed increases in mucosubstance staining, by reagents such as Alcian Blue or periodic acid-Schiff, induced by prior exposure to alkaline blocking agents (CULLING et al., 1971; LEV and SPICER, 1965; REID et al., 1973). Also the increased staining by dinitrofluorobenzene and by acid dyes of dense hair shafts, after exposure to alkaline blocking agents (STAPLE, 1969). However no critical study of such physical possibilities seems to have been made. Discussions in the literature often endeavour to explain such phenomena on the basis of reactive organic chemistry alone (REID et al., 1973), or do not attempt to explain them at all (STAPLE, 1969).

Quite *a different complication arises if the mechanism of the post-blockade staining process is assumed to be understood, and this assumption is incorrect.* Consider interpreting observations on the use of dyes after application of blockades. As discussed at length in Chapter 4 a wide variety of factors contribute to dye-tissue affinities. Also a variety of mechanisms, amongst which the role of differential affinity is only one, underlie selective staining. At present this remains unappreciated by some workers, who continue to attribute dye binding to single causes, such as coulombic attractions or hydrogen bonding or . . . Similarly the problem of selectivity is either ignored or, more commonly, regarded as totally explained by differential affinities. Consequently the effective action of a blockade is often interpreted in terms of the blocking of a specific dye-binding site when a more realistic view is that one out of several contributions to affinity has been affected. Examples where such a narrow theoretical viewpoint seems to have confused interpretation include the effects of acetylation on the staining of chromatin with Gallocyanin-chrome alum and Methylene Blue (KIEFER et al., 1968), the effects of sulphation on the staining of collagen fibres with Biebrich Scarlet (LILLIE, 1964) and the effects of a variety of blockades on the Guard stain for supposedly basic proteins (COWDEN et al., 1976).

A number of these snags are diagramatically summarised in Fig. 63.

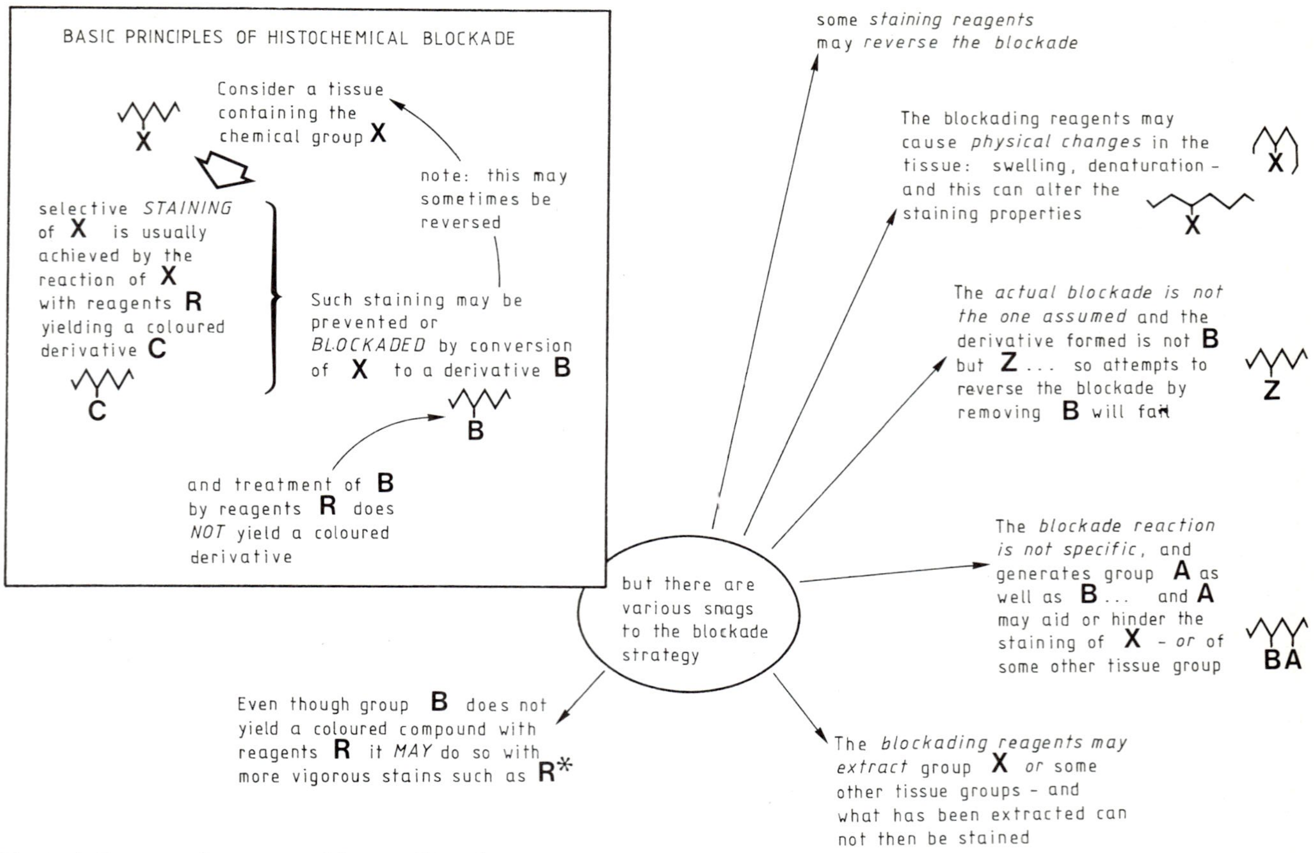

Fig. 63. A diagrammatic summary of the use of histochemical blockades, and some of the possible confusions

11.4 The use of other control procedures

One way of checking that the supposed staining reaction is the one actually taking place is to omit a reagent from the staining schedule, or to omit a step from a multi-stage procedure. An example is the treatment of tissue sections with Schiff reagent without prior periodate oxidation when using the periodic acid-Schiff procedure, or without acid hydrolysis when using the Feulgen nucleal procedure. Positive staining of such controls will indicate the presence of native, as distinct from reagent generated, aldehydes. Other instances arise when using diazonium salts as visualising reagents, such as with the disulphide-dihydroxydinaphthyl, 1-naphylmaleimide and dinitrofluorobenzene methods for protein side groups. In such methods a control step involving treatment of the tissue with the diazonium salt alone identifies native diazonium-reactive species, such as tyrosine residues. Applications of this type of control method are widespread in the enzyme histochemical and labelled antibody technologies, and are discussed in Chapters 7 and 8 respectively.

Another checking procedure is the use of test objects. This is particularly valuable when using unfamiliar or capricious staining methods. A typical test object will be a tissue preparation in which regions containing, and also regions free from, substrate X can be readily identified morphologically. An example is the use of a test-mixture of Gram-positive and Gram-negative organisms in bacteriological Gram staining. If such a test preparation is stained alongside the tissue under study a measure of positive and negative artifactual staining is available. Alternatively, and much less often, model systems are used as test objects. The use of such systems is discussed later, in Section 11.12.

11.5 Does staining intensity reflect substrate concentration?

This is a wider topic than quantitation. Consider staining a tissue section to demonstrate substrate X. On microscopic inspection adjacent cells may differ very markedly in their intensity of staining. An analogous instance is where substrate X-rich sections, derived from different animals, stain with very different intensities. In either case one may ask whether the differences in staining reflect biologically significant variation in the distribution of substrate X? Or are some other factors operating? The present discussion will consider a variety of «other factors».

Probably the simplest such factor is *variation in section thickness*. The presence of cellular or tissue structures thicker than the section, or of multiple superimposed structures thinner than the section, can lead to variations in staining intensity between serial sections. This problem has resulted in the use of a variety of techniques to measure thickness. These include mechanical procedures such as weighing or the use of a microcator; and optical methods such as absorption measurements with coloured materials, precise focussing on the top and bottom of sections, and interferometry. Electron scattering, using the electron microscope; tracer labelling; and autoradiography have also been used. Those interested may in the first instance consult reviews such as those of ERÄNKÖ (1955) and SHUGAR (1962), and later papers such as those of SILVERMAN, SCHREINER and GLICK (1969), PEARSE and MARKS (1974), WILLIAMS and MEEK (1966) and HALKJAER-KRISTENSEN and INGEMAN-HANSEN (1978). Certain procedures are destructive of the section, and many methods give only average thickness values. This is a serious limitation since thickness often varies across sections as well as between sections, as beautifully displayed by WILLIAMS (1969, figs. 3 and 4).

A more complicated case is when substrate X is distributed non-uniformly inside cells of uniform morphology. If the thickness of the tissue section is small compared to the dimensions of a cell, then a chessboard pattern may occur after staining for substrate X. Glycogen laden hepatocytes fixed in alcoholic fluids provide a clear example. Near the surface of the tissue block much of the glycogen in each cell tends to be precipitated along one side of that cell. Consequently when inspecting thin sections stained to demonstrate glycogen, some cell profiles are almost unstained whilst others are brightly coloured (ROOS, 1974).

Section thickness can also affect staining intensities by influencing *the staining process itself.* Consider a tissue section of thickness greater than the diameter of the cell nuclei of that tissue. In this case some nuclei will be intact, totally within the section, and others will be sliced through by the section's surfaces. It has been reported by several workers (e.g. LISON, 1955) that a number of common stains colour sliced nuclei more strongly than intact ones. This effect has been explained in terms of a rate control mechanism (GOLDSTEIN, 1962a).

Quite another factor is *site-dependent loss or inactivation of substrate X.* Losses can occur at any time in the processing-staining sequence, and inactivations can be due either to the events of this sequence or to native interaction in the tissues. Both these possibilities are discussed below, in Sections 11.8 and 11.9 respectively. Here it is enough to note that such effects can occur differentially in different sites. Similarly differential losses of stain, rather than of substrate, can occur. Mechanisms of such fading or diffusion processes are discussed in Section 11.10, below.

Now assume a situation in which such confusions do not occur. To what extent are subjective statements such as «the cells are intensely stained for substrate X, but the fibres are not» likely to be objectively true? This brings us to the important, though specialised, topic of *quantitation.* Though the subject is largely outside the scope of this volume some attention will be given to it. Note initially that some work described as quantitative is better described as objective. The aim of much numerical measurement is that of giving objective description of what is subjectively apparent, rather than arriving at numerical values of intrinsic significance.

There are several quite different *ways of quantifying staining-histochemistry,* that is of measuring how much stain is present in a tissue section or, better, in part of the section. Perhaps most simply the *stain can be extracted from the tissue and then assayed in some way.* Examples include spectrophotometric assay of Nitro-Blue Formazan, resulting from dehydrogenase activity (ALTMAN, 1969), and of azo-dyes resulting from acid and alkaline phosphatases (FRIEDE and KNOLLER, 1965). Though prior microscopic examination shows the stained sites, such a procedure remains an averaging one. A method which can achieve localised quantitation is *for stain to be determined in situ using microspectrophotometric or densitometric measurements* on tissue or on photomicrographs. This may be a direct measure, for example by determining the visible absorption of coloured materials such as haemoglobin, or the ultraviolet absorption of proteins and nucleic acids. However native coloured compounds are unusual, and ultraviolet work requires specialised equipment. The more commonplace assays of substrate X concentrations, by measurements of the intensity of staining of substrate X, involve two quite distinct kinds of problems. The first is the basic one of the relationship between the amount of substrate X and the amount of stain. The second concerns *the relationship between amount of stain and the measured absorbance.*

The second point is the most straightforward. Though instrumental difficulties have been substantial, both from the viewpoint of practice and of understanding, this is now in principle an area without fundamental problems. Those seeking a sound understanding of this topic may with profit first read the papers of Goldstein (BEDI and GOLDSTEIN, 1974; GOLDSTEIN, 1970, 1971, 1975).

The awkward question of *the relationship between the amount of staining and the amount of substrate X* is perhaps best approached by an initial consideration of staining stoichiometry. *Stoichiometric reactions* are those in which the products are of constant composition, regardless of the ratio of the reactants. Are any biological staining reactions likely to be stoichiometric? Not surprisingly the answer depends on the type of staining process involved. Stains such as the Feulgen nucleal, the periodic acid-Schiff and Mercury Orange involve the formation of stain-tissue covalencies and there is every reason to suppose that they are stoichiometric. Some metal impregnation processes (as for reticulin or for iron deposits) and labelled antibody stains are also in principle stoichiometric. However most other staining methods are in principle not so. For instance the binding of dyes to tissues typically involves a variety of non-stoichiometric processes, such as hydrophobic bonding and van der Waals attractions. Furthermore when dyeing procedures are rate-controlled, stoichiometry is irrelevant. Another inately non-stoichiometric process is the catalytic influence of an enzyme, as in the staining reactions of enzyme histochemistry. Again, silver stains of the type used to demonstrate nerve processes are critically influenced by rate dependent reactions, indeed this staining process is multi-step with only the initial stage involving a reagent-tissue reaction at all.

Returning to the more hopeful possibilities of the reactive stains, it is apparent that even here complications exist. Firstly there may be more than one product generated. Hence though the reaction leading to each individual product may be stoichiometric the overall process may not, for instance due to changes in the yield ratio of the products. A possible instance is the periodic acid-Schiff procedure, where the initial oxidation step gives rise to both aldehydes and periodate esters (OVADIA and STOWARD, 1971). The second step of the staining procedure, the interaction of Schiff reagent with aldehyde, provides another instance of multiple reaction products (e. g. HOROBIN and KEVILL-DAVIES, 1971). Yet another instance is in the Feulgen nucleal stain (HIRAOKA, 1973). In this case the multiplicity of products does not prevent the use of the stain for quantitative work, possibly because the same product ratios always occur? Another problem is how close to completion, to use the organic chemist's term, does the reaction go? Are 100% of the reactive tissue groups converted to coloured derivatives, or 1%? When the answer is nearer the latter than the former then any minor variation in staining circumstances could induce major changes in staining intensities. A dramatic example of this is again provided by the oxidation step in the periodic acid-Schiff procedure. In the study of OVADIA and STOWARD (1971), under conditions similar to those of histochemical usage oxidation was less than a quarter complete. The difficulty experienced when attempting to block the staining of amino groups by reactive Procion-type dyes, or the staining of thiols by Mercury Orange suggests that this is not unusual.

The discussion so far leaves few potentially stoichiometric methods unscathed. In keeping with this the Feulgen nucleal procedure is the only widely used quantitative staining procedure. However there are other, non-stoichiometric, staining systems which have been considered quantitative. Such claims are based on the proposition that even a non-stoichiometric reaction can, if the reaction conditions are carefully controlled, give rise to a *proportionality between staining intensities and substrate concentrations.* Whilst such proportionality allows a comparison of relative concentrations, to attempt absolute measurements or even to validate relative values, some external standard is required. Use of such standards is necessary for most stoichiometric quantitative processes also, with the only widely exploited exception being the Feulgen nucleal procedure. For this diploid and haploid nuclei, identifiable morphologically, provide an *internal standard* of relative DNA concentration. Analogously sperm heads have been used as internal standards for assaying thiols and disulphides in cell nuclei (WIESE and SIPPEL, 1978).

The *external standards* used have been of two kinds. A tissue can be chemically assayed for substrate X and results compared to the intensities of stains for X. Alternatively the staining intensity of a tissue can be compared to a model containing a known amount of substrate X. Of these the former is less complex, and the least used, and will be discussed first.

Examples of studies involving *comparisons between staining intensities and chemical assays* are given in Table 51. These were chosen to illustrate the range of this approach. Note an absence from the Table. There are quantitative enzyme histochemical studies in which physical-chemical characteristics of enzymes were determined and were found to be comparable to values determined biochemically on tissue homogenates. These are discussed elsewhere, in Section 7.7. Such results may at any rate be atypical, see for instance the comments of LAKE and ELLIS (1976).

The Table does contain examples ranging from real attempts at quantitation to an assay being carried out to answer the question «staining for substrate X is positive, but does a chemical assay indicate the presence of ANY substrate X in the tissue»? This latter can either be regarded as a limiting case of quantitation or as just another control procedure of the type discussed earlier in Section 11.4. When evaluating information of this sort, it is important to appreciate that chemical and biochemical assays do not themselves necessarily yield accurate results. Perusal of analytical chemistry texts demonstrates that artifact and ambiguity is not restricted to histochemistry.

Now to consider the *use of models as external standards* for histochemical quantitation. This tactic has been applied to a wide variety of staining systems, as illustrated by Table 52.

The nature of the models varies in several ways. For instance in the degree of similarity between the model and the tissue structure being modelled. Some models are merely the compound supposedly causing the staining. Compounds which have been used in this way include glycogen, a variety of proteins, and nucleic acids. Other models contain tissue, whole or in the form of fractionated homogenates. For example frozen sections of gelatin-embedded brain-homogenate have been used to calibrate a quantitative staining system for acetyl cholinesterase (WENK et al., 1973). Another type of model contains a similar chemical grouping to that under study in the tissues, but attached to a very different molecular skeleton. An example is the use of aminoethylcellulose as a model for calibrating the staining of tissue amino groups. A distinct classification of models arises from the variations in their mechanical characteristics. Some models are merely the naked substrate which is to be stained, though this may require some mechanical support. Thus microdrops of glycogen or nucleic acid solutions have been

Table 51. Examples of studies in which staining intensities have been compared to chemical assays

Staining procedure	Assay procedure	Correlation between assay and staining claimed	Reference
Examples where quantitation has been attempted			
Xiphoid cartilage was stained with Toluidine Blue, and the intensity measured by microdensitometry	The hexuronic acid content of whole cartilages was assayed	Correlation between staining intensities and hexuronic acid contents «linear» down to 42% of the normal acid content, at which point staining ceased	POOLE (1970)
Muscle glycogen stained by periodic acid-Schiff method, intensities assessed microphotometrically	Glycogen contents estimated chemically in comparable tissue samples	Correlation of staining intensities with chemical assays was 0.9	HALKJAER-KRISTENSON and INGEMAN-HANSEN (1979)
Examples where only semi-quantitation was sought			
Catecholamines in nervous tissue were converted to fluorescent derivatives by formaldehyde	The nor-adrenaline content of the tissue was assayed	The number of fluorescent variscosities was found proportional to the nor-adrenalin contents	AGNATI and FUXE (1974)
Succinate dehydrogenase (SDH) and glycolysis and L-glycerol-2-phosphate: acceptor oxidoreductase (GPOX) were demonstrated in various skeletal muscles using tetrazolium salt methods	SDH and GPOX contents were assayed biochemically in homogenates of various skeletal muscles	The areas «strongly stained» correlated, though poorly, with the assayed enzyme contents	JOBSIS and MEIJER (1973)
Example where the assay was only required to say if the substrate was/was not present			
Granuloma tissue was stained by Alcian Blue, after prior hyaluronidase and neuraminidase digestion	Carrageenan was isolated, stained with Toluidine Blue, and assayed densitometrically	Positive or negative staining correlated with the presence or absence of carrageenan as asayed	MOSS (1973)

sprayed, and trichloracetic acid precipitated proteins have been smeared (LEEMAN and RUCH, 1972), onto glass slides. Some naked substrates are mechanically self supporting, for example aminoethylcellulose films. In many model systems however the substrate is not naked but is within or attached to some matrix. Sometimes such a matrix is mechanically self-supporting. Examples include films of acrylamide gel containing tissue homogenates or proteins, and agarose or cellulose beads. Alternatively the matrix may be itself supported on glass slides. An example is the DNA-in-gelatin model (the DNA is dissolved in the gelatin, a solid block is fixed, processed through to paraffin, sectioned, mounted on slides, dewaxed . . .) used by DEITCH (1964).

There are real limits to the use of models in calibrating quantitative histochemical staining methods. Models by definition are simplified analogues of reality, hence the choice of a given model is intrinsically a limiting process. After such a choice any factors not modelled cannot be explicitly accounted for. For instance properties of real tissues not found in the models described here include such factors as local variations of

Table 52. Examples of quantitative histochemical staining systems in which models were used to provide external standards

Staining with dyestuffs	
Aldehyde Fuchsin was used to stain oxidised serum albumin and α-chymotrypsin	Mahon and Nair (1975)
Azur B was used to stain RNA	Shea (1970)
Alkaline Fast Green FCF was used to stain histones	Cohn (1973)
Naphthol Yellow S was used to stain a variety of proteins	Tas et al. (1974)
Staining by reactive reagents	
Dinitrofluorobenzene was used to stain amino groups	van Dalen et al. (1970)
The Ninhydrin-Schiff sequence was used to stain amino groups	van Dalen and van Duijn (1971)
The periodic acid-Schiff sequence was used to stain glycogen	Gahrton et al. (1975)
Glyoxal-bis (2-hydroxy anil) was used to stain calcium	Wolters et al. (1979)
Enzyme histochemical staining	
Acid phosphatase was demonstrated by a Gomori-style method	Cabrini et al. (1975)
The horseradish peroxidase (HRP) component of an anti-HRP: IgG-HRP complex (the product of an immunochemical staining process) was demonstrated by a diaminobenzidene procedure	Yokota (1973)

internal pH and of physical density. Furthermore the different models have different practical snags, though these are too rarely compared or discussed. A study making such comparisons involved cytofluorometric determination of basic and total protein, using Sulfaflavine staining (Leeman and Ruch, 1972). These workers studied model systems in which proteins were: smeared onto glass slides; or fixed, embedded in paraffin, sectioned and dewaxed; or incorporated in polyacryamide films. Protein was lost into the staining reagents very readily from the smears, but less readily from the films. The distribution of protein within the smears was rather heterogeneous, with this being even more pronounced in the sections and less so in the films. However it was the smear models which were finally used to calibrate the histochemical staining system. The films were found unsatisfactory since if thin their percentage variation in thickness was large, and if thick their high density (the reason for their good retention of the protein?) reduced staining rates far below that of the tissue being modelled. Whether any general conclusions can be drawn is improbable, a point explicitly made by Leeman and Ruch. For further discussion see Section 11.12, and also the informative and stimulating review of van Duijn (1976).

11.6 Is the stain where the substrate was? The localisation problem

Tissue substrate may in fact be mobile at any stage of processing up to and including staining, indeed the deposited stain itself may not be static. In certain instances the stain may never have been deposited at the site of the substrate. For emphasis, examples across this spectrum of ambiguity will be discussed.

It is generally assumed that once cells are removed from their usual environment then changes leading to loss of localisation of substrates will occur. This belief leads to the perennial concern of histopathologists for the state of biopsy samples, and also to the urgent processing of excised tissues by electron microscopists. In spite of this, surprisingly few studies of *post-mortem changes in biochemistry and morphology* are available. One entity which has received attention is the hepatocyte. Ultrastructural changes have been observed in the mitochondria of buffer-incubated liver slices as early as fifteen minutes after excision. The studies of Trump and his co-workers are pertinent here, and a leading reference and a source of sobering pictorial evidence of agonal changes is found in a paper of TRUMP, GOLDBLATT and STOWELL (1965). Morphological damage in fact preceded losses detected by biochemical assay. Other studies on hepatocytes, carried out under similar conditions, seem less pessimistic, perhaps because they were less extensive (ITO, 1962). ITO does however emphasise the marked differences in fragility between organelles. Plasma membranes for instance were found to quickly degenerate into myelin whorls, and the Golgi apparatus soon became vesiculated. Significant changes in mitochondria however seemed to require two hours or more, and the endoplasmic reticulum appeared intact even after two days. Nevertheless many tissue components do seem fragile and subject to dispersion, sometimes due to procedures usually ignored in such discussions. Congruent with this view, other workers have found that morphological changes occurred prior to dectable biochemical alterations. Moreover changes in low molecular weight RNA and enzymes of the cytosol preceded those in DNA, rRNA and some membrane-bound enzymes (MANN et al., 1978). An enzyme histochemical study of lysosomal hydrolases found that a major influence on localisation was the way in which experimental animals were killed (AL-AZZAWI and STOWARD, 1973).

Regardless of prior changes, fixation usually though not universally acts as an immobilising process. However the fixation process itself may induce gross ultrastructural alterations, see Fig. 2. In addition to such crude changes fixation can give rise to more selective redistributions of certain substrates. When highly directional flows of fixative pass through a tissue, cell contents may be transported and only finally immobilised on the opposite side of the cells (or nuclei) to that first entered by the fixative. Such effects are most pronounced when the fixative generates major ultrastructural damage; and where the flow of fixative is very uneven, such as will occur at the edge of a thick tissue block. The classic example is the «alcohol flight» of glycogen seen in hepatocytes and in cross sections of striated muscle fibres. Analogous phenomena may be seen in other large cells, such as the Purkinje cells of the cerebellar cortex. A widely used alternative to fixation by immersion is to perfuse with fixative. Though this reduces directional fluxes of fixative, it is however not lacking its own peculiar hazards (CAMMERMEYER, 1978).

Even when the substrate remains immobile, *the stain itself may not be deposited at the same site as the substrate.* Consider two quite different examples. Firstly metal impregnation with silver. Here the final stain comprises crystals of metallic silver which may be larger than the structures giving rise to them. At any rate the crystals may, for reasons discussed in Chapter 6, be deposited quite outside the structure. Both these possibilities, which are illustrated in Fig. 42, limit the accuracy of localisation. Large crystal size does offer the advantage of allowing demonstration of structures, such as intracellular layers of mucosubstance, too small to be themselves resolved or intensely stained by other methods. A second example of deposition of stain remote from the substrate is provided by many enzyme histochemical procedures. Here the stain, or final reaction product, is commonly in the form of precipitated crystals deposited in the zone around the enzymic site. Localisation in enzyme histochemistry is considered more fully in Chapter 7.

Another possibility is that *the stain itself may be mobile.* This topic is discussed for the case of enzyme histochemistry in Section 7.6. Other circumstances where redistribution of stain occurs include fat staining. Dyes such as the sudans, orginally molecularly dispersed inside fat droplets, may on storage be redistributed to form large crystals.

11.7 Checking the meaning of no-staining: an introduction

When a staining procedure intended to selectively colour substrate X fails to result in staining various possibilities arise:

1. That X was present *in vivo* but was lost during the pre-staining or staining procedures,
2. That X was present during staining but for some reason, for instance masking, did not stain,
3. That X was present and was stained, but that fading or diffusion of the stain took place before observations were carried out.

Much of the information needed for a consideration of these points has already been presented or alluded to elsewhere in this book. Consequently the following discussion presents only what new material is required but otherwise refers the reader to other Chapters, as appropriate.

11.8 Have there been substrate losses?

Two kinds of loss will be considered, of substance and of original nature. For example an enzyme may be physically lost from the section, and this is a loss of substance. Alternatively the enzyme may be retained in the section, but in a chemically modified form no longer possessing the original enzymic activity, and this is a loss of nature.

The classic procedure in biological staining for limiting losses is fixation, and the efficacy of fixation has already been repeatedly discussed. A general account of losses of substance was given in Section 3.6, and of loss of original nature in Sections 3.3 and 3.7. Specific attention has also been paid to the contradictions existing between preservation of substance and of nature of enzymes in Section 7.2, and of antibodies and antigens in Section 8.9. The general accounts also provide some information on losses into processing fluids subsequent to fixation. Losses have also been detected in other tissue and section preparation steps, such as the demineralisation of calcified tissue (WEATHERFORD and MANN, 1973). Changes in nature also probably occur with processes such as removal of mercury precipitates by treatment with iodine (see PEARSE, 1968, p. 171; FISCHER, 1959). Yet other processes causing losses, both of substance and nature, are the solvent-extractions and histochemical blockades used as control procedures. Such unwanted side effects have already been discussed in Sections 11.2 and 11.3 respectively.

Though this is not often discussed, *losses certainly occur during staining* in the dyebath, reagent solution, or incubating medium. Examples include losses of RNA and glycogen, already mentioned in Section 3.6; losses of mucosubstances and polysaccharides from cartilage matrix in the presence of periodate (SCOTT, TIGWELL and SAJDERA, 1972); and extraction of lipids into the peracetic and performic acids used in the peracid-Schiff reagent procedures for demonstrating carbon-carbon double bonds (ELLEDER and LOJDA, 1970 a). As a further twist, substrate X may be lost from the tissue during the staining of substrate Y. This can occur for example during the acid hydrolysis stage of the Feulgen nucleal procedure. The hydrolysis can induce loss of both DNA and glycogen fragments. Alteration in nature of the substance being demonstrated during the process of staining is important in enzyme histochemistry. Inactivation of enzymes by species such as lead ions and diazonium salts is well documented, and was discussed in Section 7.5. Unfortunate modifications of substrate nature are those where staining reagents reverse a previously applied histochemical blockade. For example tissue aldehyde groups which have been blocked by conversion to some derivative unreactive to Schiff's reagent, may be regenerated by the hydrolytic or oxidative action of periodic acid. Aldehyde blocking agents generating such acid labile groups include bisulphite, borohydride, dimidone, hydroxylamine and semi- and thiosemi-hydrazide (BAYLISS and ADAMS, 1976; STAPLE, 1958).

11.9 Is there masking of the substrate?

Consider attempting to stain substrate X without any staining resulting. Yet there may be good grounds, perhaps even a chemical assay, for supposing X to be present. One recurring explanation proffered in these circumstances is that substrate X was indeed present but was masked. Such masking is commonly regarded as the inhibition of staining of X by some other tissue constituent. When staining with acid or basic dyes masking by background materials carrying the opposite electric charge to the substrate is sometimes discussed. Proposed examples of such masking agents include proteins, when staining epithelial acid mucins with acid dyes (LEV and GERARD, 1966) or mast cell granules with basic dyes (BUTLER, 1971); lipids when staining basophilic substrates with colloidal iron (ADAMS and BAYLISS, 1968 b); and DNA and RNA when staining histones or cytoplasmic basic proteins with an acid dye (DAVENPORT and DAVENPORT, 1965; PRENTO and LYON, 1973).

The term masking is not always used with precision, and has at any rate also been applied to rather different processes. An instance is the failure of Feulgen nucleal staining of certain chromosomes due to the trapping of Schiff reagent by cellular inclusions (HEINMANN, 1969). Again, increasing the selectivity of a staining process by removing substances giving rise to interfering background staining has been described as «unmasking» (MAUNDER and ROST, 1972). Yet another example occurs when the chemical modification or denaturation of an unstained substance or complex to yield a stainable form is described as unmasking (CLAYTON, 1958). How such a process differs from a multiple-step stain such as the periodic acid-Schiff procedure is unclear.

Contrariwise many instances not described as masking could validly be so labelled. In many rate controlled staining processes the reagents have an affinity for tissue substrates but access is limited. So hydration-induced tissue swelling is often required before staining will occur, as with Alcian Blue (CEJKOVA and BRETTSCHNEIDER, 1969), Gram staining (TUCKER and BARTHOLOMEW, 1962) and the periodic acid-Schiff procedure (BEDI and HOROBIN, 1976). Other masking-of-access examples seem to reflect general differences of access existing even in fully hydrated tissues (for instance GOLDSTEIN, 1962b; HOROBIN, 1974). On the other hand access-control can also be due to differential distribution of non-polar embedding medium, as discussed for the case of plastic section staining in Section 10.4. Yet further instances are provided by methods where staining is markedly enhanced after apparently non-specific removal of tissue components, examples are given in Section 11.2.

After such qualifications it must be admitted that the masking concept, even when clearly applied, always teters on the brink of superfluity. When the mechanism of a staining process is clearly defined the influences of the chemical and physical environment of a substrate will be apparent. What need will there be then for the idea of masking? Furthermore the use of the masking concept may both reflect and propagate the mischievous oversimplification that substrates typically occur naked in the section, but just occasionally interact with their surroundings. This is contrary to the repeatedly expressed viewpoint of this book, namely that biological staining is (at least) a two-phase process and cannot be understood as merely solution biochemistry or organic chemistry or . . .

11.10 Was there staining, which has been lost?

Stain may be lost between the staining process itself and the time of observation, or of re-observation. Such losses can be considered from several points of view. For instance losses may be of substance or of nature. Stain may have diffused, either from its original site to some other as already discussed in Section 11.6, or completely out of the section. Or the stain may have changed colour, or changed into a colourless derivative (i. e. faded). The way in which such losses occur, the point in the staining-viewing sequence in which they occur, and the timescale of such losses may all be considered. Such questions have often been studied rather narrowly, with an emphasis on losses of a particular stain in a particular mode. A more general view is sought for here, as outlined in Fig. 64.

In multi-operation procedures applied stains may be lost into subsequent solutions. This influences the design of several classic procedures, such as the choice of iron-hematoxylin as the nuclear stain in many picro-trichromes. Unlike many other nuclear stains the iron-hematein co-ordination complex is stable at the low pH produced by the picric acid solution. Extraction of one stain by another occurs in electron microscopic staining systems also, as noted for example by KALIMO and colleagues (1968). These workers observed that crystals of barium and lead sulphates, deposited when demonstrating aryl sulphatases, are soluble in the lead citrate and uranyl acetate solutions subsequently used as section stains.

The terminal stage in most staining procedures is a wash or rinse, to remove unwanted stain from the section. Here we ignore differentiation methods, in which such removal results in gross change in staining pattern, and focus on the more commonplace removal on non-specific background staining. A perennial problem is how to remove such unwanted stain whilst leaving the wanted. For instance washing, in practice, does often constitute differentiation. Thus after staining collagen fibres by trichrome methods, or elastic tissue by dyes such as Gomori's Aldehyde Fuchsin or Resorcinol

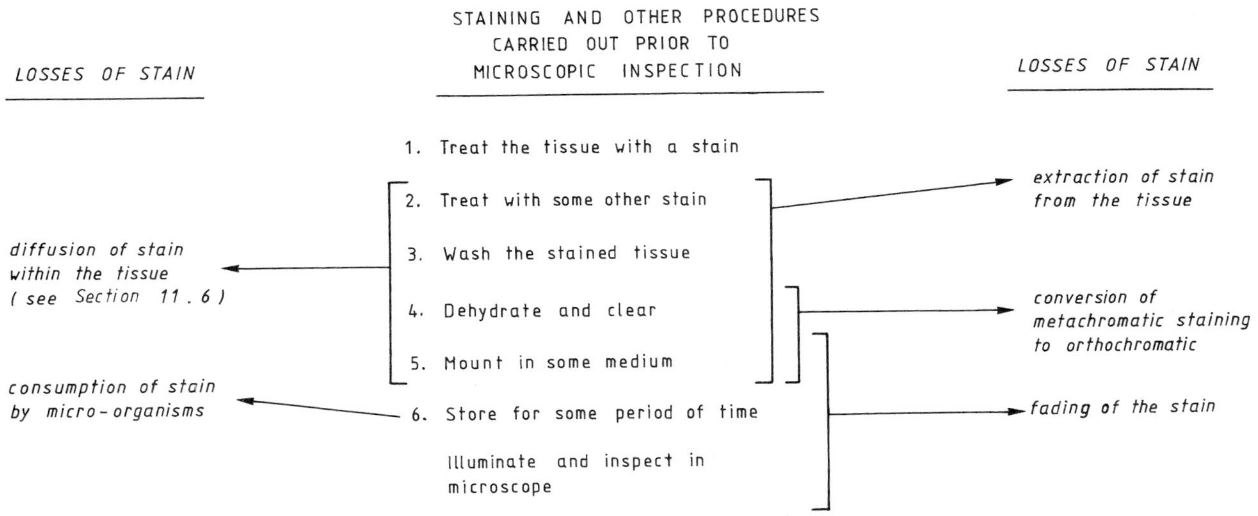

Fig. 64. An outline of ways in which stains may be lost, either in substance or nature, subsequent to their deposition and prior to their observation

Fuchsin, washing is found to remove dyes selectively from the fine connective tissue fibres whilst leaving colour in the coarser material. More quantitatively, DEMALSY and CALLEBAUT (1967) showed that in the Feulgen nucleal procedure the traditional post-Schiff sulphite rinse reduced staining both of the nuclei and of the background. Consequently they recommended the use of plain water as a rinse. Such observations raise a fundamental question: is it always possible to distinguish between «non-specific background» and «real» staining which happens to be weak and diffuse?

Other routine operations involved in preparing many stained sections are dehydration and clearing. That most basic dyes, and picric acid, are highly soluble in solvents such as ethanol is well known. It is nevertheless commonplace to find a student's Methyl Green-Pyronin preparations lacking Pyronin and his picro-trichromes lacking Picric Acid: though the beginner is often puzzled.

A dramatic change of nature of deposited stain is the *change of metachromatic staining to orthochromatic.* This usually occurs on transfer of the section from the aqueous staining solution to non-aqueous dehydrating solvents, but occasionally takes place in the embedding media. Hence metachromatic staining is best seen in non-dehydrated sections mounted in water, or second best in aqueous glycerol or syrup. Even the latter mountants however often suppress metachromasia after a few hours. Broader discussions of these topics are provided by PEARSE (1968, p. 335) and THOMPSON (1966, chapter 8). Something of the physico-chemical background may also be found in Section 4.2 of the present book. A plausible explanation for the substrate – selective retention of metachromasia, and also for the slow loss of metachromasia seen in certain mounting media, is selective retention of water by various tissue elements. It is the very ionic polysaccharides, such as the chondroitin sulphate and heparin of cartilage matrix and mast cell granules, which will be most strongly hydrated and so lose water with the greatest reluctance.

Loss of stain can also occur by *diffusion into mounting media.* Misadventures apart, such losses of stain typically occur when using permanent aqueous mountants such as sugar syrups, gum arabic, glycerol gelatin, or polyvinyl alcohol. Basic dyes are particularly easily lost. One important factor here is pH, since acidic media tend to accelerate losses. Other variables are ionic strength and sugar content (viscosity?). Addition of large amounts of salts such as potassium acetate, or sugars such as sucrose or D-sorbitol to glycerol or gum arabic greatly reduces such dye losses.

Far from trivial is the *fading of deposited stain.* Few systematic studies seem to have been carried out, and no summarising account is provided by major texts. Consequently this present account is based, when no citations are offered, largely on the rather anecdotal information offered by LILLIE (1965) plus analogous personal experience from the present author's laboratory, set against information from the textile dyeing industry.

Fading of stained preparations can be caused by light or may occur in the dark. These two processes will be discussed separately, though in practice it is often not clear what fading was due to.

Fading due to light or ultraviolet radiation involves electronically excited species as reactants. Some physico-chemical and chemical background information to such photofading is provided in Supplement 13.

Such information does rationalise many of the practical manouvers used to restrict fading. For instance choosing dye and mountant carefully, avoiding oxididants, using long wavelenght illumination, as summarised for fluorescent stains by PEARSE (1972,

Chapter 29). Similarly the poor light-fastness of dyes such as Acid Fuchsin (based upon an electron accepting chromophore), or the observations of GARCIA and IORIO (1966) that drying out of oil-mountant accelerated fading (after drying out, a preparation is exposed to atmospheric moisture) can also be understood. Other observations remain obscure, for instance the report by these latter authors and others that the rate of fading of Feulgen stained nuclei was proportional to the surface area of the nucleus. Could this have been instrumental error? C. f. BEDI and GOLDSTEIN (1976).

Bleaching of stained preparations is not limited to photofading, and *can occur with preparations stored in the dark* though often at a slower rate. Sometimes plausible explanations can be proposed. Thus synthetic mountants such as cellulose caprate or coumarone resins, natural resins such as Canada balsam, and semi-synthetic resins such as Euparal, all contain numerous reducing groups. The ready bleaching of Prussian Blue deposits in such media could be due to reduction to greenish-white ferrous ferrocyanide. Significantly, such fading does not occur so readily in non-reducing mountants. Contrariwise, precipitates of cobalt sulphide are well preserved in reducing media but decolourise in non-reducing media, perhaps due to atmospheric oxidation to cobalt sulphate. The enhancement of fading in the presence of water, noted previously with photofading, is also reported for non-photolytic bleaching (DEWSE and POTTER, 1975). Since dyes may fade by either oxidative or reductive processes, the lack of any clear correlation between type of resin and their degree of fading is not surprising. However basic dyes do appear to fade more readily than acid. This may relate to the typical electron accepting and donating character of these two broad dye classes rather than with any property of the mountants.

In keeping with chemical expectations, fading is minimised by careful choice of stain; by mounting in a polar but neutral non-aqueous medium without marked oxidising or reducing properties; by storage in the dark; and sometimes by addition of an antioxidant to the mountant (HOLLANDER and FROST, 1971).

To conclude exotically: deposited stain is sometimes *eaten.* Against the odds stained preparations, mounted in non-aqueous synthetic polymers, do sometimes provide a meal for a crop of fungi.

11.11 The inevitability, value and nature of staining theories

Many users of biological staining procedures see theorising as an unnecessary and avoidable distraction from the application of biological staining to everyday problems. Hopefully this book demonstrates that a theory is merely a label for the way in which we organise our everyday actions and thoughts and information gathering. Consequently making any kind of decision («Oh try...1%...40°...half an hour... fixing in Bouins») displays our theoretical biases. To act without a theory is not possible, though we may indeed fail to appreciate what our operational theories and assumptions are.

Though it is obvious how inadequate or unacknowledged theories could mislead, how is theorising useful in everyday work? Theories are useful:

1. When things do not go according to plan. Trouble shooting without a theory resembles looking for our own lost spectacles,

2. when events do apparently go as expected. The assessment of the significance of staining or of non-staining requires theoretical ideas on the mechanism of the methods,
3. when established staining methods are modified or new ones are devised. The plethora of minor variations and of methods difficult to repeat, reflect in part the general lack of clear theories.

Some account will now be given of the nature of the theories used in the present book. This discussion will, for illustrative convenience, be focussed on the selective staining of tissues by dyes. As described in Chapter 1 the organising questions used in this area were:

1. why does a dye colour anything?
2. Given that it does, why does a dye not colour everything?

These questions already constitute a theoretical perspective, a directive for studying dye-tissue interactions. The first problem arising is how to relate this rather personal theoretical view to more general physico-chemical ideas. In practice Question (1) was approached largely by assuming or imagining that staining with dyes may be carried to equilibrium. Responses to Question (2) either assumed equilbrium conditions, with selectivity reflecting variations in affinity of different components; or assumed non-equilibrium conditions, with selectivity reflecting variations in rates of staining or destaining of different components.

For further illustration of the theoretical style of the present book, consider equilibrium processes. An appropriate theoretical mode of describing physico-chemical equilibria is thermodynamics. Though a rigorous and abstract thermodynamic treatment is validly applicable to all systems, such a treatment yields only a minimum of information concerning the physical causes of the events described. A further limit to the practical value of rigorous thermodynamics in the biological staining area is that, in most cases, the quantities involved cannot be determined by independant measurements nor can they be predicted without using non-thermodynamic arguments. Consequently whilst the theoretical elements of Sections 4.3 to 4.5 are built around a thermodynamic skeleton, most of the detailed theoretical arguments utilized conceptual models reflecting the actions of various physical forces. Examples were the water-water hydrogen bonding forming the basis of hydrophobic bonding, the Coulombic effects arising with ionic dyes and ionic tissues, and van der Waals attractions. With such conceptual models the influences on the behaviour of staining systems of changing particular properties (such as the alcohol content or pH of an aqueous dyebath, or the size or charge of a dye) can be analysed. Remember however that each committment to a conceptual model, with specified properties, involves a deviation from thermodynamic rigour. Also models (conceptual as much as physical, compare Section 11.12) are only meaningful when the relevant properties of the actual system are in fact represented by the model chosen.

Finally note that this book does also acknowledge the value of semi-empirical theories, in particular the structure-staining correlation models described in Section 11.14.

11.12 Uses and abuses of test objects, model systems and histochemical-chemical comparisons as aids to understanding staining results

Here is discussed one of the traditional scientific responses to complex reality: the parallel study of a simpler physical system, a model system. The concept of a physical model system is somewhat blurred in the histochemical literature. Systems so described range from preparations of histochemically familiar tissues (test objects), though cell fractions and various solid preparations hopefully mimicking relevant characteristics of cells and tissues, to solutions of those compounds regarded as responsible for staining. However the systems most often described as models do seem to be solid preparations. The reasons for adopting such a range of approaches are expressed in Table 53.

Table 53. Advantages and disadvantages of various histochemical model systems

Types of model system		
COMPREHENSIBILITY AND CONTROL increase in the direction of the arrow ↓	tissue preparations test objects cell fractions solid model preparations chemical/biochemical investigations	↑ REALITY AND RELEVANCE increase in the direction of the arrow

Studies at more than one level are typically complimentary. Hence rigorous investigations may use information from several different types of models.

The use of physical model systems as external standards for histochemical quantitation studies was discussed in Section 11.5. The cautionary comments regarding the limits of model system studies offered there are more generally valid. A model is indeed not the reality we seek to understand, and by simplifying we may have excluded or altered factors significant to staining. This is so important and yet so often overlooked that several illustrative examples will be given. An early cautionary tale was told by Sibatini and Naora in 1953, who found that although addition of histone to an aqueous DNA solution altered the Feulgen staining intensity, no comparible effect was found with tissue sections. Sibatini was so impressed by this and similar observations that he later (1954) said «it is rather questionable that the effect of cellular substance on any kind of histochemical reactions can be evaluated correctly with model experiments in test-tube (conditions)... this is not only because the cellular system is far from simple and hence difficult to reproduce experimentally, but also because the effect observed with the reaction developed in solution does not necessarily hold with the precipitated system, such as fixed cells». It this sounds too sceptical, remember that Sibatini had solution models in mind. Analogous observations led van Duijn (1976) to a similarly cautious position with regard to solution models. Yet he did have an optimistic view of model systems such as controlled-porosity polyacrylamide films. Others who have emphasised the caution needed when moving from model system to section include John E. Scott who percipiently commented, after demonstrated the ability of

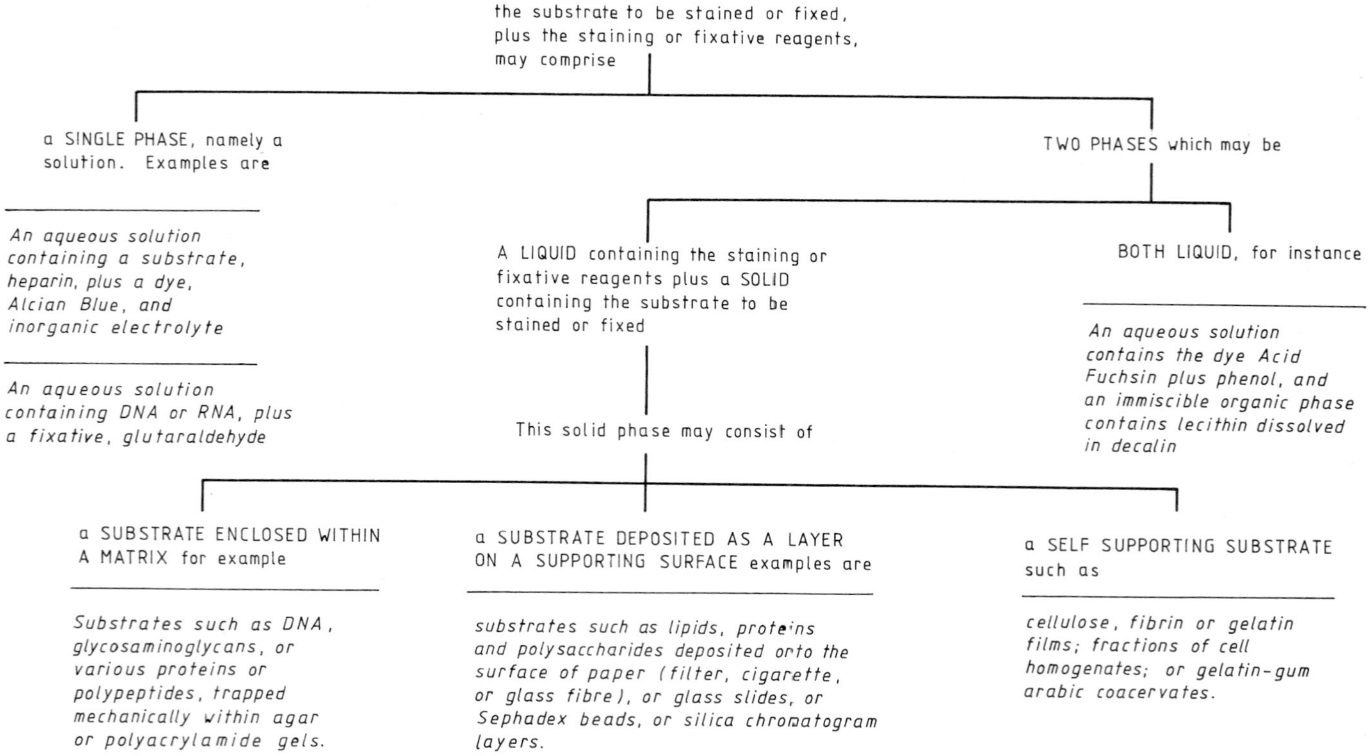

Fig. 65. Some examples of mechanical forms adopted for histochemical models used in the investigation of staining mechanisms

phosphotungstic acid to precipitate a wide range of polar polymers, that «these reactions *will* take place *unless prevented.* It remains to be seen which factors prevent some of the staining reactions taking place in prepared, embedded tissues» (SCOTT, 1971).

It is useful to consider *the range of topics studied* with the aid of model systems. Specificities and modes of action of stains have often been considered. So has the use of models as external standards in studies attempting histochemical quantitation, see Section 11.5. In these areas investigations have been carried out on staining systems both for electron and light microscopy. Various fixation procedures have also been studied with model systems. In such studies *the wide range of mechanical forms* adopted for the model systems is noteworthy, a typical selection are displayed in Figure 65.

Consideration will now be given to some uses and abuses of model systems, discussing applications of model studies to the staining process and to fixation.

There are numerous *investigations of the selectivity of histochemical stains.* Occasionally model system studies have convincingly validated supposedly selective histochemical stains. Recently for example Toluidine Blue metachromasia was shown to demonstrate heparin in the presence of other glycosaminoglycans (TAS, 1977b). More often model studies explore the limitations of staining systems. An example is the extensive investigation of the alkaline Fast Green procedure for the selective staining of histones, carried out by COHN (1973). Cohn commented that «the results of these investigations indicate that some of the traditional assumptions about Fast Green staining reaction are not valid» and that «it is essential that if any reliance is to be placed upon the use of this stain... the conditions be controlled with great care». Demonstrated limitations of methods sometimes arise from more limited causes: positive artifactual staining by a dye impurity when staining lipoproteins with Nile Blue for instance (DELCOURT 1967), or excessive solubility of certain substrates in staining reagents (ELLEDER and LOJDA, 1970a).

However a common consequence of careful model system investigations is scepticism about claims of specificity. Typical is the finding that «contrary to published observations, mixtures of Fast green and Eosin Y show no specific staining of either lysine or arginine» (MITCHELL, 1977); the demonstration that the peracid-Schiff methods designed to stain unsaturated lipids are more likely to be staining plasmalogens (ELLEDER and LOJDA, 1970a); the conclusion that the Alcian Blue-Safronin O method for mast cell granules «cannot be used for the histochemical demonstration of glycosaminoglycans» (TASS, 1977a); and the observations indicating that phosphotungstic acid may not be a selective stain for polysaccharides but rather «a ›universal‹ (non-specific) precipitant for polar polymers in acid solution» (SCOTT, 1971).

Study of the mechanisms of staining processes has also received considerable attention. The breadth and depth of such studies vary widely. Certain workers have investigated very general questions, such as the basis of acid and basic dyeing (SINGER and MORRISON, 1948; BIRKEDAL-HANSEN, 1973), and of the rates of diffusion and/or precipitation of reaction products of enzymic reactions (CORNELLISE and VAN DUIJN, 1973; DE JONG et al., 1979a, b) and of enzyme molecules (VAN DUIJN, PASCOE and VAN DER PLOEG, 1967). More common are investigations of particular staining systems, e. g. of chromation-acid hematein stains (ELLEDER and LOJDA, 1970b; ROOZEMOND, 1971), and of metal impregnations of lipids and proteins with osmium tetroxide (HAYES et al., 1963) and silver (BURR, 1973). However most investigations are concerned with details of a few widely used techniques, especially the Feulgen nucleal and the periodic acid-Schiff methods. Studies of the Feulgen procedure have considered both the hydrolysis step and the subsequent reaction of the apurinic acids with Schiff reagent (for example DUIJNDAM and VAN DUIJN, 1975; HARDONK and VAN DUIJN, 1964a, b; HIRAOKA, 1973). Similarly with the periodic acid-Schiff method, both the initial oxidation and the subsequent treatment with Schiff reagent haven been investigated (for instance HOROBIN and KEVILL-DAVIES, 1971; OVADIA and STOWARD, 1971; ROBERTS, 1977; SCOTT and HARBINSON, 1968).

After this account of applications of model systems, an account must also be given of abuses, or misapplications, where models have increased obscurity rather than enlightenment. At one extreme *model substances are sometimes adopted whose relationship to tissue contents is tenuous.* Consider a study of the Ziehl-Neelson stain for acid-fast mycobacteria (HARADA, 1976). The choice of model was influenced by conclusions drawn in the late 1940's, namely that the substrates responsible for acid-fast basic-dye staining were «unsaturated fatty acids of high molecular weight». The model system used comprised lecithin (low molecular weight, and a phospholipid) dissolved in thirty times its weight of a liquid hydrocarbon (of low molecular weight, and non-ionic in character).

A more common abuse arises because the model system may lack *characteristics relevant to the staining process.* It will be useful, if a little arbitrary, to divide such omissions into chemical and physical factors. A common chemical omission arises when a model system lacks protein, a marked deviation from most situations in the tissues. An example is the polyanion-on-paper model widely used by the Critical Electrolyte Concentration (CEC) school when studying salt-induced decreases in staining intensity. Due to this absence of protein, and so of polycations, the salt-induced increases in staining so often seen in tissue sections (HOROBIN and GOLDSTEIN, 1974) cannot be simulated by the model. This limitation has been extremely misleading, and has encouraged CEC theorists to write as if salt-induced staining increases rarely occur in tissue sections or at most are minor excentricities (SCOTT, 1973). Since this is not so, a gross over-estimate of the applicability of CEC procedures to practical histochemistry has resulted. Other instances where omission of protein has misled are the numerous lipid-on-paper model studies of the dichromate-hematein stains for «phospholipid». Absence of protein from these models has resulted in histochemists discussing chromium binding in these procedures solely as a function of lipid-dichromate reactivity. The significance of the vigorous reactivity of proteins with warm dichromate, and of the subsequent binding of chromium of lower oxidation states to the protein, has consequently been overlooked, in spite of the phenomenon being well documented (for a review see HARTLY, 1969). Ironically JOHN R. BAKER, who devised one of the more widely used dichromate-hematein stains («Baker's acid-hematein»), appears to have been quite aware of the reactivity of proteins with chromium in the Cr(VI) form (BAKER 1958, p. 56–58, 105–107), but not to have considered this significant.

The most common physical omissions are the absence of controlled variation in permeability. As discussed in Sections 4.13 and 4.9, variations in tissue porosity can and often do influence rate of staining and destaining, and under some circumstances stain-tissue affinities. Many model systems however are unable to mimic permeability variations. Although this parameter could in principle be varied with some types of model, such as the polyacrylamide gels pioneered by the Netherlands School, this has not normally been carried out in practice. As an example we will consider staining methods in which selective colouration is only achieved after differentiation, that is after a rate-controlled destaining step. Such processes normally depend on the lowered permeability of one region of the section. The binding of stain to some substrate may on the other hand initially occur throughout the tissues. Such a situation is clearly not modelled by smearing the relevant substrate onto a glass slide, or by trapping it in a single homogeneous gel. In this context consider once again the lipid-on-paper models often used for studying the dichromate-hematein «phospholipid» stains. Users of such models have usually (though not always, see ROOZEMOND, 1971) ignored the following significant observations: when staining tissue sections colouration is typically NON-specific prior to differentiation, or if the dichromate treatment is extended or is carried out at too high a temperature, or after use of certain fixatives. Such lack of correlation between differentiation-controlled staining patterns of tissues and cells, and affinity-controlled staining patterns of models has also passed with little comment in studies of the Ziehl-Neelson method (HARADA, 1976) and the Gram procedure (for example, MURRAY, 1962). Analogous confusions have arisen from model system investigations of other rate-controlled stains. These include studies of the Methyl Green-Pyronin stain for DNA-RNA (SCOTT, 1967, compare to GOLDSTEIN, 1961), and of certain trichrome acid-dye mixtures (LACOUR et al., 1958, compare to BAKER, 1958; GOLDSTEIN, 1961).

Further confusions arise when workers either fail to appreciate the complexities of their model systems or are so fascinated by these complexities that the staining of tissue sections becomes a secondary consideration. The chromation-hematein stains provide examples of overlooking significant model system complexities. Different model system studies have found: that such stains are (BOURGEOISE and HUBBARD, 1965) or are not (HOLCZINGER, 1964) selective for phospho- as compared to neutral lipids; that phosphatidyl ethanolamine is strongly stained (CHAYEN, 1968) or is hardly stained at all (ADAMS, ABDULLA, BAYLISS and WELLER, 1965); and that the degree of lipid unsaturation does (HOLCZINGER, 1964; CHAYEN, 1968) or does not (ROOZEMOND, 1971; ADAMS, ABDULLA, BAYLISS and WELLER, 1965), have a marked influence on staining intensity. These contradictions show that such systems are so complex that their value as models is doubtful. An instance which may be regarded as an over-enthusiasm for the model arises from studies by the Critical Electrolyte Concentration school of dye-polyanion binding. In a series of papers the interactions of at least fourteen dyes with a wide variety of polyanion-on-paper model systems was investigated. Yet apart from the widely used dye Alcian Blue, very few micrographs illustrating dye-tissue staining seem to have been published. As a further and quite different example consider the widespread use of formaldehyde as a model aldehyde. This disregards the fact that formaldehyde is an extremely atypical aldehyde.

In conclusion consider the *application of models to the investigation of fixation.* One topic, the degree of fixative-induced cross-linking, has been studied using solutions of such compounds as glycogen, nucleic acids or proteins. Changes in molecular size have been monitored using gel filtration chromatography and acrylamide gel electrophoresis, as described by Hopwood and his associates (HOPWOOD, 1975 and papers cited herein). The possible degradative effect of fixatives has also been studied using the same approach (HOPWOOD, 1969b). Electric charge is another physical property of tissue components important for staining processes, and is often influenced by fixation. At least one attempt has been made to study this by using a solution of protein, measuring changes in pK by means of isoelectric focussing (HOPWOOD, 1970). Other topics studied include the loss of tissue components, such as glycogen and protein, during and after fixation (SMITHERMAN et al., 1972; TAS, OUD and JAMES, 1974). Somewhat more subtle losses, namely of immunological and enzymic activities, have also been investigated (YOKOTA, 1973; BERTELOOT and HUGON, 1975). Once again there are abuses. Oversimplification for instance, such as the omission of protein from certain of the models. This precludes the possibility of observing substrate-protein crosslinking. A more substantive, and in many ways unavoidable, problem arises from the use of solution models. The essence of fixation is insolubilisation, whilst the essence of gel filtration or electrophoresis or isoelectric focussing is the study of particles in solution. Solution models are inevitably limited and quite inadequate for studying possible trapping of unreacted substrates in denatured and hence insolubilised proteins.

11.13 The baleful influence of reagent impurities on the interpretation of staining results

The term reagent impurity describes misidentification or mislabelling of the reagent, uncertain concentration, presence of more than one staining component, and presence of a non-staining compound which nevertheless influences the staining process.

Such varied impurities are responsible for an analogous array of artifacts. These may be positive or negative, indeed some impurities simultaneously generate both in different parts of the tissue. Artifacts may also be either qualitative or quantitative.

Some effects of reagent impurity have already been touched on. When discussing enzyme histochemistry, in Section 7.3, the role of substrate impurities in generating positive artifacts was noted; when considering labelled antibodies the non-specific background staining resulting from over-labelling with reagents such as fluorescent dyes was described in Section 8.7; and the effects on histochemical extraction procedures of using impure enzymes, and also of the presence of water in supposedly anhydrous extraction solvents, were discussed in Section 11.2.

It is however confusions following from the use of impure dyes, and dyestuff-derived stains such as Schiff Reagent, which have received most attention. As this has been discussed elsewhere (HOROBIN, 1969, 1980b; ROSENTHAL et al., 1965, and the numerous papers cited therein) this account will be limited to illustrating the consequences of the major types of impurity.

With some dyestuffs the name on the label gives little indication as to the content. In some cases this may be regarded as essentially the responsibility of the supplier, in other cases less so, and sometimes definitely not so.

Instances where confusion arises either from packaging errors on the part of the supplier, or perhaps from a belief that for the histochemist one pinkish-purple dye is as good as another, include: substitution of the azine dye Methylene Violet for the thiazine dye Methylene Violet Bernsthen and *vice versa* (MARSHALL and LEWIS, 1975); substitution of a mixture containing the acid dye Rhodamine B for the basic dye Pyronin Y (HOROBIN, 1969); and replacement of Stains-All by its triethylated homologue (GREEN and PASTEWKA, 1979). Such instances, even if occurring too frequently, do at least constitute anomalies. Sometimes however the structure of the dye supplied under a particular label is consistantly other than the labelled dye. One example is Toluidine Blue O. Samples labelled as Toluidine Blue were purchased from many suppliers. Thin-layer chromatography showed that all the samples contained the same major blue constituent. Nuclear magnetic resonance spectroscopy demonstrated that this contained ethyl groups. However Toluidine Blue O is methylated, and carries no ethyl groups.

At this juncture it is necessary to state the usual authorities for specifying dye structures. The major work for English-speaking workers has been *Conn's Biological Stains,* now in its ninth edition (LILLIE, 1977). There is also the massive, textile-orientated *Colour Index,* the third edition of which is produced by the Society of Colourists and Dyers (Bradford, UK) aided by the American Association of Textile Chemists (Lowell, USA). More rigorous use of such information sources would help eliminate errors described so far.

However not all confusions of this type would be eliminated. For dyes such as Astra Blue, Brilliant Cresyl Blue, Cresyl and Nile Blue, as indicated by Conn, neither the suppliers *nor* the users appear to know which dye structure is required! Astra Blue could be a phthalocyanin derivative or a triarylmethane; and Brilliant Cresyl Blue could apparently be any one of three amino-oxazines, with quite different substitution patterns and numbers of N-H, N-CH_3 and N-C_2H_5 groups. These ambiguities arise, in part at least, from the profusion of trivial names applied to industrially produced dyestuffs, with much overlap and minor variation. Again, use of available sources of information by the original authors would perhaps have pre-empted some errors. Other confusions of labelling offer no such easy answers, as confusions are in the chemistry not in the nomenclature. The Romanowsky stains are classic example. These stains contain Eosin plus dyes traditionally obtained by oxidation of Methylene Blue. These oxidation products, contrary to widely held opinions, are unlikely to be more than crude mixtures. Obtaining standardised products requires the substitution of new and effective chemical procedures such as those of MARSHALL and LEWIS (1976) or LOHR and colleagues (1974) for traditional ones. With all the labelling errors so far mentioned the original authors did at least believe they knew which dye structure was the appropriate one. Several widely used staining reagents are however admitted chemical unknowns. Aldehyde-fuchsin, Resorcin-fuchsin and, more importantly, Schiff reagent are examples of this.

Even when the label is correct, the amount of dye present is usually unspecified. Indeed dye content often varies markedly between one commerical sample of a given dye and the next. The consequences include variations both in rate of staining and in equilibrium dye uptake. These are widely acknowledged, and bench workers expect to have to adjust the staining time or the period of differentiation or the pH or... in order to obtain the required results. Though irritating, this is often trivial, except when using automatic staining machines. The situation is commonly more complicated when staining with dye mixtures, such as the trichromes of Masson, Mallory and Gomori, or the Methyl Green-Pyronin stain. In many laboratories it is common practice to prepare concentrated stock solutions, and to dilute as appropriate.

Coloured impurities, perhaps the first thought in people's minds when thinking of the impurities of dyes, will now be discussed. Such impurities may give rise to non-specific background staining, and colouration in an unwanted shade. Dyes whose contaminants are well known include Methyl Green, whose Crystal Violet impurity is usually removed by solvent extraction; and Basic Fuchsin, whose acridine impurities are removed by charcoal extraction during the preparation of Schiff reagent.

A wide variety of colourless contaminants occur in commercial dyestuffs and will be discussed. Contaminants include inorganic salts, and organic compounds such as dispersing agents and dextrin. These, as described in Chapter 4, can actively influence the patterns and intensities of staining and so are not inert diluents. Inorganic salts for instance, which have been found to comprise up to 40% by weight of certain Pyronin Y and Alcian Blue samples, can directly cause increases and decreases in staining intensities of various substrates (for illustration of this see HOROBIN and GOLDSTEIN, 1974, Figs. 1–4). Other salt effects include changes in pH (CLEMENS and TOEPFER, 1968) and, with salts such as zinc chloride present in some samples of Acridine Orange, quenching of fluorescence (STOWARD 1967). Though less well documented in the biological staining literature both dextrin and dispersing agents can give rise to marked alterations in staining intensities and patterns (HOROBIN and GOLDSTEIN, 1972, Figs. 6 & 7; BENNION, HOROBIN and MURGATROYD, 1975; MELOAN and PUCHTLER, 1978).

How are such errors to be avoided? In practical terms this question reduces to the following requirements:

1. Knowledge of which dye structure is appropriate for a given procedure,
2. Existence of an agreed nomenclature for describing these structures,
3. Use of suitable analytical methods for checking structure-name correlations,
4. Availability of pure dyes at low prices.

Requirements (1) and (2) are in part met by utilizing wherever possible the *Colour Index* (C. I.) nomenclature system. This allocates a unique number and name to each known dye structure in the *Index*. Thus although the everyday dye Eosin Y may also be sold under such labels a Bromo Acid J, Bromofluorescein, Bronze bromo ES, and Eosin DH, the *Colour Index* provides the unambiguous descriptions C. I. number 45380 and C. I. Acid Red 87.

A number of sources are available for those seeking analytical methods for biological dyes, i. e. part of requirement (3). In the first instance reviews by the present author (HOROBIN, 1969, 1980b) and the Technical Appendix on Methods for Testing Biological Stains in *Conns Biological Stains* (LILLIE, 1969a). Other sources include a major multi-author compilation *The Analytical Chemistry of Synthetic Dyes* (VENKATARAMAN, 1977), and the *Handbook of US Colourants for Food, Drugs and Cosmetics* (MARMION, 1979). Information useful for the quality control of other staining reagents, such as enzyme substrates or immunoglobulins, is of course also available in the literature.

Requirement (3) does also need someone to *apply* these methods. Should this be the suppliers of biological stains? Most certainly yes. However experience teaches that users will also require some independant check on dye purity. There are certainly some simple and useful procedures, which the user can carry out themselves, for instance:

Checking the identities of dyes, and gaining information on the coloured impurities by

- retaining samples of effective dyes to use as standards,
- chromatographing fresh dye samples, along with standards, on Sephadex LH20, using salty dimethylformahide as eluent (HOROBIN, 1971).

Checking the quantity and nature of colourless impurities by

- extracting dye samples with alcohol or chloroform, as these solvents dissolve dyes but leave most colourless impurities undissolved. Before-and-after weighings give a measure of dye content (HOROBIN, 1970),
- Making simple identity checks on colourless residues. For instance inorganic salts are not flammable, whilst dextrin does burn and leaves no residue, and dispersing agents burn but if ionic leave an ash. The traditional wet-analyses for inorganic ions, dextrin

and so on are also easy to perform. For such procedures see for instance VOGEL's texts (1978a and b).

For obtaining pure dye samples

- This is not in general a good investment of time and resources for most workers. A better procedure is to keep checking commercial samples until a satisfactory one is found.
- As a partial exception, dextrin and salt can often be fairly easily removed by scaling up the solvent extraction procedure mentioned above.

Beyond these suggestions the investment of time and of money is, for most workers, unlikely to be worthwhile. However there is an independent body which checks on the value of dyes for biological staining. This is the *Biological Stain Commission* based in Rochester, New York, USA. This organisation tests batches of dyestuffs submitted to them by suppliers for use in various routine biological staining procedures, and certifies satisfactory samples (CLARK, 1974). Unfortunately this does not ensure a supply of pure, as distinct from standardised, dyes since the analytical methods adopted by the Commission are somewhat restricted.

11.14 The uses of structure-staining correlations, practical and theoretical

The present book shows that though there are many partial theories about biological staining and histochemistry, there is no single and fundamental «theory of staining». Even if such a unified basic theory is not at this time possible, this should not preclude the development of semi-empirical, quantitative and predictive theories of a general character. However such theories too are lacking in the histochemical field, an omission all the more surprising when we consider the situation in such a related field as pharmacology. Both pharmacologists and histochemists face some of the same problems, involving the movement and binding of low molecular compounds (drugs and staining reagents respectively) through and to the tissues. Pharmacologists have developed quantitative and predictive structure-biological activity relations to aid their selection of potential drugs from amongst the prodigious variety of compounds available, and also to enhance their understanding of drug action (ALBERT, 1979). No comparable predictive correlations exist in the histochemical field. Hence an invaluable tool for facilitating selection or design of new staining reagents, for understanding staining mechanisms, and for evaluating histochemical observations remains unavailable to histochemists.

Qualitative correlations of reagent structure with reagent staining performance have a long history. The role of electric charge in determining whether a dye was acid (i. e. anionic) or basic (i. e. cationic) was clarified by EHRLICH in 1897, who also argued for the importance of dye size in certain staining systems, as subsequently did many other workers (e. g. MANN, 1902; HEIDENHAIN, 1903; PAPPENHEIM, 1918). Since then many other structural features of staining reagents have been proposed as contributing to stain-tissue affinities, or to staining selectivities. Examples include chemical groupings capable of forming charge-transfer complexes, or covalent and hydrogen bonds, or van der Waals attractions with the tissues, or of facilitating hydrophobic bonding. However, except for the role of electric charge, there is a widespread lack of agreement as to the validity of such proposals. Compare for instance the disparate views of such major figures as BAKER (1958) and LILLIE (1977) regarding the significance of the size of dyestuffs; or the conclusions of FULMER and LILLIE (1956), GOLDSTEIN (1962c) and PUCHTLER and colleagues (1961) on the one hand with those of HOROBIN and JAMES (1970) on the other concerning the role of hydrogen bonding when staining elastic fibres. Why should this be so? There is no doubt that the failure of histochemists (the present author included) to express structural data numerically is one significant cause. Consequently when structure-staining correlations were identified they could only be expressed in the form of a paragraph of prose: and prose is hard to remember, lacking the impact of a simple table or graphical display. In addition there has been an inexplicable reluctance to display structure-staining correlations graphically even when the structural data has been discussed numerically. In this context the present author can only regard some of his earlier publications (e. g. HOROBIN and JAMES, 1970; HOROBIN and MURGATROYD, 1971) with bewilderment. In passing one should note an honourable exception to these strictures, in the form of a study by Seligmans group (NACHLAS et al., 1959) of the influence of reagent structure on azocoupling.

For structure-staining correlations to be identified and used, the *choice and description of structural features of the staining reagents* must be relevant, clear, and convincing. Considering the earlier discussions in the present book it is possible to specify certain structural features of staining reagents which will influence affinities and selectivities. Thus *coulombic effects* are influenced by the electric charges of reagents, whose signs and magnitudes are readily expressed numerically. Formation of *covalent and hydrogen bonds* requires the presence of the appropriate chemical on the reagent (amino, hydroxyl and so on), which may readily be counted. A numerical estimate of the influence of adjacent structural fragments on the *reactivity* of a given chemical group may, at least in aromatic systems, be made using structural formulae and Hammett σ constants. These σ values derive from measurements of the rates of reaction of various model compounds, carrying a wide range of sustituents. Once the Hammett constants of particular fragments have been determined then the influences on a particular reactive group of such fragments may be assessed by summing the appropriate values. For a histochemically relevant example see NACHLAS, GOLDSTEIN, ROSENBLATT, KIRSCH and SELIGMAN (1959). The rate and extent of such a reaction may also depend on the concentration of the reagent inside the tissue. Hence structural features influencing both non-covalent sources of affinity and also the rate of the diffusion of the reagent into the tissue may be important.

Van der Waals attractions are influenced by such reagent features as polarity and polarisability. A simple numerical parameter giving some indication of these is provided by totalling the number of covalent bonds involved in the conjugated and /or aromatic systems of the reagent molecule, i. e. a conjugated bond number. A variant is to use a conjugated bond length (GILES et al., 1978).

Hydrophobic bonding largely depends on the overall hydrophobic-hydrophilic balance of the reagent molecule. A numerical measure of this overall character may be derived from knowledge of the structural formula by use of Hansch π values. These values are coefficients describing the hydrophobic (positive values) or hydrophilic (negative values) nature of structural fragments, e. g. methyl has a value of $+0.6$ and aromatic hydroxyl a value of -0.7. The Hansch π values were derived from measurements of the partition coefficients of large numbers of compounds between water and octanol (HANSCH et al., 1973). A critical review of the validity and usage of

Hansch π values, and of the alternative fragmental (f) values, has been given by ALBERT (1979, p. 70). *Rates of reagent diffusion* will be strongly influenced by the sizes of the reagents, often as indicated by molecular or ionic weights or, for very large and approximately spherical reagents, by particle radii. Rates may also be influenced by structural features relevant to affinity, since a high reagent-tissue affinity will tend to reduce the rate of reagent diffusion.

It will be useful at this point to show *the calculation of certain structural parameters.* Although a wide range of physico-chemical effects can be important in biological staining it is coulombic interactions, van der Waals attractions, hydrophobic bonding, and the rates of diffusion and reaction which are of repeated significance. Consequently a short-list of corresponding structural features may be constructed, namely overall electric charge, conjugated bond number, overall Hansch π values, ionic or molecular or ionic weight, and Hammett σ constant. How is the practical histochemist to go about calculating these numerical parameters? Here it will be assumed that the only sources of information concerning dye structure likely to be available are such books as *Synthetic Dyes in Biology, Medicine and Chemistry* (GURR, 1971), *Handbuch der Farbstoffe für die Mikroskopie* (HARMS, 1957), and *Conn's Biological Stains* (LILLIE, 1977). Inspection of the structural formulae contained in these texts will (after certain errors of draughtsmanship are allowed for) directly provide values of electrical charge and conjugated bond number. Inspection of formulae plus tables of atomic weights (widely available) and Hansch π values (less widely found), followed by a little arithmetic, provides molecular or ionic weights and overall Hansch π values. To minimise library work a selection of commonly required Hansch π values is provided in Table 54.

Sample calculations are shown below, using the two acid dyes Orange G and Light Green SF whose structural formulae are shown in Figs. 16 a and b respectively, as examples. Inspection of Fig. 16 immediately provides the information that under normal histochemical conditions both dyes carry an electric charge of −2. The conjugated bond numbers may also be obtained by inspection. For Orange G we may

Table 54. A collection of Hansch π values, useful for estimating the hydrophobic-hydrophilic balance of histochemical reagents and stains

Molecular Fragment	π value	Molecular Fragment	π value	Molecular Fragment	π value
CH_3	+0.6	$N(CH_3)_2$	+0.2	aromatic OH	−0.7
C_2H_5	+1.0	NHC_2H_5	+0.1	aliphatic OH	−1.2
C_6H_{11}	+2.5	$NHCH_3$	−0.5		
C_6H_5	+2.0	NH_2	−1.2	O^-	−3.9
C = C	+0.8	$= NH_2^+$	−7.5	OCH_3	0.0
$N = N\text{-}C_6H_5$	+1.7	$\text{-}N(CH_3)_2^+$	−6.0	O-β-glucose	−2.8
$C_{10}H_7$	+3.4	$NH\text{-}C_6H_5$	+1.4	-O-	−0.6
SO_3^-	−4.8	-S-	+0.1	C = O	−0.7
COO^-	−4.4	SH	+0.4	Ferrocenyl	+2.5
COOH	−3.2	NO_2	−0.3	benzothiazole	+2.1
$COOCH_3$	0.0	NO	−0.1	pyridyl	+0.3
CONH	−1.5	-N =	−0.5	$OSO_2C_6H_5$	+0.9
Cl	+0.7	N = N	−0.3		
Br	+0.9				
I	+1.1				

sum the conjugated bonds of the naphthyl and phenyl rings (11 + 6), the azo bridge (3), and the conjugated aromatic hydroxyl (1) to obtain an overall conjugated bond number of 21. For Light Green SF the same process indicates a higher value, of 34. The empirical formulae of Orange G and Fast Green SF can be seen to be $C_{16}H_{10}N_2O_7S_2Na_2$ and $C_{37}H_{34}N_2O_9S_3Na_2$ respectively. So with the aid of a table of atomic weights, the anionic weights may be calculated as 406 and 747 respectively. Hansch π values are calculated as follows. For instance Light Green SF has various hydrophobic elements, namely five phenyl rings, an unsaturated bridging carbon, two ethyl and two methylene groups. Table 54 enables us to assign Hansch π values to them as follows: (5X+2.0)+(1X+0.4)+(2X+1.0)+(2X+0.6). The hydrophilic groups of this dye comprise three sulphonates and two aromatic amines, one of which is ionised. Assigned Hansch π values are: (3X−4.8)+(1X−7.5)+(1X−1.2). The overall Hansch π value for this hydrophilic dye is thus −9.2. A similar calculation for Orange G indicates that it is somewhat less hydrophilic, with an overall Hansch π value of −5.2.

The ways of *describing staining results* when studying structure-staining correlations require comment. Quantitative stain-tissue affinity data would be preferred, as this would allow to some extent for variations in reagent concentration. However such an approach would be extremely restrictive, since there is very little data of this type available. We will usually wish to study practical procedures for which the available information concerning staining takes the form of statements such as «selective staining occurred/staining did not occur/overstaining took place». Sometimes information may be given as subjective assessments of the −/+/+++ variety, or of objective measurements of staining intensity with however no allowance made for effects of reagent concentration.

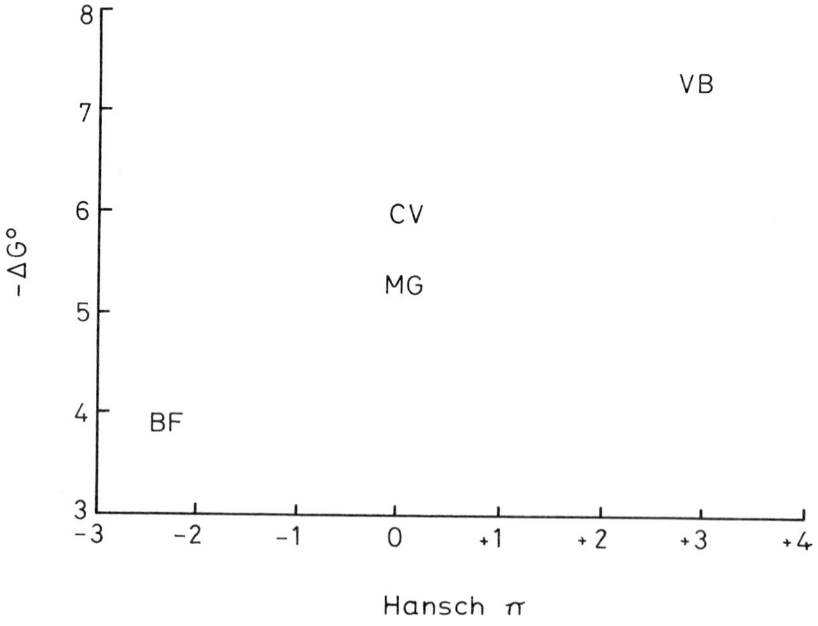

Fig. 66. A plot of dye-mucin affinity *versus* Hansch π value, for the staining of goblet cell mucin by monocationic arylmethane basic dyes. B, C, M and V correspond to Basic Fuchsin, Crystal Violet, Malachite Green, and Victoria Blue 4R respectively. Affinity data from MARSHALL and HOROBIN (1973)

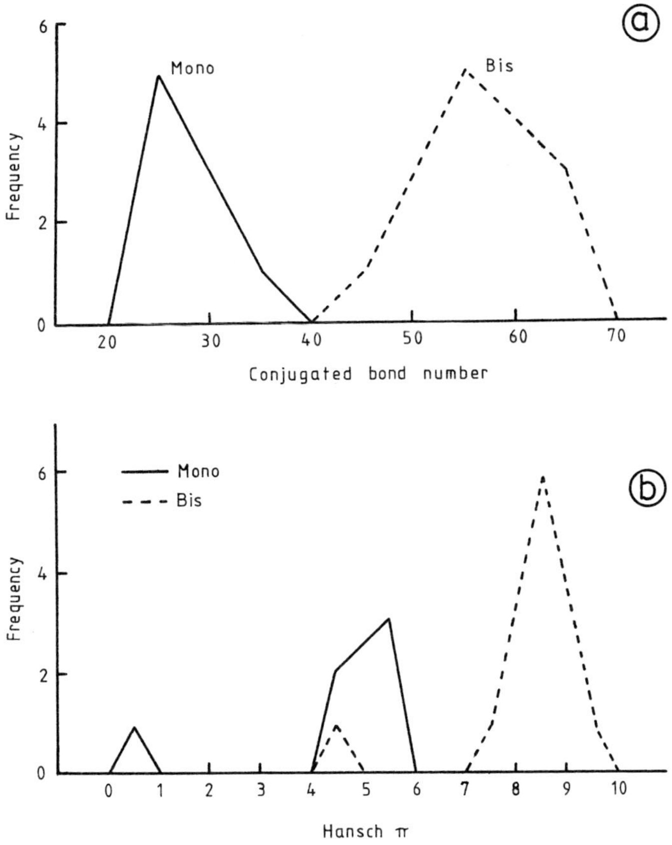

Fig. 67. Frequency diagrams illustrating the differences in conjugated bond numbers (diagram a) and Hansch π values (diagram b) between mono- and bis-formazans. The formazans derive from the tetrazolium salts listed in Conn's Biological Stains (LILLIE, 1977)

In spite of the limitations of such reports they are the raw data of practical histochemistry and must be dealt with.

We can now discuss *ways in which structure-staining data may be displayed* to best disclose correlations.

The simplest modes of display consider a single structural feature at a time. For instance we may plot stain-tissue affinity against a numerical parameter describing some structural feature. An illustration is provided in Fig. 66, which shows a plot of dye-mucin affinity against Hansch π value for several monocationic basic dyes of fundamentally similar structure. A positive correlation is seen between high affinity and hydrophobic character.

Another mode of data display is to use frequency diagrams. We may plot the frequency of occurrence of staining reagents within specified intervals of the specified parameter describing some structural feature. It is often useful to plot different reagent populations on the same diagram. Differences may be staining versus non-staining, or may be further structural characteristics. An example of the latter is in Fig. 67. Here the hydrophobic-hydrophilic characters of mono- and bis-formazans, as indicated by their

Hansch π values, are displayed in (a), and the conjugated bond numbers of the same compounds in (b). In this sample of formazans (which derive from those tetrazolium salts listed in *Conn's Biological Stains* (LILLIE, 1977) as being, or having been, used histochemically) the bis-formazans are of noticably higher conjugated bond number and also have more positive Hansch π values than the mono-formazans.

Of other ways of displaying data concerned with single structural parameters, Fig. 68 illustrates the influence of relative dye size in certain staining procedures. Figure 68a shows that when pairs of acid dyes are applied to tissue sections it is predominantly the larger dye of a pair which stains collagen fibres, with the smaller dye colouring cytoplasms. The sample of dye-pairs in Fig. 68a is taken from a standard textbook of histopathology (LILLIE and FULMER, 1976). Figure 68b shows that when pairs of basic dyes of different sizes are applied to tissue sections it is the larger of each pair which predominates in the chromatin, with the basophilic cytoplasms and nucleoli being coloured by the smaller dyes. The sample of dye pairs is taken from GOLDSTEIN (1961). It is the clarity and impact of Fig. 68 which is so valuable. Compare Fig. 68a to the prose arguments concerning the significance of relative dye size in trichromes (e. g. BAKER, 1958; LILLIE, 1977). Or compare Fig. 68b to the same data in tabular form, e. g. Table 31 of *this* book.

As an alternative to considering one structural feature at a time, reagents may be plotted on graphs whose axes are parameters describing two different structural features. It will now be different areas of the plot which are significant. Figure 69 provides an example of this, showing various basic dyes displayed on a plot of conjugated bond number *versus* cationic weight. This particular diagram resulted from staining for 10 minutes with 0.5% w/v solutions of basic dyes made up in 3% v/v aqueous acetic acid solution, the tissue being fixed in Carnoy's fluid (HOROBIN and FLEMMING, unpublished results). It can be seen that non-staining, non-selective staining, and selective staining dyes for goblet cell mucins fall into different areas of the plot. For the conditions specified and on the basis of the dye sample adopted, it seems that selective staining of goblet cell mucin often occurs if basic dyes with conjugated bond numbers of over 35, and cationic weights of over 600 are used.

Possible *applications of structure-staining correlations* are varied. For instance correlations can aid choice of new staining reagents, and avoid choosing reagents likely to give rise to high background staining (i. e. «positive artifacts»). Correlations may also be used to increase our understanding of staining mechanisms.

Several examples of the development of *criteria for choosing new staining reagents* are available. For instance HOROBIN and FLEMMING (1980) found that with the Taenzer-Unna elastic fibre staining system (i. e. basic dyes in acid alcoholic solutions) dyes with conjugated bond numbers of 25 or over were required for the fibres to be coloured. With staining systems using acid dyes from alkaline aqueous-formamide or -dimethyl-sulphoxide solution, dyes with conjugated bond numbers of 35 or over were needed. In another paper HOROBIN (1981) reported that for really effective lipid staining using non-ionic dyes from saturated solutions in alcohol or aqueous-alcohol, dyes with Hansch π values of $+7$ or over are used.

The use of the correlations for *avoidance or clarification of artifactual staining* is another potentially valuable field of application. The paper concerned with lipid staining discussed above did also consider artifactual staining. Thus many of the monotetrazolium salts and naphthyl substrates used in enzyme histochemistry give rise to monoformazans and to naphthols with Hansch π values of $+1$ or over (HOROBIN,

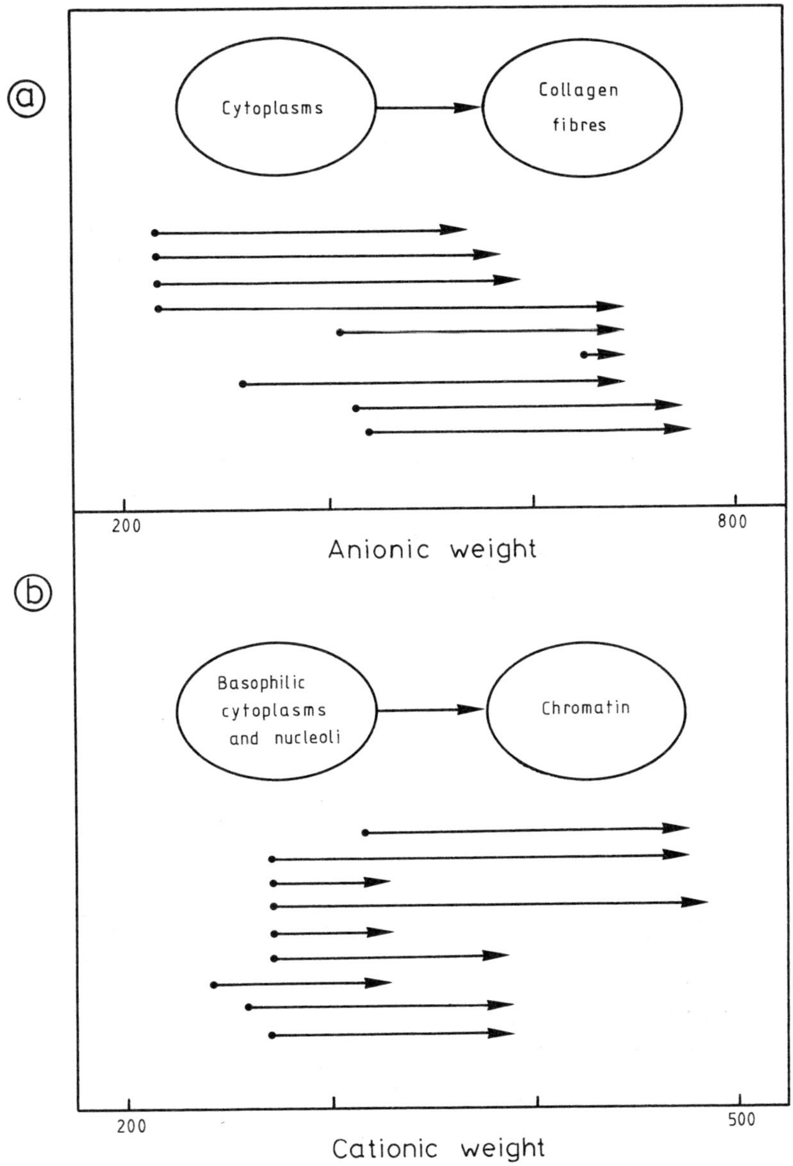

Fig. 68. Diagrams illustrating the influence of relative dye size on certain staining results. Diagram (a) demonstrates that when tissues are stained with pairs of acid dyes it is consistantly the larger dye which colours the collagen fibres. Dye pairs are those noted in Lillie and Fulmer (1976). Diagram (b) shows that with pairs of basic dyes it is constantly the smaller dyes which colour basophilic cytoplasms and nucleoli. Dye pairs are those studied by Goldstein (1961)

1981, Fig. 4). Consequently these compounds are expected to be lipophilic, which is in keeping with many workers observations. In a recent paper Altman and his colleagues provide a dramatic visual demonstration of the importance of hydrophilic character in this context (Altman et al., 1979). The preferred formazans, which give no artifactual

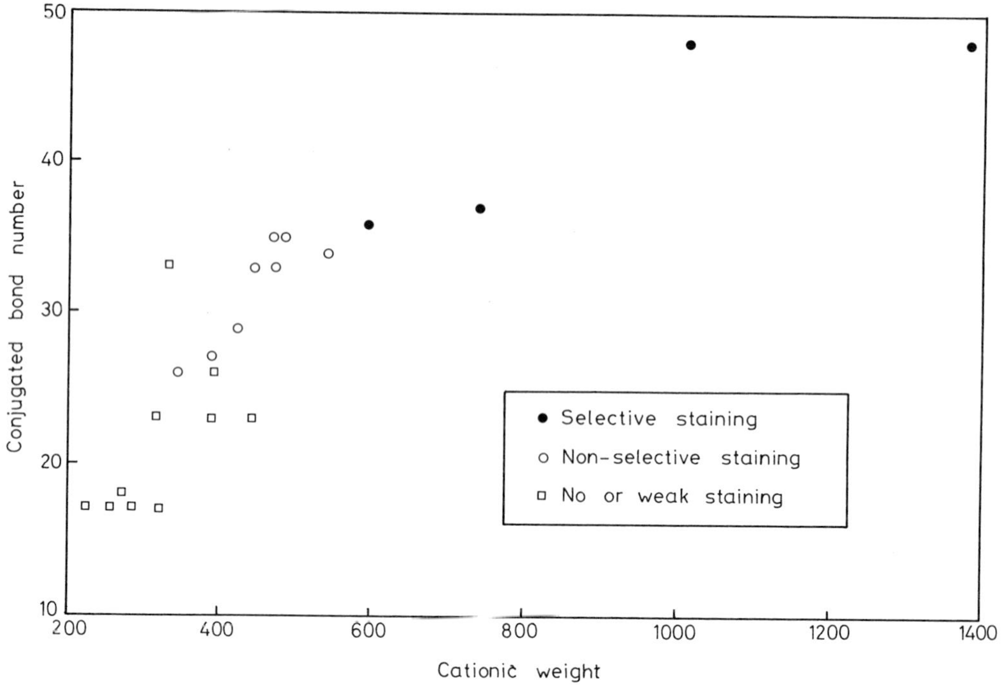

Fig. 69. Basic dyes plotted on a conjugated bond number-cationic weight diagram. The non-nucin staining, non-selective, and selective nucin staining dye populations are seen to fall in different areas of the diagram

lipid staining, were metal complexes of BSPT. These hydrophilic compounds will have Hansch π values in the region of -6.1, assuming that they are 2:1 formazan-metal complexes. Formazans giving pronounced lipid staining have Hansch π value of $+1.3$ or more. Uncomplexed BSPT, which gives less artifactual lipid staining than the other formazans, has a Hansch π value of $+0.7$. Both positive and negative artifactual staining also occurs when using elastic fibre stains, as noted in the study already mentioned (Horobin and Flemming, 1980). When using acid dyes under alkaline conditions it was noted that non-specific staining of proteins additional to elastin required the use of dyes with anionic charges of -3 or more. When using the Taenzer-Unna basic dye in acid alcohol system, staining was limited to the surfaces of the elastic fibres if the dyes used were too large, e. g. with cationic weights of 900 or more. A further structural parameter sometimes correlating with non-specific or background staining is conjugated bond number. For instance in studies of stains for glycogen (Horobin and Murgatroyd, 1971) and calcium (Speirs, 1970) the intensity of unwanted colouration of the background rose with conjugated bond number of the dyes concerned. Such effects were either not noticed by these authors, or if noticed not expressed quantitatively or graphically.

Structure-staining correlations can also help us in *understanding staining mechanisms,* and some examples follow illustrating this. A study of elastic fibre staining (Horobin and Flemming, 1980) showed that staining correlated well with conjugated bond number; with Hansch π values having only a minor significance (their Fig. 2). This suggests that elastic fibre-dye affinity is largely due to van der Waals attractions, with some contribution from hydrophobic bonding. On the other hand, involvement of dye charge (their Fig. 3) demonstrates that for progressive staining to be highly selective, coulombically based acid dyeing effects must be minimal. Another example can be seen by taking the information concerning selective goblet cell mucin staining contained in Fig. 69 above. There we can see that for mucin staining to occur at all a dye must have a conjugated bond number above a certain minimum value, suggesting that mucin-dye affinity involves a significant contribution from van der Waals attractions. Note however than Hansch π values also correlate, see Fig. 66, and so hydrophobic bonding also probably makes some contribution. However for mucin staining to be selective further requires that the dyes are above a minimum size, supporting the view (c. f. GOLDSTEIN, 1962b) that selectivity of mucin staining involves rate-control. As a final example consider the formazans generated as final reaction products in certain enzyme histochemical procedures. Fig. 67 shows that the bis-formazans which have been used in histochemistry have higher conjugated bond numbers and more positive Hansch π values than the mono-formazans. This suggests that the bis-compounds will both be better at binding by van der Waals attraction to, say, tissue proteins and also will be more likely to dissolve in tissue lipids than the mono-compounds. This does support the practice of referring to bis-formazans as «more substantive» but contradicts another commonly stated view, that the bis-formazans are «less lipophilic». The marked reduction of artifactual staining of lipid following use of bis-tetrazolium salts in thus apparently due to the high formazan-protein affinity (substantivity) rather than low lipid affinity. Another aspect of the use of formazans as final reaction products is the conversion of certain formazans into formazan-metal complexes as *final* final reaction products. Such a manouver succeeds in both increasing the conjugated bond number and also making the Hansch π values more negative, as shown in Table 55. Hence these metal complexes are likely to bind to protein by van der Waals attractions more strongly than the uncomplexed compounds, but the metal complexes are also likely to be less lipid soluble.

Table 55. A comparison between certain monoformazans and their metal co-ordination complexes

Formazan[1]	Hansch π value	conjugated bond number[2]
BSPT	+0.7	34
BSPT-metal complex	−6.1	68
MTT	+4.4	22
MTT-metal complex	+1.3	44

[1] Full names: BSPT = 2(2-benzothiazolyl)5-styryl-3(4-phthalhydrazidyl) tetrazolium; MTT = methylthiazolyldiphenyl tetrazolium

[2] For the metal complexes these values assume 2:1 formazan-metal compounds

Finally we should consider some *limitations and complications* of the structure-staining approach. One possibly commonplace possibility is the confusion of rate and affinity control, since many everyday procedures employ staining times far short of equilibrium times. It will be simplest to model those situations where either rates or affinities dominate the staining pattern, but a mixed situation can be studied. For instance a plot of size *versus* some affinity-related numerical parameter may provide a clear cut correlation. An example is found above in Fig. 68, where cationic weight is plotted against conjugated bond number. Alternatively an affinity model may be utilised and any anomalously weakly staining reagents scrutinised. If they are of large size or liable to aggregate, rate effects may well be occurring. An example is found in the use of basic dyes for staining elastic fibres, where very large dyes (i. e. above 900 daltons cationic weight) gave only superficial staining (HOROBIN and FLEMMING, 1980). Another complication occurs when the values of one structural parameter adventitiously parallel another. Thus if in a particular sample of stains all the hydrophilic groups were ionic, then Hansch π values would be inversely proportional to overall electric charge. In practise this means that convincing and clear graphical displays can only be expected if the population of reagents under consideration cover a large area of the plot.

11.15 A summary of some practical strategies

Throughout Chapter 11 a variety of ways of testing the significance of staining and of non-staining have been discussed. Too many trees may have obscured the wood, so as a finale Fig. 70 summarises practical strategies used as significance-tests in the biological staining field.

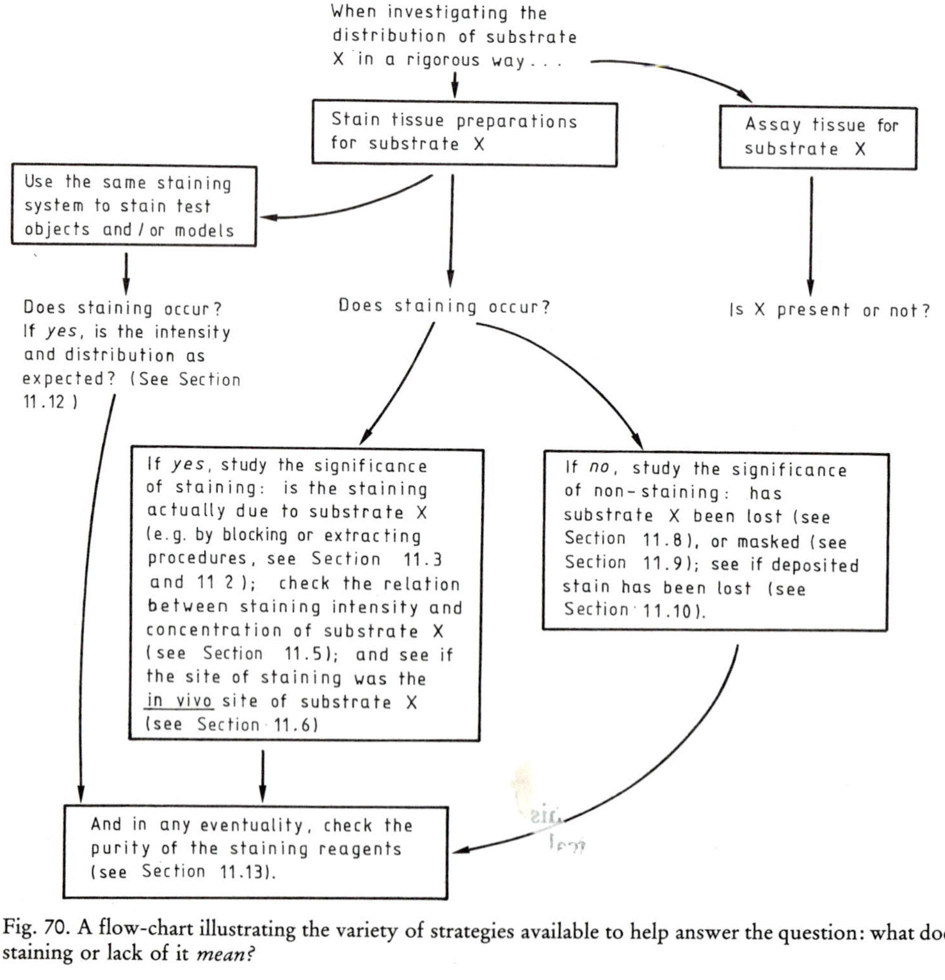

Fig. 70. A flow-chart illustrating the variety of strategies available to help answer the question: what does the staining or lack of it *mean?*

12 Supplements

12.1 Nomenclature of traditional colour theory: chromagen, chromophore and auxochrome

These terms originated in early colour theories which described dyestuffs as comprising *chromophoric* groups attached to unsaturated molecules giving rise to the colour, plus *auxochromic* groups which intensified the colour produced by the chromophore and also gave the coloured compounds affinity for substrates. A compound containing a chromophore, not necessarily a dye since an auxochrome could be absent, was termed a chromogen.

Examples of unsaturated groupings considered to be *chromophores* included:

$>C=O$	carbonyl group, as in anthraquinone dyes
$-N=N-$	azo group
$-NO_2$	nitro group
	quinoid group

Examples of groupings considered to be *auxochromes* included:

$-NH_2$ $-NHCH_3$ $-N(C_2H_5)_2$	often occuring as cations such as $-N(C_2H_5)_2H^+$
$-COOH$ $-OH$ $-SO_3H$	often occuring as anions such as $-SO_3^-$

Though the theoretical backing for this nomenclature is obsolete, the terms live on in many texts concerned with biological staining. The term chromophore still has a practical value, as the structural elements used when classifying dyes according to chemical constitution are often the chromophores of this traditional colour theory.

Bibliographic note: An interesting historical sketch of the development of colour theories can be found in vol. 1 of VENKATARAMAN's *Chemistry of Synthetic Dyes* (1952).

12.2 π bonds and σ bonds: a glance at the valency of carbon

The traditional structural formulae for organic compounds show carbon-carbon single bonds and double bonds, aromatic rings, and so on. Such bonding may be described in molecular orbital (MO) terms. To achieve this we must first consider the electron distribution around a carbon atom. This can be visualised in terms of those regions (atomic orbitals or AOs) having a high probability of containing an electron.

The orbital arrangement around carbon can take different forms. For instance the sp^3 hybridised situation, involving four equivalent AOs whose major lobes are directed tetrahedrally away from the carbon nucleus (Figure 71 – I). Overlap of two such orbitals on two carbon atoms gives rise to a bonding MO (Figure 71 – II), corresponding to the traditional carbon-carbon single bond, (Figure 71 – III). In another electronic arrangement, sp^2 hybridisation, the AOs take the form illustrated below (Figure 70 – IV). Three major orbital lobes lie in a plane 120° apart with the lobes of an unhybridised p_z orbital situated above and below this plane. If two such atoms interact, two bonding MOs arise, (Figure 71 – V), corresponding to the traditional carbon-carbon double bond (Figure 71 – VI). One MO arises from overlap of sp^2 orbitals and is centred along the C-C axis (a sigma [σ] bond). The other MO arises from the overlap of the p_z AOs and comprises two lobes, one above and one below the bond skeleton (a pi [π] bond). Note that the traditional formulae representing carbon-carbon double bonds (Figure 71 – VI) misleadingly implies that the two bonds are equivalent.

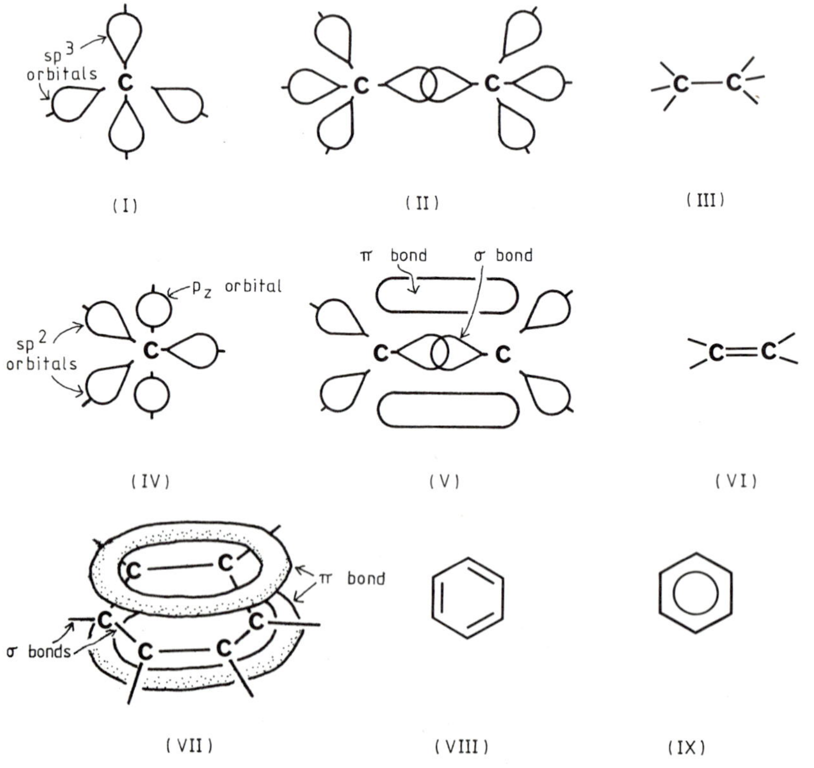

Fig. 71. Various representations of the valency of carbon. For details see text

Finally we can consider the MO representation of interacting, that is conjugated, carbon-carbon double bonds. If several suitably oriented sp^2 hybridised carbon atoms are allowed to interact, the formation of a sigma bonded skeleton occurs due to overlap of sp^2 orbitals, as above. The pi bonds formed by p_z orbital overlap however can have MOs distributed over the whole interacting system and not localised between particular carbon-carbon atom pairs. This is illustrated for the aromatic benzene ring (Figure 71 – VII), showing the lowest energy pi orbital. Since the electron pair in such a MO is distributed over the whole ring the electrons are said to be delocalised. For this reason the traditional structural representation of a benzene ring (Figure 71 – VIII) is sometimes replaced by (Figure 71 – IX).

Electron delocalisation does not stop at a six carbon ring. Much more extensive pi orbitals are found in dyestuffs, with the electron pair distributed over a dozen or more atoms.

Bibliographic note: Accounts, at all levels, of the bonding of organic molecules may be found in many modern chemistry texts (e. g. MURRELL et al., 1978) though some accounts will not make use of the MO formulations.

12.3 Relation of affinity to other thermodynamic quantities

So far affinity has been related to the thermodynamic quantity $-\Delta G^\circ$. This increases the precision of our usage but for the concept to be of further value $-\Delta G^\circ$ must be related to other thermodynamic quantities. In fact Gibbs free energy (G) may be *defined* as:

$$G = H - TS$$

(where H is the enthalpy, or heat content, T the temperature, and S the entropy). Thus at constant temperature, pressure and defined standard state one can write:

$$-\Delta G^\circ = -\Delta H^\circ + T\Delta S^\circ$$

So the influence of enthalpy and entropy on affinity can be set out as follows:

ΔH° &	ΔS°	
−	+	Gives $-\Delta G^\circ$, so staining can occur spontaneously.
+	−	Gives $+\Delta G^\circ$, so staining will never occur spontaneously.
− +	− +	The sign of ΔG°, and thus the possibility of spontaneous staining, will depend on precise conditions.

Considered the latter two situations in the table, we can say that when the temperature is low (minimising the $T\Delta S^\circ$ term) then exothermic reactions (i. e. $-\Delta H^\circ$) could occur spontaneously. At higher temperatures the $T\Delta S^\circ$ term must dominate.

Thermodynamics does not require the molecular meaning of H° and S° to be known. One could say though that processes in which bonds are formed are often exothermic (i. e. $-\Delta H^\circ$), and processes in which bonds have to be broken by energy from the surrounding are often endothermic (i. e. $+\Delta H^\circ$). Processes in which the reactants achieve a more random, less ordered state are often associated with positive entropy changes (i. e. $+\Delta S^\circ$).

12.4 The nature and characteristics of van der Waals attractions

The phrase van der Waals forces describes various polar attractions which, though weak and of extremely short range, occur between all atoms and molecules.

Fig. 72. Examples of dipole-dipole (I) and dipole-induced dipole (II) attractions

These polar forces include dipole-dipole interactions (Figure 72 – I), and dipole-induced dipole attractions (Figure 72 – II) where the presence of a permanent dipole moment induces a dipole in a neighbouring polarizable, though more symmetric, species. The third of the polar forces is the London or dispersion force. To understand this consider that any neutral atom or molecule can be described as a positive centre surrounded by a cloud of negative charge. This positive centre, displaced from its equilibrium position, will oscillate. According to the quantum theory an oscillator, even when in its lowest possible energy state, possesses energy and a fluctuating dipole moment. This dipole, finite at most instants though zero if time averaged, will induce similar dipoles in adjacent atoms or molecules and so give rise to attractive forces. Though these dispersion forces occur between all species their magnitude is low; the other dipolar forces will be of greater magnitude if the molecule contains a permanent dipole moment.

What are the important properties of van der Waals attractions? The interaction energies for the various forms of dipolar forces vary with r^{-6} (where r is the distance between the interacting molecules) and also with some function of the molecules' permanent dipole moments or with polarisabilities for symmetrical molecules. Thus for van der Waals interactions to occur molecules must be extremely close together; unsymmetrical, so as to possess a permanent dipole of some magnitude; and readily polarizable.

Common polarizable systems include pi-electron systems and high atomic number atoms such as bromine and iodine. Examples of groupings with large dipoles are amines, nitro groups and sulphonic acids. The use of conjugated bond number counts of staining reagents to estimate the significance of van der Waals attractions is discussed in Section 11.14.

Extensive and vigorous acounts of these intermolecular forces are available, for example the monograph of KIHARA (1978).

12.5 The hydrogen bond

An atom of hydrogen can form a normal covalent bond with only one other atom. Sometimes however a hydrogen atom can form an additional weak bond (with a dissociation energy of only some 20 KJ $mole^{-1}$ compared to some hundreds of KJ $mole^{-1}$ for a covalent bond), termed a *hydrogen bond.*

Some examples of hydrogen bonds (shown dotted ···) are given below. Note that both inter- and intra-molecular hydrogen bonds are possible.

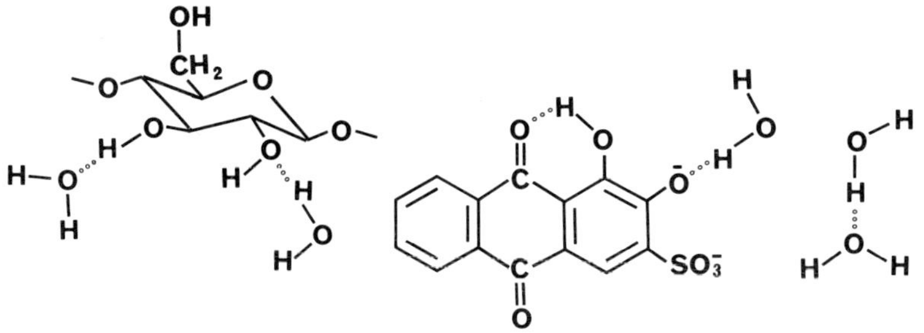

These bonds form, in general, between hydrogen and atoms of electro-negative elements such as oxygen and nitrogen.

If the bonds are not of the typically covalent type, how may their origins be pictured? Qualitatively by realising that as the proton is extremely small its electrostatic field is extremely intense. So a bonding can occur due to the attraction of the positive proton for the electrons of the bonded atom. This is represented diagrammatically below, by extreme electrostatic and covalent forms.

What factors favour the formation of hydrogen bonds between two species? Linearity of the two electronegative atoms and the hydrogen atom, and an absence of a third efficiently hydrogen bonding species such as that common solvent water.

12.6 The coordinative bond

The coordinative bond forms between metal cations and electron donating groups or ligands. The ligands can be neutral species such as $-NH_2$, $-OH$, C=O and N=N, or anions such as $-O^-$, $-COO^-$ or $-OPO_3^{3-}$.

The coordinative bond is essentially a localised, directed covalent bond, though with appreciable ionic character. Qualitatively the bond can be considered as arising from the interaction of positively charged metal ions with the dipoles or ions of the ligands.

The arrangement of the ligands around the metal ion varies with the nature of both the metal and the ligand. A common arrangement is octahedral (Figure 73 – I), found for instance with Cr (III) and Fe (III).

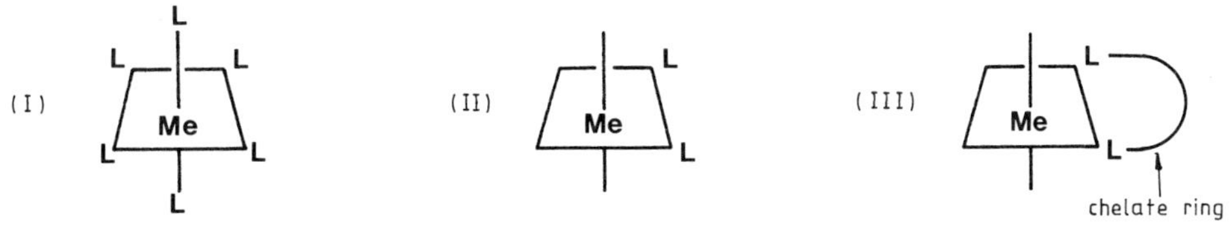

Fig. 73. Stereochemical features of certain metal co-ordination complexes. Details in the text. L: ligand, Me: metal cation

The electron-donor groups may all be on different molecules (unidentate ligands) or on the same molecule. For example the structure of Gallocyanin-chrome alum may be represented as in Figure 20. Here the Cr (III) ion is coordinated octahedrally by two unidentate water molecules (not shown) plus two bidentate Gallocyanin molecules.

The stability of the coordinate bond varies very widely. Characteristics of metal ions favouring stable coordination complexes include high charge and small ionic radius: the higher the electrostatic field intensity the greater the interaction with the electron donor. Ligand characteristics favouring complex formation include high base strength and the ability to form chelate rings. Thus the complex involving only unidentate ligands illustrated in (Figure 73 – II) will be less stable than the complex (Figure 73 – III) in which two similar coordinating groups are present on one ligand molecule. The size of such a chelate ring is another factor. Five or six membered rings are the most stable (see Figures 17 and 18). The overall size of the ligands is also important. Because of steric repulsions, bulky ligands form less stable complexes than do otherwise similar, smaller ligands.

The stability of coordination complexes also depends on the particular combination of metal and ligand. The more electropositive metals such as aluminium and iron form their most stable complexes with ligands in which the donor atoms are such as nitrogen or oxygen. Metals such as mercury or silver form their more stable complexes if the donor atoms are such as sulphur.

Finally back to kinetics. The reactivity of a complex is of no consequence unless the rate of reaction is appreciable. Thus dissolved in water Cr $(H_2O)_6^{3+}$ complexes exchange water for other ligands extremely slowly, whilst Fe $(H_2O)_6^{3+}$ complexes react very rapidly.

12.7 Hydrophobic bonding

When dissolved in water, molecules containing large hydrocarbon groupings, such as detergents, most dyes, proteins and nucleic acids, tend to aggregate, or to coil if flexible, so as to bring the hydrocarbon groups into contact. This pushing together of hydrocarbon groups is termed hydrophobic bonding and is illustrated below (Figure 74) for organic cations.

To understand hydrophobic bonding, an understanding of the structure of liquid water is needed. At room temperature water is comprised of islands of highly structured hydrogen-bonded water molecules surrounded by a sea of single water molecules. As the ordered water's structure is relatively ice-like, and as any given volume of water is continually switching between the ordered and disordered forms, this description of liquid water has been called the flickering iceberg model.

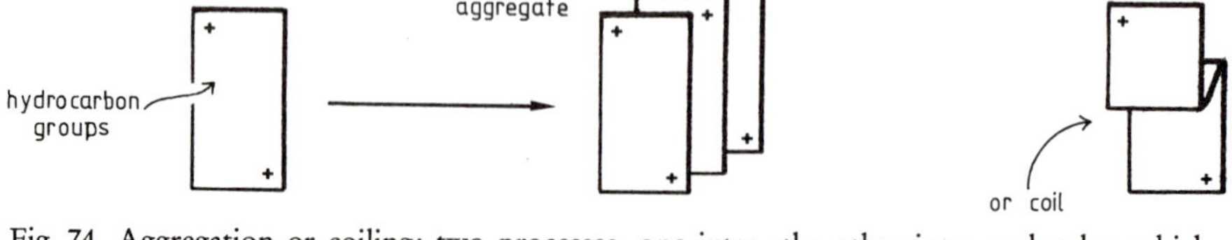

Fig. 74. Aggregation or coiling: two processes, one inter- the other intra-molecular, which can maximise contact between hydrophobic groupings of organic molecules dissolved in water

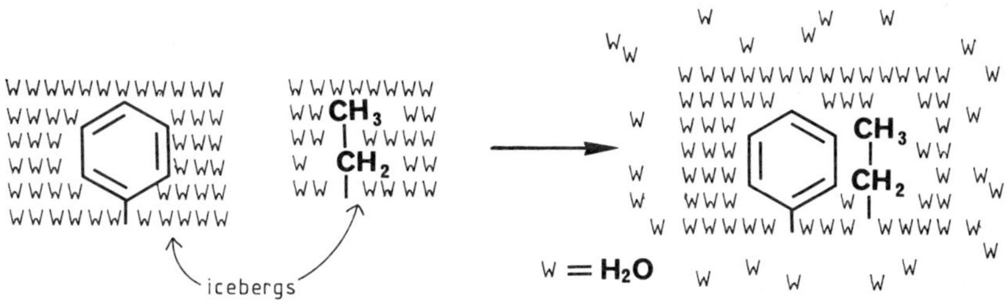

Fig. 75. An illustration of the molecular events underlying hydrophobic bonding. It can be seen that as the hydrophobic groups come together the solvent molecules (W) become less ordered

The relevance of this to hydrophobic bonding is that the ice-like water is stabilised, relative to the disordered form, by the presence of hydrocarbon groupings. Consequently in an aqueous dyebath both the hydrocarbon groupings of dye molecules and the hydrophobic residues on the biopolymers of the tissue section will be surrounded by ice-like water. However when dye and tissue are in contact much of the icebergs will of necessity be excluded and so will revert to disordered single water molecules as shown diagrammatically in Figure 75.
So the movement of dye onto the tissue is accompanied by a decrease in order of the solvent and an increase in entropy: consequently the overall process will take place spontaneously.

When will hydrophobic bonding occur? Most readily if the solute molecules have large hydrocarbon groupings uninterrupted by hydrophilic substituents, e. g. with staining reagents with positive Hansch π values, see Section 11.14. The hydrophobic effect also entails loss of structure in water. If the structured form of water is destroyed however then such a change cannot occur, and neither can hydrophobic bonding. For instance destruction of the icebergs by heat or by molecules, such as alcohol or urea, capable of competitively forming hydrogen bonds with water will prevent hydrophobic effects.

12.8 Quantitative estimation of affinity: use of Freundlich and Langmuir isotherms

When the dye can be regarded as being dissolved in or diffusely adsorbed onto the substrate then the distribution of dye can be described by the ratio

$$a_t/a_b = K$$

where the equilibrium constant K is the usual distribution coefficient for the partitioning of a soluble between two immiscible solvents. If the dye is assumed to form an ideal solution in both phases the activities may be replaced by the concentrations D, where subscripts t and b stand for tissue and dyebath respectively. So the standard affinity $(-\Delta G^\circ)$ is

$$-\Delta G^\circ = RT \ln D_t/D_b$$

This relationship holds even if the dye does not form ideal solutions, as long as the deviation from ideality is similar in both phases.

In practice such a simple relation is not observed. Consider for example deviations due to aggregation of dye. Suppose the dye forms aggregates in both solution and substrate, with the mean aggregates containing respectively v and w dye molecules. The dyeing process may then be written:

$$\begin{array}{ccc} \textit{dyebath} & & \textit{substrate} \\ w\,(\text{dye})_v & \rightleftharpoons & v\,(\text{dye})_w \end{array}$$

so that one can then write an equilibrium constant in the usual way

$$K = D_t^v / D_b^w$$

This is mathematically equivalent to the Freundlich adsorption isotherm

$$D_t = k\,D_b^x$$

where k and x correspond to $K^{1/v}$ and w/v respectively in the previous equation. Some dyeing systems are fitted by this model, with various values of x and k.

With acid and basic dyes the requirement for electrical neutrality has encouraged workers to regard adsorption as being onto a finite number of sites in the substrates, namely ionic groups such as ammonium and phosphate. Consequently the *Langmuir adsorption isotherm* has been applied, which assumes all binding sites to be similar and non-interacting and also assumes that activities may be replaced by concentrations. Hence we can write the following:

$$D_t = KSD_b/(1 + KD_b)$$

Here S is the saturation value of the dye in the substrate, that is the value of D when all available sites are occupied, with other symbols as before. Note that in this case, unlike that of the Freundlich isotherm, a plateau of dye uptake is reached at high D_b values.

The expression above may be rearranged, so:

$$1/D_t = 1/S + 1/KSD_b$$

so that a plot of $1/D_t$ against $1/D_b$ should be linear, allowing evaluation of K, and so a determination of the affinity.

12.9 Methods of determining half staining times and activation energies for the staining of tissue sections

Half staining times ($t_{1/2}$). As long as the uptake of dye follows either a Freundlich or a Langmuir isotherm (see Supplement 8) the amount of dye adsorbed at equilibrium from a fairly dilute dyebath by a moderately stained substrate is proportional to the concentration of the dye in the bath. So the half staining time can be assessed by staining sections for different times, including equilibrium, in a given dyebath and then either; (i) matching them with the aid of a comparison eyepiece with a section stained to equilibrium in a dyebath of half the given dye concentration; or (ii) measuring the absorbances directly using a microspectrophotometric system.

The time taken for a given staining system to achieve equilibrium should be found by staining a section in the dyebath together with a further section previously overstained in a more concentrated bath. The first section takes up dye from the bath whilst the second section loses dye into it. At equilibrium the final staining intensities of both are similar. Examples of equilibrium times when staining 6 μm sections at 4 C are 24 hours for elastic fibres using a 0.01 w/v solution of Orcein in alcohol, and 30 min for mucin using a 0.008% w/v aqueous Azur A solution.

To compensate for increases or decreases in stainability due to the action of the solvent (both occur), sections are first treated for varying times with dye-free solvent. The time in the pure solvent plus the time in the staining bath is arranged to be constant for all sections, and equal to the time of staining of the equilibrium controls.

Activation energies. If half staining times are measured at two temperatures, say T_1 and T_2, then the activation energy of staining, E, may be calculated from the expression:

$$E = R\,T_1\,T_2\,t_{1/2\,(1)}/(T_2 - T_1)\,t_{1/2\,(2)}$$

where $t_{1/2\,(1)}$ and $t_{1/2\,(2)}$ are the half staining times at T_1 and T_2 and R is the Gas Constant.

This expression is based on the fact that if k_1 and k_2 are the rate constants of a process at temperatures T_1 and T_2 then,

$$\ln k_1/k_2 = E\,(T_2 - T_1)/R\,T_1\,T_2$$

and for both first and second order reactions k is inversely proportional to $t_{1/2}$.

For a full account of these topics see the papers of GOLDSTEIN: leading references are FRIEDBERG and GOLDSTEIN, 1969, and GOLDSTEIN, 1963.

12.10 Aggregation of dyes in solution: some physical chemistry

How do we know that dyes do aggregate? Most directly by ultra-filtration studies, either through membranes or using gel-filtration media. Other investigations which have supported the idea of aggregation usually involved optical methods (metachromatic effects, both in absorption and emission spectra; also light scattering), osmotic phenomena, conductivity measurements and diffusion effects (directly, or using polarography).

Though such investigations have produced some conflicting results, the following points seem generally agreed. Aggregation occurs with dyes of a wide variety of structural types. At room temperature in aqueous solutions of low ionic strength with dye concentrations of 10^{-4}–10^{-10} M or less, aggregation is commonly restricted to dimerisation. As dye concentration or ionic strength rises, or temperature falls, the size of the aggregate tends to increase. Micelles containing tens or even hundreds of dye molecules sometimes occur. The rates of formation-dissociation of dye aggregates vary widely. Dimerisation, and its reversal, is usually extremely rapid but formation of larger aggregates may occur only over a period of days. Most dyes flocculate at sufficiently high ionic strength, with the formation of increasingly large aggregates proceeding until precipitation occurs.

Why do dyes aggregate? What enables dyes to come together, with such an ordering process apparently being entropically unfavourable and, in the case of acid and basic dyes, occurring in spite of coulombic repulsions? (An aside concerning such repulsions: the less symmetrical a dye the more readily such repulsions can be minimized by suitable molecular orientation. For instance Congo Red aggregates more readily than

Chrysophenine G). The current view is that van der Waals attractions (see Supplement 4) play a major role in all solvents, with other phenomena such as hydrogen bonding (see Supplement 5) or charge transfer complexation being of occasional importance. Thus the more extensive the area of polarisable, delocalised bonds (see Supplement 2), and so the larger the conjugated bond number (see Section 11.14), the stronger the dye-dye interactions. For instance huge micelles are formed by dyes possessing large chromophoric groups such as the phthalocyanines and the polyazo dyes.

In aqueous solutions another important factor favouring dye aggregation is hydrophobic bonding, see Supplement 7. In this case also the larger the number of hydrocarbon groups the greater the dye-dye interaction, but now saturated sidechains as well as the delocalised chromophoric structures are significant, see Section 11.14. The importance of hydrophobic effects helps to rationalise the disaggregating effects of co-solutes such as aniline, ethanol, phenol and urea. All these co-solutes will act to reduce the possible positive entropy change occurring during dye aggregation. The role of neutral salts in promoting dye aggregation is more complex since various ions can either increase or degrease the degree of structure in the solvent water. However, apart from such specific effects a high salt concentration stabilises large dye micelles by providing a high concentration of counter ions.

A useful review of dyc aggregation, though limited to aqueous solutions, has been provided by Coats (1969).

12.11 The precipitation of sparingly soluble («insoluble») salts

The initial step in the precipitation of a sparingly soluble salt, such as lead phosphate, is nucleation. This entails the formation of ionic clusters, termed crystallisation nuclei, of above a certain critical size. Such clusters form spontaneously due to the random fluctuations in the density of the solution. If the cluster size exceeds a certain minimum it can start to grow (homogeneous nucleation), whereas smaller clusters will redissolve.

With many salts this process is trivially slow even at the equilibrium solubility concentration. To exceed the energy barrier against nucleation the concentration of the salt must then be increased, so the solution will become supersaturated. The minimum dimensions of the crystallisation nuclei decrease as the supersaturation ratio (that is, concentration of precipitants/equilibrium solubility concentration) increases, as consequently does the rate of homogenous nucleation.

Often however nucleation occurs with the catalytic aid of additional species: impurities in the solution, or certain substrates in the solid phase. This process is termed heterogeneous nucleation. This commonly predominates at low supersaturation ratios, with homogeneous nucleation becoming the important nucleation mechanism at higher supersaturations. The basis of heterogeneous nucleation is essentially that the energy barrier to nucleation is lowered by a catalytic surface. One commonly favourable situation is the presence of a surface onto which strong adsorption of precipitant ions can occur.

Crystallisation nuclei grow mostly by repeated formation of new nuclei on the crystal surface, this being a rate-limiting step. At very high supersaturation ratios,

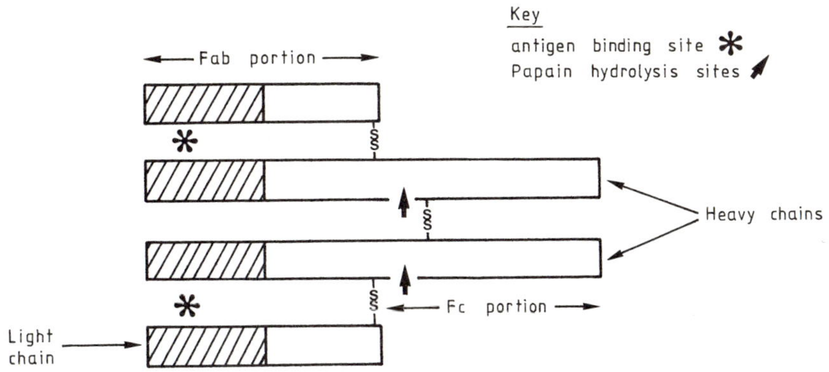

Fig. 76. Diagrammatic representation of an IgG-type immunoglobulin molecule

however, the rate-limiting process for crystal growth may become the diffusion of ions through the solution onto the crystal surface.

The overall rate of precipitation depends both on the total number of nuclei formed and on their subsequent growth rate. Both processes may be strongly inhibited by the presence in solution of surfactants (for instance certain dyes, enzyme substrates or detergents) which can become adsorbed onto nuclei or crystal faces.

12.12 Some structural characteristics of immunoglobulins

There are a number of major structural types of immunoglobulins in the human, of which immunoglobulin G (IgG), is the most common. The IgG molecule is built up from four peptide chains, two «heavy» and two «light», linked by disulphide bridges (Figure 76).

The molecular weight of the IgG molecule is about 150,000 Daltons. When IgG is subjected to enzymolysis with the protease papain the heavy chains are split at the arrows. This fission generates two Fab fragments and one Fc fragment, with only the former still able to bind antigen.

The great range of IgGs, each able to bind selectively a given hapten, arises due to differences in the amino acid composition of the «variable regions» of the four peptide chains, shaded above.

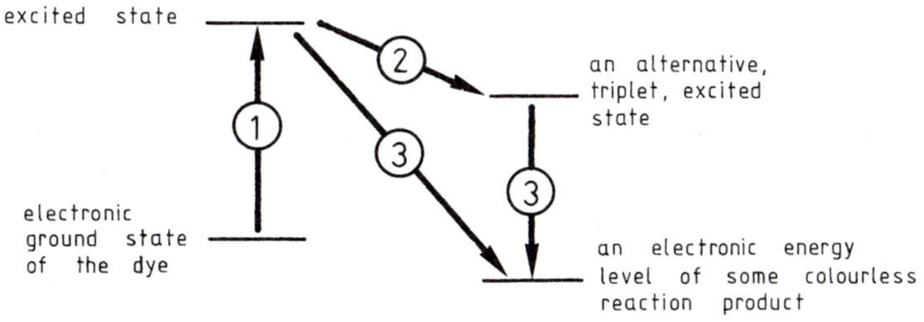

Fig. 77. The fundamental electronic events of photofading

The grossly different classes of immunoglobulin arises from variations in the «constant regions» of the peptide chains (well, they *are* less variable!), giving rise to IgG, IgD, IgE and certain IgA types. Other major structural varients involve «polymerisation» of the basic unit. Thus IgM is effectively a pentamer of the basic IgG unit with a molecular weight of around 900,000, and certain IgA sub-classes are dimers.

12.13 Photofading: some physico-chemical background to the fading of stained preparations following exposure to visible or ultra violet radiation

The *basic electronic processes involved in photofading* is illustrated in Fig. 77, where the arrows represent electronic transitions.

Photofading comprises an initial absorption of a quantum of radiation (arrow 1), after which the dye is in an electronically excited state. Transfer to an alternative, triplet, excited state may now occur (arrow 2). Dye molecules in such electronically excited states may loose energy in a variety of ways (see Section 4.2). Here however we are concerned only with the processes represented by arrow 3. These entail the formation of products, usually colourless, with new chemical structures.

Thus photofading involves chemical reactions of electronically excited states. In photofading of stained preparations *several quite different reaction types can occur* as listed in the equations below. Dye and non-dye reactants are represented below by D and R respectively, with ground and excited states distinguished by adding an asterisk in the latter instance (so: D^* and R^*). Quanta of radiation are denoted by hν.

Photoreduction of excited dye molecules by the substrate (such as histidine and tryptophane residues of protein), or by some other substance in the system.

$$D + h\nu \rightarrow D^*$$
$$D^* + R \rightarrow D_{reduced} + R_{oxidised}$$

Examples are the reduction of nitro dyes to azoxy compounds, and of azo dyes to amines.

Photooxidation of the excited dye, either directly by the substrate or indirectly by energy transfer from excited dye to oxygen or water followed by attack of the resulting active species:

$$D + h\nu \rightarrow D^*$$
then variously, e. g.
$$D^* + R \rightarrow D_{semioxidised} + R_{reduced}$$
and
$$D^* + O_2 \rightarrow D^{\bullet+} + O_2^{\bullet-} \rightarrow \text{colourless products}$$

Azo dyes are oxidised to azoxy compounds in this way.

Photodecomposition of dyes by excitation of non-dyes including hydrogen donors such as amino, carboxyl and keto groups present in substrate or additives:

$$R + h\nu \rightarrow R^*$$
$$R^* + D \rightarrow \text{colourless products}$$

An example is the reduction of azo dyes to amines by mandelic acid.

Photocatalytic processes often involve the excitation of some substance (C) which then reacts with a reducing agent, and is reoxidised by the dye:

$$C + h\nu \rightarrow C^*$$
$$C^* + R \rightarrow C_{reduced} + R_{oxidised}$$
$$C_{reduced} + D \rightarrow D_{reduced} + C$$

An example is the reductive bleaching of Thionin by ethanol, catalysed by $(FeCl_4)^-$.

So these are the possible varieties of photo-reactions. But which reaction, in a particular dye-substrate combination, will in fact take place? To what extent? And which dyes or dye-substrate combinations are most prone to photofading?

First consider the influence of the chemical structure of the dye. Dyes differ intrinsically in their reactivities, for instance towards oxidants and reductants. Many acridine, thiazine and xanthene dyes, and some anthraquinones (and riboflavin) have a higher reactivity towards oxygen than towards reducing agents. The reverse is so for many anthraquinones, azines and indanthrenes. Some generalisations on the effects of various substituent groups can be made. For a given dye class and a given substrate, substituents which are electron donors (such as amino and methoxy groups) tend to accelerate oxidative and retard reductive photofading. The reverse holds for electron accepting groups such as nitro and halogeno.

However these dye-structure effects are commonly obscured by influences of the environment, that is influences originating both externally and internally to the dye-substrate system.

External environmental factors include availability of oxygen and water vapour, both of which often accelerate photofading especially on protein substrates. Temperature also has a small influence. Photofading usually falls slightly as the temperature rises, probably due to loss of substrate water. An important external effect is the wavelength of the incident illumination. The action spectrum of the fading process is that of the activated substance, and this may be dye or substrate or catalyst.

Of the internal environmental factors some are physical. For instance the surface area available to interact with oxygen, radiation, substrate and water. Hence the less aggregated the dye the more photofading occurs. Dye aggregation has been discussed earlier, in Section 4.12. The growth of dye particles and the diffusion of water depend on the porosity of the substrate, another important variable.

An important internal chemical factor is the chemical nature of the substrate. A useful empirical generalisation is that dyes are often oxidised on non-protein substrates such as cellulose and derivatives, and reduced on proteins, such as keratin and fibroin.

Readers with physico-chemical background will find useful information in a review by Meier (1971).

References

Abramczuk, J.: Effect of formalin fixation on the dry mass of isolated rat liver nuclei. – Histochemie *29*, 207–212 (1972).

Accini, L., Hus, K. C., Spiele, H., De Martino, C.: Picric acid-formaldehyde fixation for immunoferritin studies. – Histochemistry *42*, 257–264 (1974).

Adams, C. W. M.: Lipid Histochemistry. – In: Adv. Lipid Res. – Academic Press, New York 1969.

Adams, C. W. M., Abdulla, Y. H., Bayliss, O. B., Weller, R. O.: The reaction of Bakers chromate reagent with lipids. – J. Histochem. Cytochem. *13*, 410–411 (1965).

Adams, C. W. M., Bayliss, O. B.: Histochemistry of myelin. V. Trypsin-digestible and trypsin-resistant proteins. – J. Histochem. Cytochem. *16*, 110–114 (1968a).

–: Histochemistry of myelin. VII. Analysis of lipid-protein relationships and absence of acid mucopolysaccharide. – J. Histochem. Cytochem. *16*, 119–127 (1968b).

Adamo, N. J., Daigneault, E. A.: Silver staining versus dye staining of decalcified epoxy-embedded spiral ganglia. – Stain Technol. *48*, 89–91 (1973).

Agnati, L. F., Fuxe, K.: Quantitative comparisons of amine fluorescence in cortical nor-adrenaline terminals using smear preparations. – J. Histochem. Cytochem. *22*, 1122–1127 (1974).

Ahlqvist, J.: The affinity of some naphthols for tissues before and after extraction of lipids. – Acta path. microbiol. scand. *59*, 171–180 (1963).

Ainsworth, S. K., Karnovsky, M. J.: An ultrastructural staining method for enhancing the size and electron opacity of ferritin in thin sections. – J. Histochem. Cytochem. *20*, 225–229 (1972).

Al-Azzawi, H. T., Stoward, P. J.: The effects of lethal procedures on the histochemical localisation of acid hydrolases. – Histochemie *37*, 187–195 (1973).

Albert, A.: Selective Toxicity, 6th edition. – Chapman and Hall, London 1979.

Aldridge, W. G., Watson, M. L.: Perchloric acid extraction as a histochemical technique. – J. Histochem. Cytochem. *11*, 773–781 (1963).

Alexander, P., Stacey, K. A.: The colloidal behavior of dyes. – Proc. roy. Soc. *A 212*, 274–290 (1952).

Alfert, M., Das, N. K., Eastwood, J. M.: Effects of ribonuclease on living cells and on cells fixed by different methods. – J. Histochem. Cytochem. *10*, 681–682 (1963).

Alroy, J., Temaura, K., Davidsohn, I., Weinstein, R. S.: A method for demonstrating blood group isoantigens in permanent tissue sections. – Stain Technol. *53*, 53–56 (1978).

Alsop, D. W.: Rapid single-solution polychrome staining of semithin epoxy sections using polyethylene glycol 200 (PEG 200) as a stain solvent. – Stain Technol. *49*, 265–272 (1974).

Altman, F. P.: The quantitative elution of Nitroblue Formazan from tissue sections. – Histochemie *17*, 319–326 (1969).

–: The use of a new grade of polyvinyl A for stabilizing tissue sections during histochemical incubations. – Histochemie *28*, 236–242 (1971).

–: An introduction to the use of Tetrazolium salts in quantitative enzyme cytochemistry. – Koch Light, Colnbrook, U.K. 1972.

Altman, F. P., Hoyer, P. E., Andersen, H.: Dehydrogenase histochemistry of lipid-rich tissues: a tetrazolium-metal chelation technique to improve localisation. – Histochem. J. *11*, 485–488 (1979).

Amsterdam, A., Schramm, M.: Rapid release of zymogen granule protein by osmium tetroxide and its retention during fixation by glutaraldehyde. – J. Cell Biol. *29*, 199–207 (1966).

Anderson, W. A., André, J.: The extraction of some cell components with pronase and pepsin from thin sections of tissue embedded in an epon-araldite mixture. – J. Microscopie *7*, 343–354 (1968).

ANSELL, P. L., MARSHALL, J.: A method for staining incompletely dehydrated Araldite embedded biological material for light microscopy. – J. Microsc. *97,* 385–387 (1973).

APARICIO, S. R., MARSDEN, P.: A rapid Methylene Blue-Basic Fuchsin stain for semi-thin sections of periferal nerve and other tissues. – J. Microsc. *89,* 139–141 (1969).

ASQUITH, R. S., KWOK, W. F., OTTERBURN, M. S.: Effect of pretreating wool and urea solutions on the rate of dyeing with acid dyes. – J. Soc. Dyers Colour. *94,* 212–213 (1978).

AVRAMEAS, S.: Coupling of enzymes to proteins with glutaraldehyde. Use of the conjugates for the detection of antigens and antibodies. – Immunochemistry *6,* 43–52 (1969).

–: Enzyme markers: their linkage with proteins and use in immuno-histochemistry. – Histochem. J. *4,* 321–330 (1972).

BACSY, E., RAPPAY, Gy.: Microspectrophotometric data on the kinetics of acid phosphatase activity in cells removed from the peritoneal cavity of the rat. – Histochem. J. *3,* 193–199 (1971).

BAKER, J. R.: Principles of Biological Microtechnique: a study of fixation and dyeing. – Methuen, London 1958.

–: Experiments on the action of mordants. 1. ‹Single bath› mordant dyeing. – Quart. J. micr. Sci. *101,* 255–272 (1960).

–: Experiments on the action of mordants. 2. Aluminium hematein. – Quart. J. micr. Sci. *103,* 493–517 (1962).

–: The fine structures produced in cells by fixatives. – J. roy. micr. Soc. *84,* 115–131 (1965 a).

–: The fine structure produced in cells by primary fixatives. 2. Potassium dichromate. – Quart. J. micr. Sci. *106,* 15–21 (1965 b).

–: Cytological Techniques, 5th edition. – Methuen, London 1966.

BAKER, J. R., LUKE, B. M.: The fine structure produced in cells by primary fixatives. 1. Mercuric chloride. – Quart. J. micr. Sci. *104,* 101–106 (1963).

BAKER, J. R., MCCRAE, J. M.: The fine structure resulting from fixation by formaldehyde: the effects of concentration, duration and temperature. – J. roy. micr. Soc. *85,* 391–399 (1966).

BALLANTYNE, B.: Artifacts in monamine oxidase histochemistry. – Proc. anat. Soc. *104,* 592 (1969).

BARBOSA, P., PETERS, T. M.: The effects of vital dyes on living organisms with special reference to Methylene Blue and Neutral Red. – Histochem. J. *3,* 71–93 (1971).

BARER, R., JOSEPH, S.: Refractometry of living cells. Part 1. Basic Principle. – Quart. J. micr. Sci. *95,* 399–423 (1954).

BARRNETT, R. J., ROTH, W. D.: Effects of fixation on protein histochemistry. – J. Histochem. Cytochem. *6,* 406–415 (1958).

BARRNETT, R. J., SELIGMAN, A. M.: Histochemical demonstration of sulphydryl and disulphide groups of protein. – J. nat. Cancer Inst. *14,* 769–804 (1954).

BARTHOLOMEW, J. W.: Variables influencing results, and the precise definition of steps in Gram staining as a means of standardizing the results obtained. – Stain Technol. *37,* 139–155 (1962).

BAUGH, P. J., LAWTON, J. B., PHILLIPS, G. O.: Polyanions and their complexes. VIII. Interaction in heterogeneous systems. – J. phys. Chem. (Ithaca) *76,* 688–695 (1972).

BAUMAN, J. G. J., DUIJN, P. VAN: Fluorescence-microscopical localisation of specific DNA sequences by cytochemical hydridisation with fluorochrome-labelled RNA. – Abstr. VIth Int. Congr. Histochem. Cytochem. 1980 (LAKE, B. D., BAYLISS-HIGH, O., HOLT, S. J., STOWARD, P. J., eds.). – The Royal Microscopical Society, Oxford 1980.

BAYER, E. A., SHUTELSKY, E., WYNNE, D., WILCHEK, M.: Preparation of ferritin-avidin conjugates by reductive alkylation for use in electron microscopic cytochemistry. – J. Histochem. Cytochem. *24,* 933–939 (1976).

BAYLISS, O. B., ADAMS, C. W. M.: Resistance of aldehyde blockades to periodic acid. – Histochem. J. *8,* 71–75 (1976).

BEADLE, D. J., DAWSON, A. L., JAMES, D. J., LIVINGSTONE, D. C.: The avidity of heavy metal diazotates for animal lysosomes and plant vacuoles during the ultrastructural localisation of acid hydrolases. – Histochem. J. *6,* 57 (1974).

BEDI, K. S.: Quantitative histochemistry of DNA with special reference to the hypothesis of DNA constancy. – Ph. D. Thesis, University of Sheffield, U.K. 1976.

BEDI, K. S., GOLDSTEIN, D. J.: Cytophotometric factors causing apparent differences between Feulgen DNA contents of different leucocyte types. – Nature (Lond.) *251,* 439–440 (1974).

–: Apparent anomalies in nuclear Feulgen-DNA contents. Role of systematic microdensitometric errors. – J. Cell Biol. *71,* 68–88 (1976).

BEDI, K. S., HOROBIN, R. W.: An alcohol-soluble Schiff's reagent: A histochemical application of the complex between Schiff's reagent and phosphotungstic acid. – Histochemistry *48,* 153–159 (1976).

BENJAMINSON, M. A.: Conjugates of chitinase with Fluorescein Isothiocynate or Lissamine Rhodamine as specific stains for chitin *in situ*. – Stain Technol. *44*, 27–31 (1969).

BENNION, P. J., HOROBIN, R. W.: Non-electrostatic factors in the dyeing of tissue sections. – Proc. roy. micr. Soc. *8*, 36 (1973).

–: Some effects of salts on staining: Use of the Donnan equilibrium to describe staining of tissue sections with acid and basic dyes. – Histochemistry *39*, 71–82 (1974a).

–: The aggregation of dyes in solution: Some preliminary estimates using gel-filtration chromatography, and its importance in biological staining. – Proc. roy. micr. Soc. *9*, 111 (1974b).

BENNION, P. J., HOROBIN, R. W., MURGATROYD, L. B.: The use of a basic dye (Azure A or Toluidine Blue) plus a cationic surfactant for staining of RNA: a technical and mechanistic study. – Stain Technol. *50*, 307–313 (1975).

BERG, G. G., LYON, D., CAMPSBELL, M.: Faster histochemical reaction for ATPase in the presence of chloride salts, with studies of the mechanism of precipitation. – J. Histochem. Cytochem. *20*, 39–55 (1972).

BERGMAN, R. A., UENO, H., MORIZONO, Y., HANKER, J. S., SELIGMAN, A. M.: Ultrastructural demonstration of Acetylcholinesterase activity of motor endplates via osmiophilic diazothioethers. – Histochemie *11*, 1–12 (1967).

BERNARD, G. R.: Microwave irradiation as a generator of heat for histological fixation. – Stain Technol. *49*, 215–224 (1975).

BERTELOOT, A., HUGON, J. S.: Effect of glutaraldehyde and lead on the activity of hepatic glucose-6-phosphatase. A biochemical and cytochemical study. – Histochemistry *43*, 197–214 (1975).

BETRABET, S. M., BAGWE, V. B., DARUWALLA, E. H.: Behaviour of bifunctional and polyfunctional reactive dyes applied to cotton cellulose. – J. Soc. Dyers Colour. *93*, 338–345 (1977).

BIGBEE, J. W., KOSEK, J. C., ENG, L. F.: Effects of primary antiserum dilution on staining of «antigen rich» tissues with the peroxidase antiperoxidase technique. – J. Histochem. Cytochem. *25*, 443–447 (1977).

BIRD, C. L., BOSTON, W. S.: The Theory of Colouration of Textiles. – Dyers Company Publication Trust, Bradford, U.K. 1975.

BIRKEDAL-HANSEN, H.: Eosin staining of gelatine. – Histochemie *36*, 73–87 (1973).

BISWAS, B. B., BASU, P. S., PAL, M. K.: Gram staining and its molecular mechanism. – Int. Rev. Cytol. *29*, 1–27 (1970).

BJERRUM, O. L., JOHNSEN, S. G.: Two types of human erythrocytes demonstrated by differential dye uptake. – Scand. J. Haematol. *8*, 476–480 (1971).

BLOCH, D. P.: Histone differentiation and nuclear activity. – Chromosoma *19*, 317–339 (1966).

BLOCH, D. P., HEW, H. Y. C.: Schedule of spermatogenesis in the pulmonate snail Helix Aspersa, with special reference to histone transition. – J. biophys. biochem. Cytol. *7*, 515–532 (1960).

BOADLE, M. C., BLOOM, F. E.: A method for the fine structural localisation of monamine oxidase. – J. Histochem. Cytochem. *17*, 331–340 (1969).

BOCKING, A., GROBARTH, C., VAN DEIMLING, O.: Esterase. XVIII. Biochemical and histochemical investigations of the subcellular localisation of the non-specific esterase. – Histochemistry *42*, 359–375 (1974).

BOHN, W.: A fixation method for improved antibody penetration in electron microscopical immunoperoxidase studies. – J. Histochem. Cytochem. *26*, 293–297 (1978).

BOORSMA, D. M., STREEFKER, J. G., KORS, N.: Peroxidase and Fluorescein isothiocyanates as antibody markers. A quantitative comparison of two peroxidase conjugates prepared with glutaraldehyde or periodate and a fluorescein conjugate. – J. Histochem. Cytochem. *24*, 1017–1025 (1976).

BORGERS, M.: Levamisole and its analogue R8231: new potent inhibitors of alkaline phosphatase. – In: Electron Microscopy and Cytochemistry, Proc. Second Int. Symp., Drienerlo, 1973 (WISSE, E., DAEMS, W. Th., MOLENAR, I., VAN DUIJN, P., eds.), pp. 51–54. – North-Holland, Amsterdam 1974.

BOURGEOIS, C., HUBBARD, B.: A method for the simultaneous demonstration of choline-containing phospholipids and neutral lipids in tissue sections. – J. Histochem. Cytochem. *13*, 571–578 (1965).

BOWES, J. H., CATER, C. W.: The reaction of glutaraldehyde with proteins and other biological materials. – J. roy. micr. Soc. *85*, 193–200 (1966).

BRADBURY, E. M.: Current ideas on the structure of chromatin. – Trends biochem. Sci. *1*, 7–9 (1976).

BRONNER, R.: Simultaneous demonstration of lipids and starch in plant tissue. – Stain Technol. *50*, 1–4 (1975).

BROOKER, L. G. S., WHITE, F. L., HESELTINE, D. W., KEYS, G. H., DENT, S. G., VAN LARE, E. J.: Spatial configuration, light absorption, and sensitizing effects of cyanine dyes. – J. photogr. Sci. *1*, 173–183 (1953).

BROWN, D. J.: The Pyrimidines. – Wiley, New York 1962.

Brown, J. E., Brown, P. K., Pinto, L. H.: Detection of light-induced changes of intracellular ionised calcium concentration in *Limulus* ventral photoreceptors using Arseno III. – J. Physiol. (Lond.) *267*, 299–300 (1977).

Brozman, M.: Immunohistochemical analysis of formaldehyde and trypsin- or pepsin-treated material. – Acta histochem. (Jena) *63*, 251–260 (1978).

Bruchhaus, H., Geyer, G.: Studies on the alkaline hydrolysis of nucleic acids in tissues. V. Cytophotometric evaluation of the effect of an alkaline hydrolysis on the Feulgen reaction. – Histochem. J. *6*, 579–581 (1974).

Bruchhausen, D. R., Hofermann, Krieg, H., von Mayersbach, H.: Stabilisation and lyophilisation of protein during freezing-drying for histological purposes. – Histochemie *20*, 215–224 (1969).

Buffa, R., Solcia, E., Fiocca, R., Crivelli, O., Pera, A.: Complement-mediated bonding of immunoglobulins to some endocrine cells of the pancreas and gut. – J. Histochem. Cytochem. *27*, 1279–1280 (1979).

Bulmer, D.: Observation on histological methods involving the use of phosphotungstic and phosphomolybdic acids, with particular reference to staining with phosphotungstic acid-haematoxylin. Quart. J. micr. Sci. *103*, 311–323 (1962).

Burns, J., Stoward, P. J.: Studies in fluorescence histochemistry. VIII. Evidence that N-acetyl groups do not take part in the acetic anhydride-salicylhydrazide-zinc (or fluorescent ketone) method for demonstrating protein C-terminal carboxyl groups. – Histochemie *26*, 266–271 (1971).

Burr, F. A.: Staining of a protein-polysaccharide model with silver methinamine. – J. Histochem. Cytochem. *21*, 386–387 (1973).

Burstone, M. S.: An evaluation of histochemical methods for protein groups. – J. Histochem. Cytochem. *3*, 32–49 (1955).

Burton, K., Taylor, A. B.: Oxidation of pyrimidine nucleosides and nucleotides by osmium tetroxide. – Biochem. J. *104*, 686–694 (1967).

Buschmann, R. J.: Extraction of absorbed lipid (Linoleic acid-l-^{14}C) from rat intestinal epithelium during processing for electron microscopy. – J. Cell Biol. *38*, 252–255 (1968).

Butcher, R. G.: Tissue stabilisation during histochemical reactions. – Histochemie *28*, 231–235 (1971).

Butler, W. F.: Variations in the staining of cutaneous mast cells. – Histochem. J. *3*, 365–370 (1971).

Butterworth, J.: Histochemistry of non-specific phosphatase: the use of p-nitrophenyl phosphate and glycerol phosphate as substrates. – Histochem. J. *3*, 477–487 (1971).

Cabrini, R. L., Frasch, A. C. C., Itoiz, M. E.: A quantitative microspectrophotometric study of the lead precipitation reaction for the histochemical demonstration of acid phosphatase. – Histochem. J. *7*, 419–426 (1975).

Cammermeyer, J.: Is the solitary Dark Neurone a manifestation of postmortem trauma to the brain inadequately fixed by perfusion? – Histochemistry *56*, 97–115 (1978).

Campbell, D. S. E., Cathcart, D., Giles, C. H., Rahmman, S. M. K.: Optical investigations into the physical state of dyes adsorped at solid surfaces. Part 2. – Trans. Faraday Soc. *55*, 1631–1643 (1959).

Cardno, S. H., Steiner, J. W.: Improvements of staining technics for thin sections of epoxy-embedded tissue. – Amer. J. clin. Path. *43*, 1–8 (1965).

Cascarano, J., Zweifach, B. W.: Physical and metabolic factors in diffusion of solutes into tissue slices. – Amer. J. Physiol. *200*, 1285–1292 (1961).

Casselman, W. G. B.: Cytological fixation by chromic acid and dichromates. – Quart. J. micr. Sci. *96*, 203–222 (1955).

–: Some applications of interference microscopy in enzyme histochemistry, with particular reference to alkaline phosphatase. – Exp. Cell Res. *7*, 28–39 (1959).

Cater, C. W.: The evaluation of aldehyde and other bifunctional compounds as cross-linking agents for collagen. – J. Soc. Leath. Trades Chem. *43*, 203 (1963).

Cejkova, J., Brettschneider, I.: The effect of hydration of the rabbit cornea on the histochemical demonstration of acid mucopolysaccharides. Histochemie *17*, 327–336 (1969).

Celio, M. R.: Immunohistochemistry on Bouin fixed fetal tissue, stored for thirty years in ethanol. – Histochemistry *61*, 347–350 (1979).

Chambers, R. W., Bowling, M. C., Grimley, P. M.: Glutaraldehyde fixation in routine histopathology. – Arch. Path. *85*, 18–30 (1968).

Chandra, S., Skelton, F. R.: Staining juxtaglomerular cell granules with Toluidine Blue or with Basic Fuchsin for light microscopy embedding. – Stain Technol. *39*, 107–110 (1964).

Chauvet, J., Nouvel, G., Archer, R.: Structure primaire d'un inhibiteur pancreatique de la trypsine (inhibieur de Kunitz et Northdrop). I. Enchainements N- et C-terminaux. – Biochim. biophys. Acta (Amst.) *115*, 121–129 (1966).

Chayen, J.: Chapter 13 – In: Cell Structure and its Interpretation (McGee-Russell, S. M., Ross, K. F. A., eds.). – Arnold, London 1968.

The Chemical Society, London: The Chemical Society Specialist Reports, Amino-acids, Peptides, and Proteins. A chapter on «Metal derivatives of amino-acids, peptides, and proteins» occurs in odd numbered volumes.

Chou, J. T. Y.: A note on the preservation of adipose tissue. – Quart. J. micr. Sci. *99*, 229–230 (1958).

Christie, K. N.: Celloidin as a protective film for improving the histochemical localisation of succinate dehydrogenase. – Histochem. J. *5*, 473–477 (1973).

Churg, A.: Carbonic anhydrase histochemistry: evidence for non-enzymatic reaction and artifact production. – Histochemie *36*, 293–301 (1973).

Cibis, P. A., Yamashita, T.: Staining for fungi with saccharated iron oxide. – Stain Technol. *34*, 129–133 (1959).

Clark, G.: Diamond Jubilee. A history of quality assurance in biological dyes. – Amer. Soc. Microbiol. News *40*, 252–259 (1974).

Clayton, B. P.: Unmasking of sudanophilic lipid in the testis of house-cricket. – Quart. J. micr. Sci. *99*, 453–457 (1958).

–: The action of fixatives on the unmasking of lipid. – Quart. J. micr. Sci. *100*, 269–274 (1959).

Clemens, H. J., Toepfer, K.: Physikalisch-chemische Eigenschaften von kommerziellen Thiazinfarbstoffen. 2. Qualitative und quantitative Untersuchungen über die Verunreinigungen der Farbstoffe und das «Umlosen» zum Erhalt eines höheren Reinheitsgrades. – Acta histochem. (Jena) *31*, 126–134 (1964).

Clyne, D. H., Norris, S. H., Modesto, R. R., Pesce, A. J., Pollak, V. E.: Antibody enzyme conjugates. The preparation of intermolecular conjugates of horseraddish peroxidase and antibody and their use in immunohistology of renal cortex. – J. Histochem. Cytochem. *21*, 233–240 (1973).

Coalson, R. E., Schwartz, M. J.: Boron trifluoride-methanol for histochemical methylation. – Acta histochem. (Jena) *36*, 227–237 (1970).

Coats, E.: Aggregation of dyes in aqueous solutions. – J. Soc. Dyers Colour. *85*, 355–368 (1969).

Cohn, N. S.: A model system analysis of the parameters in histone staining. 1. Alkaline Fast Green. – Histochem. J. *5*, 529–545 (1973).

Coleman, R. A., Pratt, L. H.: Electron microscopic localisation of phytochrome in plants using an indirect antibody-labelling method. – J. Histochem. Cytochem. *22*, 1039–1047 (1974).

Conger, K. A., Garcia, J. H., Lossinsky, A. S., Kauffman, F. C.: The effect of aldehyde fixation on selected substrates for energy metabolism and amino acids in mouse brain. – J. Histochem. Cytochem. *26*, 423–433 (1978).

Conklin, J. L.: Staining reactions of mucopolysaccharides after formalin-containing fixatives. – Stain Technol. *38*, 56–59 (1963).

Cope, G. H., Williams, M. A.: Quantitative studies on the preservation of choline and ethanolamine phosphatides during tissue preparation for electron microscopy. I. Glutaraldehyde, osmium tetroxide, Araldite methods. – J. Microsc. *90*, 31–46 (1969).

–: Quantitative studies on the preservation of choline and ethanolamine phosphatides during tissue preparation for electron microscopy. II. Other preparative methods. – J. Microsc. *90*, 47–60 (1969).

–: Biochemical and histochemical studies on p-nitrophenyl phosphate and β-glycerophosphate hydrolysis at acid pH by rabbit parotid tissue. – Histochem. J. *9*, 685–701 (1977).

Cornelisse, C. J., van Duijn, P.: A new method for the investigation of the kinetics of the capture reaction in phosphatase cytochemistry. II. Theoretical and experimental study of phosphate diffusion from thin polyacrylamide films. – J. Histochem. Cytochem. *21*, 614–622 (1973).

Cottell, D. C., Campbell Livingstone, D.: Fluorescent reagent for the periodic acid-Schiff and Feulgen reaction for cytochemical studies. – J. Histochem. Cytochem. *24*, 956–958 (1976).

Coupland, R. E., Hopwood, D.: The mechanism of the differential staining reaction for adrenaline- and noradrenaline-storing granules in tissues fixed in glutaraldehyde. – J. Anat. (Lond.) *100*, 227–243 (1966).

Cowden, R. R., Curtis, S. K.: Demonstration of protein-bound sulphydryl and disulphide groups with fluorescent mercurials. – Histochemie *22*, 247–255 (1970).

Cowden, R. R., Rasch, E. M., Curtis, S. K.: Cytochemical evaluation of the Guard procedure for a regressive staining method for demonstrating chromosomal basic proteins. I. Effects of fixation, blocking reactions, selective extractions and polyacid «differentiation». – Histochemistry *48*, 81–92 (1976).

Culling, C. F. A., Reid, P. E., Dunn, W. L.: The effect of saponification upon certain histochemical reactions of the epithelial mucins of the gastrointestinal tract. – J. Histochem. Cytochem. *19*, 654–662 (1971).

DeBari, V. A., Needle, M. A.: Mechanism for transport of Nitro Blue tetrozolium into viable and non-viable leucocytes. – Histochemistry *56*, 155–163 (1978).

DeBruijn, W. C., McGee-Russell, S. M.: Bridging a gap in histology. – J. roy. micr. Soc. *85*, 77–90 (1965).

DeJong, A. S. H., Hak, T. J., van Duijn, P., Daems, W. Th.: A new dynamic model system for study of capture reactions for diffusable compounds in cytochemistry. II. Effect of the composition of the incubation medium on the trapping of phosphate ions in acid phosphatase cytochemistry. – Histochem. J. *11*, 145–161 (1979b).

–: A new dynamic model system for the study of capture reactions for diffusable compounds in cytochemistry. III. Influence of the matrix composition on the lead phosphate precipitation process in acid phosphatase cytochemistry. – Histochem. J. *11*, 163–171 (1979a).

DeMartino, C., Natali, P. G., Brown, C. B. Axxinni, L.: Influence of plastic embedding media on staining and morphology of lipid bodies. – Histochemie *16*, 350–360 (1968).

Davenport, R., Davenport, J. C.: A cytochemical study of cytoplasmic basic proteins in the ascidian oocyte. – J. Cell Biol. *25*, 319–326 (1965).

Dallam, R. D.: Determination of protein and lipid lost during osmic acid fixation of tissues and cellular particulates. – J. Histochem. Cytochem. *5*, 178–181 (1957).

Davie, F.: Reference list to literature on histological stains applicable to sections of resin-embedded tissue. – Sci. Tools *17*, 59–61 (1970).

Davis, H. G., Wilkins, M. H. F., Chayen, J., La Cour, L. F.: The use of the interference microscope to determine dry mass in living cells and as a quantitative cytochemical method. – Quart. J. micr. Sci. *95*, 271–304 (1954).

Davies, P., Tabor, B. E.: Kinetic study of the cross-linking of gelatin by formaldehyde and glyoxal. – J. Polym. Sci. *A1*, 799–815 (1963).

Defendi, V.: Observations on naphthol staining and the histochemical localisation of enzymes by naphthol-azo dye technique. – J. Histochem. Cytochem. *5*, 1–10 (1957).

Defendi, V., Pearse, A. G. E.: Significance of coupling rate in histochemical azo dye methods for enzymes. – J. Histochem. Cytochem. *3*, 203–211 (1955).

Deitch, A. D.: A method for the cytophotometric estimation of nucleic acids using Methylene Blue. – J. Histochem. Cytochem. *12*, 451–461 (1964).

Delcourt, R.: Démonstration par l'électrophorèse en gel d'acryle de la spécificité du bleu de Nil en tant que réactif des triglycérides. C. R. Soc. Biol. (Paris) *161*, 1146–1148 (1967).

Demalsy, P., Callebaut, M.: Plain water as a rinsing agent preferable to sulfurous acid after the Feulgen Nucleal reaction. – Stain Technol. *42*, 133–136 (1967).

Denton, M. D., Ginsburg, A.: Conformational changes in glutamine synthetase from *Eschrichia coli*. I. The binding of M2+ in relation to aspects of the enzyme structure and activity. – Biochemistry *8*, 1714–1725 (1969).

Dewse, C. D., Potter, C. G.: Influence of light and mounting medium on the fading of the Feulgen stain. – Stain Technol. *50*, 301–306 (1975).

Dodge, J. D.: Cytochemical staining of sections from plastic-embedded flagellates. – Stain Technol. *39*, 381–386 (1964).

Douglas, W. H. J.: Perchloric acid extraction of desoxyribonucleic acid from thin sections of epon-araldite material. – J. Histochem. Cytochem. *18*, 510–514 (1970).

Douglas, W. H. J., Ripley, R. C., Ellis, R. A.: Enzymatic digestion of desmosome and hemidesmosome plaques performed on ultrathin sections. – J. Cell Biol. *44*, 211–214 (1970).

Dubois-Dalcq, M., McFarland, H., McFarlin, D.: Protein A-Peroxidase: a valuable tool for the localisation of antigens. – J. Histochem. Cytochem. *25*, 1201–1206 (1977).

Duijndam, W. A. L., van Duijn, P.: The interaction of apurinic acid aldehyde groups with Pararosanilin in the Feulgen-Schiff and related staining procedures. – Histochemistry *44*, 67–85 (1975).

Eagland, D.: Water, A comprehensive treatise (Franks, F. ed.). *4*, Chapter 5: Nucleic acids, peptides and proteins, pp. 305–518. – Plenum Press, New York 1975.

Ehrlich, P.: The Collected Papers of Paul Ehrlich, reprinted English translation (Himmelweit, F., ed.), *1*, 117–123. – Pergamon Press, London 1879.

Ekfors, T. O., Hopsu-Havu, V. K.: Immunofluorescent localisation of trypsin-like esteropeptidases in the mouse sub-mandibular gland. – Histochem. J. *3*, 415–420 (1971).

ELLEDER, M., LOJDA, Z.: Remarks on the «OTAN» Reaction. – Histochemie *14*, 47–64 (1968).
–: Studies on lipid histochemistry. I. Criticism of performic and peracetic acid-Schiff reactions. – Histochemie *24*, 7–20 (1970a).
–: Studies in lipid histochemistry. II. The nature of the material stained with acid hematein test and with OTAN reactives in red blood cells. – Histochemie *24*, 21–28 (1970b).
–: Studies in lipid histochemistry. III. Problems of extraction with acetone in lipid histochemistry. – Histochemie *28*, 68–87 (1971).
ERÄNKÖ, O.: Quantitative Methods in Histology and Microscopic Histochemistry. – Karger, Basel 1955.
ERÄNKÖ, O., PALKAMA, A.: Improved localisation of phosphorylase by the use of polyvinylpyrrolidone and high substrate concentration. – J. Histochem. Cytochem. *9*, 585 (1961).
ERÄNKÖ, O., RECHARDT, L., HANNINEN, L.: Electron microscopic demonstration of cholinesterases in nervous tissue. – Histochemie *8*, 369–376 (1967).
ERLANDSEN, S. L., PARSONS, J. A., RODDING, C. B.: Technical parameters of immunostaining of osmicated tissue in epoxy sections. – J. Histochem. Cytochem. *27*, 1286–1289 (1979).
EVANS, E. E., KENT, S. P.: The use of basic polysaccharides in histochemistry and cytochemistry: I. The demonstration of acidic polysaccharides in tissue sections with Fluorescein-labelled *Aspergillus* polysaccharide. – J. Histochem. Cytochem. *10*, 8–13 (1962).
EVANS, H. C.: Fibre swelling and detergent adsorption in detergent textile fibre systems. – J. Colloid Sci. *13*, 537–552 (1958).
EVERETT, D. H.: An Introduction to the study of Chemical Thermodynamics. – Longmans, London 1962.
FAHIMI, H. D.: Diffusion artifacts in cytochemistry of catalase. – J. Histochem. Cytochem. *21*, 999–1009 (1973).
FEDER, N., O'BRIEN, T. P.: Plant microtechnique: some principles and new methods. – Amer. J. Bot. *55*, 123–142 (1968).
FISCHER, R.: Histochemical model experiments on factors influencing dye-protein interaction. – J. Histochem. Cytochem. *7*, 350–351 (1959).
FLITNEY, F. W.: The time course of the fixation of albumin by formaldehyde, glutaraldehyde, acrolein and other higher aldehydes. – J. roy. micr. Soc. *85*, 353–364 (1966).
FRAENKEL-CONRAT, H.: The action of epoxides on proteins. – J. biol. Chem. *154*, 227–238 (1944).
FRANCIS, R.: Chapter 10 in: Theory and Practice of Histological Techniques (BANCROFT, J. D., STEVENS, A., eds.). – Churchill Livingstone, Edinburgh 1977.
FRIEDBERG, S. H., GOLDSTEIN, D. J.: Thermodynamics of orcein staining of elastic fibres. – Histochem. J. *1*, 361–376 (1969).
FRIED, R. L., KNOLLER, M.: Quantitative tests of histochemical methods for phosphomonoesterases in brain tissue. – J. Histochem. Cytochem. *13*, 125–132 (1965).
FUKUYAMA, K., EPSTEIN, W. L.: Sulphur-containing proteins and epidermal keratinisation. – J. Cell Biol. *40*, 830–838 (1969).
FULLMER, H. M., LILLIE, R. D.: Some aspects of the mechanism of Orcein staining. – J. Histochem. Cytochem. *4*, 64–68 (1956).
–: The staining of collagen with elastic tissue stains. – J. Histochem. Cytochem. *5*, 11–14 (1957).
GABE, M.: Histological Techniques. – Masson, Paris and Springer, New York 1976.
GAHRTON, G., OLSSON, I., DAHLQVIST, A.: Determination of the glycogen content in single neutrophil leucocytes using micromodel of leucocyte glycogen. – J. Histochem. Cytochem. *23*, 59–64 (1975).
GALLAGHER, I. H.: Mechanism of action of a two-tone plaque disclosing agent. – J. Periodont. *48*, 395–396 (1977).
GALLYAS, F.: Kinetics of formation of metallic silver and binding of silver ions by tissue components. – Histochemistry *64*, 87–96 (1979a).
–: Factors affecting the formation of metallic silver and the binding of silver ions by the tissue components. – Histochemistry *64*, 97–109 (1979b).
GARCIA, A. M., IORIO, R.: Introduction to Quantitative Cytochemistry (WIED, G. L. ed.), p. 215. – Academic Press, New York 1966.
GASIC, G. J., BERWICK, L., SORRENTINO, M.: Positive and negative colloidal iron as cell surface electron stains. – Lab. Invest. *18*, 63–71 (1968).
GERSH, I., VERGARA, J., ROSSI, G. L.: Use of anhydrous vapours in post-fixation and in staining of reactive groups of proteins in frozen-dried specimens for electron microscopic studies. – Anat. Rec. *138*, 445–453 (1960).
GEYER, G., STIBENZ, D.: Quantitative aspects of the colloidal iron reaction. – Acta histochem. (Jena) *50*, 264–272 (1974).

GILES, C. H., DUFF, D. G., SINCLAIR, R. S.: Relations between the molecular structures of dyes and their technical properties. – The Chemistry of Synthetic Dyes (Venkataraman, K., ed.) *8*, Academic Press, New York 1978.

GILLIS, J. M., PAGE, S. G.: Localisation of ATPase activity in striated muscle and probable sources of artifact. – J. Cell Sci. *2*, 113–118 (1967).

GLAUERT, A., REID, N.: Practical Methods in Electron Microscopy, *3*, Part 1 Fixation, Dehydration, and Embedding of Biological Specimens, Part II Ultramicrotomy. – North-Holland, Amsterdam 1974.

GLAZER, A. N., DELANGE, R. J., SIGMAN, O. S.: Laboratory Techniques in Biochemistry and Molecular Biology, *4*, p. 205. – North-Holland, Amsterdam 1976.

GOLAND, P., GRAND, N. G., KATELE, K. V.: Cynauric chloride and N-methyl morpholine in methanol as a fixative for polysaccharides. – Stain Technol. *42*, 41–51 (1967).

GOLDBLATT, P. J., TRUMP, B. F.: The application of Del Rio Hortega's silver method to epon-embedded tissue. – Stain Technol. *40*, 105–115 (1965).

GOLDSTEIN, D. J.: Mechanism of differential staining of nucleic acids. – Nature (Lond.) *191*, 407–408 (1961).

–: Differential staining of nucleic acids? – Nature (Lond.) *193*, 999–1001 (1962a).

–: Correlation of size of dye particle and density of substrate, with special reference to mucin staining. – Stain Technol. *37*, 79–93 (1962b).

–: Ionic and non-ionic bonds in staining, with special reference to the action of urea and sodium chloride on the staining of elastic fibres and glycogen. – Quart. J. micr. Sci. *103*, 477–492 (1962c).

–: An approach to the thermodynamics of histological dyeing, illustrated by experiments with Azur A. – Quart. J. micr. Sci. *104*, 413–439 (1963).

–: Relation of effective thickness and refractive index to permeability of tissue components in fixed sections. – J. roy. micr. Soc. *84*, 43–54 (1965).

–: In Cell Structure and its Interpretation (MCGEE-RUSSELL, S. M., ROSS, K. F. A., eds.), chapter 6, p. 67–77. – Arnold, London 1968.

–: The fluorescence of elastic fibres stained with eosin and excited by visible light. – Histochem. J. *1*, 187–198 (1969a).

–: Determination of dichroism with the microscope. – J. Microsc. *89*, 19–36 (1969b).

–: Aspects of scanning microdensitometry. I. Stray light (glare). – J. Microsc. *92*, 1–16 (1970).

–: Aspects of scanning microdensitometry. II. Spot size, focus and resolution. – J. Microsc. *93*, 15–42 (1971).

–: Aspects of scanning microdensitometry. III. The monochromator system. – J. Microsc. *105*, 33–56 (1975).

–: A microdensitometric method for the analysis of staining kinetics. – J. Microsc. *119*, 331–343 (1980).

GOLDSTEIN, D. J., HOROBIN, R. W.: Rate factors in staining by Alcian Blue. – Histochem. J. *6*, 157–174 (1974a).

–: Surface staining of cartilage by Alcian Blue, with reference to the role of microscopic dye aggregates in histological staining. – Histochem. J. *6*, 175–184 (1974b).

GOMORI, G.: A rapid one-step trichrome stain. – Amer. J. clin. Path. *20*, 661–664 (1950).

GONATAS, N. K., STIEBER, A., GONATAS, J., GAMBETTI, P., ANTOINE, J. C., AVRAMEAS, S.: Ultrastructural autoradiographic detection of intracellular immunoglobulins with iodinated Fab fragments of antibody. The combined use of ultrastructural autoradiography and peroxidase. – J. Histochem. Cytochem. *22*, 999–1009 (1974).

GREEN, M. R., PASTEWKA, J. V.: The cationic carbocyanine dyes Stains-all, DBTC, and Ethyl-Stains-all, DBTC-3,3′, 9-Triethyl. – J. Histochem. Cytochem. *27*, 797–799 (1979).

GRIFFITH, W. P., ROSETTI, R.: Oxo complexes of osmium and ruthenium with organic ligands. – J. chem. Soc., Dalton Trans. 1449–1453 (1972).

GRILLO, T. A. I., OGUNNAIKE, P. O., FAOYE, S.: Effects of histological and electron microscopical fixatives on the insulin content of the rat pancreas. – J. Endocr. *51*, 645–649 (1971).

GRIMLEY, P. M.: A tribasic stain for thin sections of plastic embedded OsO_4-fixed tissues. – Stain Technol. *39*, 229–233 (1964).

GRONBLAD, M., ERÄNKÖ, O.: Fine structure of dense-cored vesicles in glomus cells of rat carotid body after fixation with permanganate or glutaraldehyde. – Histochemistry *57*, 305–312 (1978).

GUILLOUZO, A., BELANGER, L., BEAUMONT, C., VALET, J.-P., BRIGGS, R., CHUI, JEN-FU: Cellular and subcellular immunolocalisation of α_1-fetoprotein and albumin in rat liver. Re-evaluation of various experimental conditions. – J. Histochem. Cytochem. *26*, 948–949 (1978).

GURD, F. R. N., WILCOX, P. E.: Complex formation between metallic cations and proteins, peptides and amino acids. – Adv. Protein Chem. *11*, 312–427 (1956).

GURR, E.: Rational Use of Dyes in Biology. – Leonard Hill, London 1965.

–: Biological Staining Methods, 8th edition. – Searle Diagnostic Gurr Products, High Wycombe 1973.

Gustavson, K. H.: Some protein-chemical aspects of tanning processes. – Adv. Protein Chem. *5,* 353–421 (1949).

Habeeb, A. F. S. A., Hiramoto, R.: Reaction of proteins with glutaraldehyde. – Arch. Biochem. *126,* 16–26 (1968).

Halkjaer-Kristensen, J., Ingeman-Hansen, T.: Microphotometric determination of glycogen in single fibres of human quadriceps muscle. – Histochem. J. *11,* 629–638 (1979).

–: Thickness measurements of skeletal muscle sections using the light microscope. – Histochem. J. *10,* 497–504 (1978).

Hammerling, U., Aoki, T., Wood, H. A., Old, L. J., Boyse, E. A., DeHarven, E.: New visual markers of antibody for electron microscopy. – Nature (Lond.) *223,* 1158 (1968).

Hanker, J. S., Anderson, W. A., Bloom, F. E.: Osmiophilic polymer generation: catalysis by transition metal compounds in ultrastructural cytochemistry. – Science *175,* 991–993 (1972).

Hanker, J. S., Kasler, F., Bloom, M. G., Copeland, J. S., Seligman, A. M.: Coordination polymers of osmium: the nature of osmium black. – Science *156,* 1737–1738 (1967).

Hanker, J. S., Kusyk, C. J., Bloom, F. E., Pearse, A. G. E.: The demonstration of dehydrogenase and monamine oxidases by the formation of osmium blacks at the sites of Hatchetts Brown. – Histochemie *33,* 205–230 (1973).

Hansch, C., Leo, A., Unger, S. H., Kim, K. H., Lien, E. J.: "Aromatic" substituent constants for structure-activity correlations. – J. med. Chem. *16,* 1207–1216 (1973).

Harada, K.: The nature of mycobacterial acid-fastness. – Stain Technol. *51,* 255–260 (1976).

Hardonk, M. J., van Duijn, P.: The mechanism of the Schiff reaction as studied with histochemical model systems. – J. Histochem. Cytochem. *12,* 748–751 (1964a).

–: A quantitative study of the Feulgen reaction with the aid of histochemical model systems. – J. Histochem. Cytochem. *12,* 752–757 (1964b).

Harms, H.: Handbuch der Farbstoffe für die Mikroskopie: Stauffen Verlag, Kamp-Lintfort 1957.

Hartley, F. R.: The chemistry of chrome mordanting wool. – J. Soc. Dyers Colour. *85,* 66–71 (1969).

Hayashi, M.: Comments on the cytochemical techniques using Naphthol AS-BI substrates and hexazonium pararosaniline for the demonstration of Acid phosphatase, β-glucuronidase, and N-acetyl-β-glucosaminidase activities. – J. Histochem. Cytochem. *25,* 1021–1023 (1977).

Hayat, M. A.: Principles and Techniques of Electron Microscopy: Biological Applications, *1.* – Van Nostrand Reinhold, New York 1970.

–: Electron Microscopy of Enzymes, *1.* – Van Nostrand Reinhold, New York 1973.

–: Positive staining for electron microscopy. – Van Nostrand Reinhold, New York 1975.

Haurowitz, F.: The Chemistry and Function of Proteins, 2nd edition. – Academic Press, London 1963.

Hayashi, M., Shirahama, T., Cohen, A. S.: Combined cytochemical and electron microscopic demonstration of β-glucuronidase activity in rat liver with the use of a simultaneous coupling azo dye techniques. – J. Cell Biol. *36,* 289–297 (1968).

Haydon, G. B., Smith, S. O., Seligman, A. M.: Observations of mitochondrial staining for succinic dehydrogenase. – J. Histochem. Cytochem. *15,* 752–754 (1968).

Hayes, T. L., Lindgren, F. T., Gofman, J. W.: A quantitative determination of the osmium tetroxide-lipoprotein interaction. – J. Cell Biol. *19,* 251–255 (1963).

Heath, E.: Ultrastructural effects of performic acid oxidation of epoxy-embedded tissue with and without early uranyl staining. – J. Microsc. *110,* 65–70 (1977).

Heidenhain, M.: Encyklopädie der mikroskopischen Technik (Ehrlich, P., Krause, R., Masse, M., Rosin, H. eds.), pp. 327–360. – Urban and Schwarzenberg, Berlin 1903.

Heineman, R. L.: Inhibition of Schiff-reagent staining in chromosomes of ovigerous tissues of anoetids; whole mites and sections. – Stain Technol. *44,* 273–277 (1969).

Helfferich, F.: Ion Exchange. – McGraw Hill, New York 1962.

Henderson, B., Loveridge, N., Robertson, W. R.: A quantitative study of different grades of polyvinyl alcohol on the activities of certain enzymes in unfixed tissue sections. – Histochem. J. *10,* 453–463 (1978).

Henry, J. B.: Vital staining by Procion Navy Blue M3RS in laboratory animals. – Stain Technol. *43,* 297–302 (1968).

Hess, R., Pearse, A. G. E.: Enzyme immobilisation. Substantive diazonium salts in enzyme histochemistry. – Enzymol. biol. clin. *1,* 87–113 (1961/2).

Hiraoka, T.: Feulgen nucleal reaction. II. Approaches to the understanding of mechanism *in vitro* reaction. – Histochemie *35,* 283–296 (1973).

HJERTQUIST, S. O., ENGFELDT, B.: Studies on the epiphysial growth zone. 1. The preservation of acid glycosaminoglycans in tissues in some histotechnical procedures for electron microscopy. – Virchows Arch. path. Anat. (B, Zellpathologie) *1*, 222–229 (1968).
HODSON, S., MARSHALL, J.: Tyrosine incorporation into the rabbit retina. – J. Cell Biol. *35*, 722–726 (1967).
HOFF, H. F., RAYBURN, C.: A modified Alcian Blue-Periodic acid Schiff stain for epoxy-embedded atherosclerotic arteries. – Stain Technol. *49*, 241–243 (1974).
HOFSTÄDTER, F., VON MAYERSBACH, H., PEARSE, A. G. E., POLAK, J. M.: Diethylpyrocarbonate and benzoquinone vapour fixation for immuno-fluorescence studies of non-hormone protein in frozen dried tissues. – Acta histochem. (Jena) *59*, 177–181 (1977).
HOLCZINGER, L.: The reaction of unsaturated fats with the acid hematein test. – Histochemie *4*, 120–122 (1964).
HOLL, A.: Vital staining by Trypan Blue; its selectivity for olfactory receptor cells of the Brown Bullhead *Ictalurus Natalis*. – Stain Technol. *40*, 269–273 (1965).
HOLLANDER, D. H., FROST, J. K.: Antioxidant inhibition of stain fading and mounting medium crazing. – Acta cytol. (Philad.) *15*, 419 (1971).
HOLT, S. J.: Factors governing the validity of staining methods for enzymes, and their bearing upon the Gomori acid phosphatase technique. – Exp. Cell Res. Suppl. *7*, 1–27 (1959).
HOLT, S. J., O'SULLIVAN, D. G.: Studies in enzyme cytochemistry. 1 Principles of cytochemical staining methods. – Proc. roy. Soc. *B148*, 465–480 (1958).
HOOGE-PETERS, E. L., FOWLKES, B. J., HOOGE, R. J.: A new neuronal marker identified by phosphorylcholine-binding myeloma proteins. – Nature (Lond.) *281*, 376–378 (1979).
HOPSU, V. K., MCMILLAN, P. J.: Quantitative characterisation of a histochemical enzyme system. – J. Histochem. Cytochem. *12*, 315–324 (1964).
HOPWOOD, D.: The effect of formaldehyde fixation and dehydration on ox adrenal medulla with regard to the chromaffin reaction and post chroming. – Histochemie *10*, 98–106 (1967).
–: Fixatives and fixation: a review. – Histochem. J. *1*, 323–360 (1969a).
–: A comparison of the cross-linking abilities of glutaraldehyde, formaldehyde and 2-hydroxyadipaldehyde with BSA and casein. – Histochemie *17*, 151–61 (1969b).
–: Fixation of proteins by osmium tetroxide, potassium dichromate and potassium permanganate. Model experiments with BSA and BSG. – Histochemie *18*, 250–260 (1969c).
–: The reactions between formaldehyde, glutaraldehyde and osmium tetroxide, and their fixation effects on bovine serum albumin and on tissue blocks. – Histochemie 24, 50–64 (1970).
–: Theoretical and practical aspects of glutaraldehyde fixation. – Histochem. J. *4*, 267–303 (1972).
–: The reactions of glutaraldehyde with nucleic acids. – Histochem. J. *7*, 267–276 (1975).
HORISBERGER, M., VONLANTHEN, M.: Fluorescent colloidal gold: a cytochemical marker for fluorescent and electron microscopy. – Histochemistry *64*, 115–118 (1979).
HOROBIN, R. W.: The impurities of biological dyes: their detection, removal, occurrence and histological significance – a review. – Histochem. J. *1*, 231–265 (1969).
–: Dextrin and salt as impurities of histological dyes. – Histochemie *22*, 39–44 (1970).
–: Analysis and purification of biological stains by gel filtration. – Stain Technol. *46*, 297–304 (1971).
–: A preliminary quantitative study of the «rate of staining» model for rationalising various mixed acid dye stains («trichromes»). – Proc. roy. micr. Soc. *9*, 110 (1974).
–: Structure-staining relationships in histochemistry and biological staining. I. Theoretical background and a general account of correlation of histochemical staining with the chemical structure of the reagents used. – J. Microsc. *119*, 345–355 (1980a).
–: The Use of Pure Dyes for Biological Staining. – Koch Light, Colnbrook (U. K.) 1980b.
–: Structure-staining relationships in histochemistry and biological staining. – III. Some comments on the intentional and artifactual staining of lipids. – Acta Histochem. (Jena) Suppl. *24*, 237–246 (1981).
HOROBIN, R. W., BENNION, P. J.: The interrelation of the size and substantivity of dyes: the role of van der Waals attractions and hydrophobic bonding in biological staining. – *33*, 191–204 (1973).
HOROBIN, R. W., FLEMMING, L.: Structure-staining relationships in histochemistry and biological staining. II. Mechanistic and practical aspects of the staining of elastic fibres. – J. Microsc. *119*, 357–372 (1980).
HOROBIN, R. W., GOLDSTEIN, D. J.: Impurities and staining characteristics of Alcian Blue samples. – Histochem. J. *4*, 391–399 (1972).
–: The influence of salt on the staining of tissue sections with basic dyes: an investigation into the general applicability of the critical electrolyte concentration theory. – Histochem. J. *6*, 599–609 (1974).
HOROBIN, R. W., HAGUE, A.: Periodic acid-Schiff-phosphotungstic acid as an electron histochemical method. – J. Microsc. *94*, 265–271 (1971).

Horobin, R. W., James, N. T.: The staining of elastic fibres with Direct Blue 152. A general hypothesis for the staining of elastic fibres. – Histochemie *22,* 324–336 (1970).

Horobin, R. W., Kevill-Davies, I.: A mechanistic study of the histochemical reactions between aldehydes and Basic Fuchsin and acid alcohol used as as simplified substitute for Schiff's reagent. – Histochem. J. *3,* 371–378 (1971).

Horobin, R. W., Murgatroyd, L. B.: The composition and properties of Gallocyanin-chrome alum stains. – Histochem. J. *1,* 36–54 (1968).

–: The indentification and purification of Pyronin and Rhodamine dyes. – Stain Technol. *44,* 297–302 (1969).

–: The staining of glycogen with Best's Carmine and similar hydrogen bonding dyes. A mechanistic study. – Histochem. J. *3,* 1–9 (1971).

Horobin, R. W., Tomlinson, A.: The influence of the embedding medium when staining for electron microscopy: the penetration of stains into plastic sections. – J. Microsc. *108,* 69–78 (1976).

Hoshi, T., Tanizaki, Y., Ando, N.: Absorption spectra of dyes. IX. Electronic spectra of a 1:1 and a 1:2 complex, and CT-bands. – Bull. chem. Soc. *38:* 725–729 (1965).

Hsu, K. D., Rifkind, R. A., Zabriskie, J. B.: Fluorescent, electron microscopic and immuno-electrophoretic studies of labelled antibodies. – Science *142,* 1471–1473 (1963).

Huber, J. D., Parker, F., Odland, G. F.: A basic fuchsin and alkalinised Methylene Blue rapid stain for epoxy-embedded tissue. – Stain Technol. *43,* 83–87 (1968).

Huxley, H. E., Zubay, G.: Preferential staining of nucleic acid-containing structures for electron microscopy. – J. biophys. biochem. Cytol. *11,* 273–296 (1961).

Iren, von, F., Bange, G. G.: Localisation of inorganic ions by precipitate freeze dissolution. – Histochemistry *55,* 81–96 (1978).

Ito, S.: The Interpretation of Ultrastructure (Harris, R. J. C. ed.), p. 129. – Academic Press, New York 1962.

Irving, J. T.: A histological staining method for sites of calcification in teeth and bone. – Arch. oral Biol. *1,* 89–96 (1959).

Janigan, D. T., Williams, M. A., Tyler, H. M., Demster, W. J.: A biochemical approach to the study of rejection of canine renal homotransplants. III. Histochemical studies of transplanted kidney. – Brit. J. exp. Path. *45,* 347–356 (1964).

Jaubert, F., Barbey, S., Grune, M., Danel, C., Chretien, J., Nezelof, C.: Human alveolar macrophages. Spontaneous reduction of BSPT salt. – Histochemistry *64,* 163–169 (1979).

Jobsis, A. C., Meijer, A. E. F. H.: Evaluation of enzyme histochemical observations for metabolic studies. A combined histochemical and biochemical investigation of experimentally induced skeletal muscle changes. II. The biochemical investigation and comparison with the histochemical observations. – Histochemie *36,* 63–73 (1973).

Johnson, P. L., Hoffman, H., Rolle, G. K.: The Gram staining mechanism of cat tongue keratin. – J. Histochem. Cytochem. *5,* 84–90 (1957).

Jones, D.: The reaction of formaldehyde with unsaturated fatty acids during histological fixation. – Histochem. J. *1,* 459–491 (1969).

Jones, G. R. N.: Quantitative histochemical studies on the succinate-neotetrazolium reductase system. – Exp. Cell Res. *43,* 268–280 (1965).

–: Losses of nitrogenous material which occur from unfixed frozen sections during incubation. – Proc. roy. micr. Soc. *7,* 108 (1972).

Jordan Lloyd, J., Dempsey, M., Garrod, M.: The swelling of proteins fibres in organic solvents. – Trans. Faraday Soc. *42B,* 228–244 (1946).

Jordon, D. O.: The Chemistry of Nucleic Acids. – Butterworths, London 1960.

Jozsa, L., Szederkenyi, G.: Über Verluste der Gewebsmukopolysacharide während der Fixierung. – Acta histochem. (Jena) *26,* 255–260 (1967).

Kalimo, H. O., Helminen, H. J., Arstila, A. U., Hopsu-Havu, V. K.: The loss of enzyme reaction products from ultra-thin sections during the staining for electron microscopy. – Histochemie *14,* 123–130 (1968).

Kashiwa, H. K., House, C. M.: The glyoxal bis (2-hydroxyanil) method modified for localised insoluble calcium salts. – Stain Technol. *39,* 359–367 (1964).

Kasten, F. H.: The chemistry of Schiff's reagent. – Int. Rev. Cytol. *10,* 1–100 (1960).

–: The Feulgen Reaction. – Acta histochem. (Jena) *17,* 88–99 (1964).

–: Loss of RNA and protein and changes in DNA during a 30-hour cold perchloric acid extraction on cultured cells. – Stain Technol. *40,* 127–135 (1965).

KASTEN, F. H.: Cytophotometric investigations on the influence of fixation on Feulgen hydrolysis: evidence for a random loss of purines and pyrimidines from DNA during acid hydrolysis. – Acta histochem. (Jena) Suppl. *9,* 637–647 (1971).

KASTEN, F. H., LALA, R.: The Feulgen reaction after glutaraldehyde fixation. – Stain Technol. *50,* 197–201 (1975).

KATER, S. B., NICHOLSON, C.: (eds.) Intracellular staining in Neurobiology. – Springer-Verlag, Berlin 1973.

KAWASHIMA, T., MURATA, F.: Electron microscopic demonstration of cholinesterase in the human intercostal muscle by an indoxyl acetate esterase method. – Acta Histochem. Cytochem. *2,* 54–60 (1969).

KENT, S. P., EVANS, E. E.: The use of basic polysaccharides in histochemistry and cytochemistry. II: The demonstration of acidic polysaccharides in tissue sections using Fluorescein-labelled deacetylated chitin. – J. Histochem. Cytochem. *10,* 14–18 (1962).

KERPEL-FRONIUS, S., HAJOS, F.: A method for the electron microscopic demonstration of Cytochrome oxidase in fresh and formalin-fixed tissues. Histochemie *10,* 216–223 (1967).

KHAN, M. A.: Heterogeneous staining of rat hepatocyte nuclei by HBFP technique. – Histochem. J. *10,* 641–644 (1978).

KHAN, M. A., HOLT, P. G., PAPADIMITRIOU, J. M., KNIGHT, J. O., KAKULAS, B. A.: Creatine kinase, histochemical study by the gelatin film-lead precipitation technique. – Histochemie *32,* 49–58 (1972).

KIEFER, G., KIEFER, R., SANDRITTER, W.: Cytophotometric determination of the binding of basic dyes by DNA following acetylation. – Histochemie *14,* 65–71 (1968).

KIERNAN, J. A., STODDART, R. W.: Fluorescent-labelled Aprotinin: a new reagent for the histochemical detection of acid mucosubstances. Histochemie *34,* 77–84 (1973).

KIHARA, T.: Intermolecular Forces. – Wiley, Chichester 1978.

KILPATRICK, D. J., RATTEE, I.: Studies of urea-assisted wool dyeing. I. The swelling of wool fibres in aqueous urea solution. – J. Soc. Dyers Colour. *93,* 424–427 (1977).

KITCHENER, J. A.: Ion Exchange Resins. Methuen, London 1961.

KITCHING, W.: Recent aspects of the mercuration reaction. – Organometalic Revs. *3,* 35–60 (1968).

KLESSEN, C.: On the cytochemical demonstration of glycogen in neutrophil granulocytes: periodic acid-Schiff reaction and diastase (amylase) digestion test. – Histochemistry *62,* 191–206 (1979).

KLOHS, W. D., GOFF, C. W., BERNAKI, R. J.: Glycosyltransferase in plant and animal tissues. Effects of fixation and lead on enzyme activity. – J. Histochem. Cytochem. *26,* 772–781 (1978).

KOCHETKOV, N. K., BUDOWSKY, E. I.: The chemical modification of nucleic acids. – Progress in Nucleic Acid Research and Molecular Biology, *9,* 403–438, Academic Press: New York 1969.

KOELLE, G. B., GROMADZIKI, C. G.: Comparison of the gold-thiocholine and gold-thioacetic acid methods for the histochemical localisation of acetylcholinesterase and cholinesterases. – J. Histochem. Cytochem. *14,* 443–454 (1966).

KOELLE, G. B., DE T. VALK, A.: The physiological implications of the histochemical localisation of monoamine oxidase. – J. Physiol. *126,* 434–447 (1954).

KORHONEN, L. K., LINDHOLM, K.: Effect of histological fixation on carbohydrate-rich tissue compounds and their periodate consumption. – Histochemie *18,* 87–96 (1969).

KORN, E. D., WEISMAN, R. A.: I. Loss of lipids during preparation of amoebae for electron microscopy. – Biochim. biophys. Acta (Amst.) *116,* 309–316 (1966).

KOSAKAI, H.: Epoxy embedding, sectioning and staining of plant material for light microscopy. – Stain Technol. *48,* 111–115 (1973).

KRAEHENBÜL, J. P. , GALARDY, R. E., JAMIESON, J. D.: Preparation and characterisation of an immunoelectron microscope tracer consisting of a heam-octapeptide coupled to Fab. – J. exp. Med. *139,* 208–223 (1974).

KRONMAN, M. J., ROBBINS, F. M.: Fine Structure of Proteins and Nucleic Acids (FASMAN, G. D., TIMASHEFF, S. N. eds.), p. 271. – Dekker, New York 1970.

KUGLER, J. H., WILKINSON, W. J. C.: Quantitative analysis with the solutions used for the fixation of glycogen. – Acta anat. (Basel) *56,* 184–195 (1964).

KUGLER, P., WROBEL, K.-H.: Studies on the optimalisation and standardisation of the light microscopic succinate dehydrogenase histochemistry. – Histochemistry *57,* 47–60 (1978).

KURZON, R. M., STERNBERGER, L. A.: Estrogen receptor immunocytochemistry. – J. Histochem. Cytochem. *26,* 803–808 (1978).

LACOUR, L. F., CHAYEN, J., GAHAN, P. S.: Evidence for lipid material in chromosomes. – Exp. Cell Res *14,* 469–485 (1958).

LAGERSTEDT, S.: The release of ribonucleic acid from Carnoy fixed sections during incubation in Mellvaines buffer. – Experientia (Basel) *12,* 425 (1956).

LAKE, B. D.: The histochemical evaluation of the glycogen storage diseases. A review of techniques and their limitations. – Histochem. J. *2,* 444–450 (1970).

LAKE, B. D., ELLIS, R. B.: What do you think you are quantifying? An appraisal of histochemical methods in the measurement of the activities of lysosomal enzymes. – Histochem. J. *8,* 357–366 (1976).

LANE, B. P., EUROPA, D. L.: Differential staining of ultrathin sections of epon-embedded tissues for light microscopy. – J. Histochem. Cytochem. *13,* 579–582 (1965).

LAURILA, P., VIRTANEN, I., WARTIOVAARA, J., STENMAN, S.: Fluorescent antibodies and lectins stain intracellular structures in fixed cells treated with nonionic detergent. – J. Histochem. Cytochem. *26,* 251–257 (1978).

LECHAGO, J., SUN, N. C. J., WEINSTEIN, W. M.: Simultaneous visualisation of two antigens in the same tissue section by combining immunoperoxidase with immunofluorescence techniques. – J. Histochem. Cytochem. *27,* 1221–1225 (1979).

LEDINGHAM, J. M., SIMPSON, F. O.: Intensification of osmium staining by p-phenylene diamine: paraffin and epon embedding; lipid granules in renal medulla. – Stain Technol. *45,* 255–260 (1970).

LEDUC, E. H., AVRAMEAS, S., BOUTEILLE, M.: Ultrastructural localisation of antibody in differentiating plasma cells. – J. exp. Med. *127,* 109–118 (1968).

LEDUC, E. H., BERNHARD, W., TOURNIER, P.: Cyclic appearance of a typical mitochondria containing DNA fibres in cultures of an adenovirus 12-induced hamster tumour. – Exp. Cell Res. *42,* 597–616 (1966).

LEEMAN, U., RUCH, F.: Cytofluorimetric determination of basic and total proteins with sulfaflavine. – J. Histochem. Cytochem. *20,* 659–671 (1972).

LEMIN, D. R., VICKERSTAFF, T.: Some physical-chemical properties of damaged wools. – In: Fibrous Proteins. – Society of Dyers and Colourists, Bradford (U. K.) 129–141, 1946.

LEV, R., GERARD, A.: The histochemical demonstration of protein in epithelial mucins. – J. roy. micr. Soc. *87,* 361–373 (1966).

LEV, R., SPICER, S. S.: A histochemical comparison of human epithelial mucins in normal and hypersecretory states including pancreatic cystic fibrosis. – Amer. J. Path. *46,* 23–47 (1965).

LEWIS, G. N.: Rules for the absorption spectra of dyes. – J. Amer. chem. Soc. *67,* 770–775 (1945).

LEWIS, P. R., KNIGHT, D. P.: Staining Methods for Sectioned Material, Part 1 of Practical Methods in Electron Microscopy, *5,* (GLAUERT, A. ed.) – North-Holland, Amsterdam 1977.

LILLIE, R. D.: Histochemical acylation of hydroxyl and amino groups. Effect on the periodic acid Schiff reaction, anionic and cationic dye and van Gieson stains. – J. Histochem. Cytochem. *12,* 821–841 (1964).

–: Histopathologic Technic and Practical Histochemistry, 3rd edition. – McGraw-Hill, New York 1965.

–: H. J. Conn's Biological Stains, 8th edition. – Williams and Wilkins, Baltimore 1969a.

–: Mechanism of chromation hematoxylin stains. – Histochemie *20,* 338–354 (1969b).

–: Conn's Biological Stains, 9th edition. – Williams and Wilkins, Baltimore 1977.

LILLIE, R. D., BERGEAUX, G., HENDERSON, R.: Hot chromation oxyphilia. – Histochemie *11,* 332–349 (1967).

LILLIE, R. D., FULLMER, H. M.: Histopathologic Technic and Practical Histochemistry, 4th edition – McGraw-Hill, New York 1976.

LILLIE, R. D., JIRGE, S. K.: Methylation or lactoneisation of acid mucins by acid methanolic reagents in histochemistry. – Histochemistry *41,* 249–256 (1975).

LILLIE, R. D., PIZZOLATO, P.: Mechanisms of Iron (II) and Iron (III) sequence hematoxylin stains. – J. Histochem. Cytochem. *20,* 116–129 (1972).

LILLIE, R. D., PIZZOLATO, P., HENDERSON, R., DONALDSON, P.: The influence of formaldehyde fixation time on the rate of nitrous acid deamination of erythrocytes and spinal cord proteins. – Histochemie *24,* 156–158 (1970).

LIPSETT, M. N.: Aggregation of guanine oligoribonucleotides and the effect of mercuric salts. – J. biol. Chem. *239,* 1250–1255 (1964).

LISON, L.: Staining differences in cell nuclei. – Quart. J. micro. Sci. *96,* 227–237 (1955).

LISTER, J. H.: Fused Pyrimidines, Part II, Purines. – Wiley, New York 1971.

LISY, V., KOVARU, H., LODIN, Z.: *In vitro* effects of polyvinylpyrolidine and sucrose on the acetylcholine esterase, succinate dehydrogenase and lactate dehydrogenase activities in the brain. – Histochemie *26,* 205–211 (1971).

LIVINGSTON, D. C., COOMBS, M. M., FRANKS, L. M., MAGGI, V., GAHAN, P. B.: Lead phthallocyanine method for the demonstration of acid hydrolases in plant and animal tissues. – Histochemie *18,* 48–60 (1969).

LIVINGSTON, D. C., FISHER, S. W., GREENOAK, G. E., MAGGI, V.: A new procedure for the ultrastructural localisation of acid phosphatases. – Histochemie *24,* 159–167 (1970).

Llewellyn, B. D.: Improved nuclear staining with Mordant Blue 3 as a hematoxylin substitute. – Stain Technol. *53,* 73–77 (1978).
Locke, M., Krishman, N.: Hot alcoholic phosphotungstic acid and uranyl acetate as routine stains for thick and thin sections. – J. Cell Biol. *50,* 550–557 (1971).
Lohr, W., Sohmer, I., Wittekind, D.: The azur dyes: their purification and physicochemical properties. I. Purification of Azur A. – Stain Technol. *49,* 359–366 (1974).
Lojda, Z.: Some remarks concerning the histochemical detection of disaccharidases and glucosidases. – Histochemie *5,* 339–360 (1965).
Lojda, Z., Elleder, M.: Remarks on the detection of osmium derivatives on tissue sections. – Histochemie *13,* 276–282 (1968).
Lojda, Z., Malis, F.: Histochemical demonstration of enterokinase. – Histochemie *32,* 23–29 (1972).
Lojda, Z., Vecerek, B., Pelichova, H.: Some remarks concerning the histochemical detection of acid phosphatase by azo-coupling reactions. – Histochemie *3,* 428–454 (1964).
Luft, J. H., Wood, R. L.: The extraction of tissue protein during and after fixation with osmium tetroxide in various buffer systems. – J. Cell Biol. *19,* 46A (1963).
Lukas, Z., Dvorak, K., Jurankova, Z.: Improved localisation of thiamine pyrophosphatase activity in mounted cryostat sections using gel fixation and gel incubation media. – Histochemistry *56,* 345–347 (1978).
Luskin, L. S., Myers, R. J.: Acrylic ester polymers. – In: Encyclopedia of Polymer Science and Technology, Wiley-Interscience, New York, *1,* 246–328, 1964.
McGee-Russel, S. M., Smale, N. B.: On colouring epon-embedded tissue sections with Sudan Black B or Nile Blue A for light microscopy. – Quart. J. micro. Sci. *104,* 109–115 (1963).
McMillan, P. J.: Differential demonstration of muscle and heart type lactic dehydrogenase of rat muscle and kidney. – J. Histochem. Cytochem. *15,* 21–31 (1967).
Mahon, D. C., Nair, K. K.: Stoichiometry of the aldehyde fuchsin staining reaction, for proteins. – J. Histochem. Cytochem. *23,* 652–659 (1975).
Maltha, J. C., Bex, J. H. M., Nottet, S. J. A. M.: Procion Brilliant Orange, a counterstain for decalcified sections vitally labelled with Procion Brilliant Red H-8 BS. – Stain Technol. *52,* 211–215 (1977).
Mann, D. M. A., Barton, C. M., Davies, J. S.: Post-mortem changes in human central nervous tissue and the effects on quantitation of nucleic acids and enzymes. – Histochem. J. *10,* 127–135 (1978).
Mann, G.: Physiological Histology. – Clarendon Press, Oxford 1902.
Marmion, D. M.: Handbook of US Colorants for Food, Drugs, and Cosmetics. – Wiley, New York 1979.
Marmur, J., Rownd, R., Schildkraut, C. L.: Denaturation and renaturation of DNA. – Progr. Nucleic Acid Res. mol. Biol. *1,* 231–300 (1963).
Marshall, P. N., Horobin, R. W.: The chemical nature of the gallocyanin-chrome alum staining complex. – Stain Technol. *47,* 155–161 (1972).
–: The mechanism of action of «mordant» dyes – a study using preformed metal complexes. – Histochemie *35,* 361–371 (1973a).
–: Measurements of the affinities of basic and «mordant» dyes for various tissue substrates. – Histochemie *36,* 303–312 (1973b).
–: The influence of inorganic salts when staining with preformed metal complex dyes. – Histochemie *37,* 299–311 (1973c).
Marshall, P. N., Lewis, S. M.: A technique for distinguishing between Methylene Violet and Methylene Violet Bernsthen. – Stain Technol. *50,* 51–53 (1975).
–: The purification of Methylene Blue and Azure B by solvent extraction and crystallisation. – Stain Technol. *50,* 375–381 (1976).
Martinez-Hernandez, A., Merrill, D. A., Naughton, M. A., Geczy, C.: Acrylamide affinity chromatography for immunohistochemistry. Purification of specific antibodies. – J. Histochem. Cytochem. *23,* 146–148 (1975).
Mason, S. F.: The electronic spectroscopy of dyes. – J. Soc. Dyers Colour. *84,* 604–612 (1968).
Maunder, C., Rost, F. W. D.: A microfluorimetric study of masked metachromasia in cells of the endocrine polypeptide (APUD) series. – Histochem. J. *4,* 145–153 (1972).
Maunsbach, A. B.: Absorption of I^{125}-labelled homogeneous albumin by rat kidney proximal tubule cells. – J. Ultrastruct. Res. *15,* 197–241 (1966).
Mayor, H. D., Hampton, J. C., Rosario, B.: A simple method for removing the resin from epoxy-embedded tissue. – J. Biophys. & Biochem. Cytol. *9,* 909–910 (1961).
Meier, H.: In: The Chemistry of Synthetic Dyes, *4,* p. 389 (Venkatarman, K. ed.). – Academic Press, New York 1971.

MEIER, W.: Untersuchungen zur Theorie der Fettfärbung. – Z. wiss. Mikrosk. *64*, 193–208 (1959).
MEIJER, A. E. F.: Histochemical method for the demonstration of myosin adenosine triphosphatase in muscle tissue. – Histochemie *22*, 51–58 (1970).
–: Semipermeable membranes for improving the histochemical demonstration of enzyme activities in tissue sections. 1. Acid phosphatase. – Histochemie *30*, 31–39 (1972).
–: Semipermeable membranes for improving the histochemical demonstration of enzyme activities in tissue sections. III. Lactate dehydrogenase. – Histochemie *35*, 165–172 (1973).
MELOAN, S. N., PUCHTLER, H.: Demonstration of amyloid with Mesitol WLS-Congo Red: Application of a textile auxillary to histochemistry. – Histochemistry *58*, 163–166 (1978).
MENKART, J., SPEAKMAN, J. B.: The production of unshrinkability by crosslinkage formation in wool. Part II. The rate of reaction of mercuric acetate with animal fibres. – J. Soc. Dyers Colour. *63*, 322–324 (1947).
MEPHAM, B. L., FRATER, W., MITCHELL, B. S.: The use of proteolytic enzymes to improve immunoglobulin staining by the PAP technique. – Histochem. J. *11*, 345–357 (1979).
MERRIAM, R. W.: Standard chemical fixations as a basis for quantitative investigations of substances other than desoxyribonucleic acid. – J. Histochem. Cytochem. *6*, 43–51 (1958).
METZ, C.: Mahons myelin stain for celloidin-embedded nervous tissue. – Stain Technol. *51*, 59–61 (1976).
MEYBECK, J., GALAFASSI, P.: The effect of hydrocarbon substituents in azo dyes in wool dyeing. – Appl. Polym. Symp. No. *18*, 463–472 (1971).
MEYER, F. A., GELMAN, R. A., SILVERBERG, A.: Hydrogels for Medicine and Related Applications. In: American Chemical Society Symposium Series, *31*, Chapter 3. – American Chemical Society, Washington 1976.
MILAS, N. A., ILLIOPULOS, M. I.: Organic osmiates. – J. Amer. chem. Soc. *81*, 6089 (1959).
MILLER, H. R. P.: Fixation and tissue preservation for antibody studies. – Histochem. J. *4*, 305–320 (1972).
MILLICAN, G. F.: The use of Kiton Fast Green to measure the viability of the head of the femur after fracture of the neck of the femur. – Injury *10*, 235–238 (1979).
MILLONIG, G., MARINOZZI, V.: Fixation and embedding in electron microscopy. – Adv. Optical Electron Micr. *2*, 251–273 (1968).
MITCHELL, J. P.: Model system studies of staining procedures for lysine and arginine residues. – Histochemistry *52*, 151–157 (1977).
MITRENGA, D., ARNOLD, W., MAYERSBACH, H. V.: Freeze drying and embedding in glycol methacrylate (GMA). The results of morphological, histochemical and immunohistological investigations. – Histochemistry *39*, 313–326 (1974).
MOE, H., BEHNKE, O., ROSTGAARD, J.: Staining of osmium fixed Vestopal embedded tissue sections for light microscopy. Acta anat. (Basel) *48*, 142–148 (1962).
MOLENAAR, I., SIXMA, J. J., LINSSEN, W. H.: Immune ferritin labelling of ultrathin sections. – J. Histochem. Cytochem. *14*, 766 (1966).
MÖLLENDORFF, W. v., MÖLLENDORFF, M. v.: Untersuchungen zur Theorie der Färbung fixierter Präparate. III. Durchtränkungs- und Niederschlagsfärbung als Haupterscheinungen bei der histologischen Färbung. – Ergeb. Anat. Entwickl.-Gesch. *25*, 1–66 (1924).
MONNERON, A. MOULÉ, Y.: Critical evaluation of specificity in electron microscopical radioautography in animal tissues. – Exp. Cell Res. *56*, 179–193 (1969).
MOORE, W. J.: Physical Chemistry, 4th edition. – Longmans, London 1963.
MORAWETZ, H.: Macromolecules in Solution. – Interscience, New York 1965.
MORIARTY, G. C., MORIARTY, C. M., STERNBERGER, L. A.: Ultrastructural immunocytochemistry with unlabelled antibodies and the peroxidase-antiperoxidase complex. A technique more sensitive than the radioimmunoassay. – J. Histochem. Cytochem. *21*, 825–833 (1973).
MOSES, H. L., ROSENTHAL, A. S., BEAVER, D. L., SCHUFFMAN, S. S.: Lead ion and phosphatase histochemistry. II. Effect of ATP hydrolysis by lead iron on the histochemical localisation of ATPase activity. – J. Histochem. Cytochem. *14*, 702–710 (1966).
MOSS, C. A.: Glycosaminoglycans of disaggregated foetal mouse brain tissue cultures. – Histochem. J. *5*, 547–566 (1973).
MOVAT, H. Z.: Silver impregnation methods for electron microscopy. – Amer. J. clin. Path. *35*, 528–537 (1961).
MOWRY, R. W.: A comparison of technical procedures for the histochemical preservation of alkaline phosphatase. – Bull. int. Ass. med. Mus. *30*, 95–98 (1949).
MOWRY, R. W., SCOTT, J. E.: Observations on the basophilia of amyloids. – Histochemie *10*, 8–32 (1967).
MUCKERGEE, P., GHOSH, A. K.: The effect of urea on methylene blue, its self-association, and interaction with polyelectrolytes in aqueous solution. – J. phys. Chem. *67*, 193–197 (1963).

Muckergee, P., Ghosh, A. K.: The «isoextraction» method and the study of the self association of methylene blue in aqueous solutions . – J. Amer. chem. Soc. *92,* 6403–6407 (1970).

Munger, B. L.: Staining methods applicable to sections of osmium-fixed tissue for light microscopy. – J. biophys. biochem. Cytol. *11,* 502–506 (1961).

Munz, K., Meves, C.: Digestability of elastic fibres by elastase in semifine and fine sections as a function of fixation mode and embedding media. – Histochemistry *40,* 181–190 (1974).

Murray, M.: The significance of Grams stain in mammalian tissue. – J. Histochem. Cytochem. *10,* 36–38 (1962).

Murrell, J. N., Kettle, S. F. A., Tedder, J. M.: The Chemical Bond. – Wiley, Chichester 1978.

Nachlas, M. M., Crawford, D. T., Seligman, A. M.: The histochemical demonstration of leucine amino-peptidase. – J. Histochem. Cytochem. *5,* 264–278 (1957).

Nachlas, M. N., Goldstein, T. P., Rosenblatt, D. H., Kirsch, M., Seligman, A. M.: Influence of chemical structure on the rate of azo coupling and its significance in histochemical methodology. – J. Histochem. Cytochem. *7,* 50–65 (1959).

Nachlas, M. M., Young, A. C., Seligman, A. M.: Problems of enzymatic localisation by chemical reactions applied to tissue sections. – J. Histochem. Cytochem. *5,* 565–583 (1957).

Nakane, P. K.: Simultaneous localisation of multiple tissue antigens using the peroxidase-labelled antibody method: a study on the pituitary glands of the rat. – J. Histochem. Cytochem. *16,* 557–560 (1968).

Nakane, P. K., Kawaoi, A.: Peroxidase-labelled antibody. A new method of conjugation. – J. Histochem. Cytochem. *22,* 1084–1091 (1974).

Nakane, P. K., Pierce, G. B.: Enzyme-labelled antibodies for the light and electron microscopical localisation of tissue antigens. – J. Cell Biol. *33,* 307–318 (1967).

Neil, M. W., Horner, M. W.: The acid-p-nitrophenyl phosphatase activity of adult guinea-pig liver subcellular fractions. – Biochem. J. *84,* 32 (1962).

Nielson, A. J., Griffith, W. P.: Tissue fixation by osmium tetroxide. A possible role for proteins. – J. Histochem. Cytochem. *27,* 997–999 (1979).

Norton, W. T., Korey, S. R., Brotz, M.: Histochemical demonstration of unsaturated lipids by a bromine-silver method. – J. Histochem. Cytochem. *10,* 83–88 (1962).

Novelli, A.: A short method for chondriome. – J. Histochem. Cytochem. *10,* 102–103 (1962).

Novikoff, A. B.: Electron transport enzymes: biochemical and tetrazolium staining studies. – In: Histochemistry and Cytochemistry, Proceedings of the First International Congress of Histochemistry and Cytochemistry, Paris 1960, p. 465 (Wegmann, R. ed.) – Pergamon Press, Oxford 1963.

Novikoff, A. B., Albala, A., Biempica, L.: Ultrastructural and cytochemical observations on B-16 and Harding-Passey mouse melanomas. The origin of premelanosomes and compound melanosomes. – J. Histochem. Cytochem. *16,* 299–319 (1968).

Novikoff, A. B., Novikoff, P. M., Quintana, N., Davis, C.: Diffusion artifacts in 3,3'-diaminobenzidine cytochemistry. – J. Histochem. Cytochem. *20,* 745–749 (1972).

Opitz, M., Hahn von Dorsche, H.: Identification of glycogen with Alizarin Blue. – Acta histochem. (Jena) *45,* 219–223 (1973).

Ostrowski, K., Komender, K., Kwarecki, K.: Quantitative investigations on the solubility of proteins extracted from tissues fixed by different chemical and physical methods. – Ann. histochem. *6,* 501–506 (1961).

Ovadia, M., Stoward, P. J.: Some quantitative histochemical studies of the periodic acid oxidation of glycogen in situ. – Histochem. J. *3,* 233–240 (1971).

Palladini, G., Solheid, C.: Adsorption of free and antibody-conjugated horseradish peroxidase to fixed glial cells in tissue culture and possible source of error in immunohistochemistry. – Histochemistry *50,* 87–92 (1976).

Papadimitriou, J. M., van Duijn, P.: Effects of fixation and subtrate protection on the isoenzymes of aspartate aminotransferase studied in a quantitative cytochemical model system. – J. Cell Biol. *47,* 71–83 (1970).

Pappenheim, A.: Die morphologische Hämatologie. – Folia haemat. (Lpz.) *22,* 1–532 (1918).

Pearlmutter, N. L., Lembi, C. A.: Localisation of chitin in algal and fungal cell walls by light and electron microscopy. – J. Histochem. Cytochem. *26,* 782–791 (1978).

Pearse, A. D., Marks, R.: Measurement of section thickness in quantitative microscopy with special reference to enzyme histochemistry. – J. clin. Path. *27,* 615–618 (1974).

Pearse, A. G. E.: Histochemistry, Theoretical and Applied, *1,* (3rd edition). – Churchill, London 1968.

–: Histochemistry, Theoretical and Applied, *2,* (3rd edition) – Churchill–Livingstone, Edinburgh 1972.

PEARSE, A. G. E., HESS, R.: Substantivity and other factors responsible for formazan patterns in dehydrogenase histochemistry. – Experimentia (Basel) *17*, 136–141 (1961).

PEARSE, A. G. E., POLAK, J. M.: Bifunctional reagents as vapour- and liquid-phase fixatives for immunohistochemistry. – Histochem. J. *7*, 179–186 (1975).

PEARSON, R. G.: Hard and soft acids and bases, HSAB, Part 1, Fundamental principles. – J. chem. Educ. *45*, 581–587 (1968).

PENTTIHA, A., McDOWELL, E. M., TRUMP, B. F.: Effects of fixation and post-fixation treatments on volume of injured cells. – J. Histochem. Cytochem. *23*, 251–270 (1975).

PEPE, F. A., FINCK, H.: The use of specific antibody in electron microscopy. II. The visualisation of mercury-labelled antibody in the electron microscope. – J. biophys, biochem. Cytol. *11*, 521–531 (1961).

PETERS, A.: Experiments on the mechanism of silver staining. Part I. Impregnation. – Quart. J. micr. Sci. *96*, 84–102 (1955a).

–: Experiments on the mechanism of silver staining. Part II. Development. Quart. J. micr. Sci. *96*, 103–115 (1955b).

PETERS, A., HINDS, P. L., VAUGHN, J. E.: Extent of stain penetration in sections prepared for electron microscopy. – J. Ultrastruct. Res. *36*, 37–45 (1971).

PETERS, R. H.: In Diffusion in Polymers (J. Crank and G. S. Park, eds.), p. 315. – Academic Press, New York and London 1968.

PETERS, T., ASHLEY, C. A.: An artifact of radioautography due to binding of free amino acids to tissues by fixatives. – J. Cell Biol. *33*, 53–60 (1967).

PETRERE, J. A., SCHARDEIN, J. L.: Microwave fixation of fetal specimens. – Stain Technol. *52*, 113–114 (1977).

POOL, C. R.: Prestaining oxidation by acidified H_2O_2 for revealing Schiff-positive sites in epon-embedded sections. – Stain Technol. *48*, 123–126 (1973).

POOLE, A. R.: The relationship between toluidine blue staining and hexauronic acid content of cartilage matrix. – Histochem. J. *2*, 425–430 (1970).

PRATTEN, M., WILLIAMS, M. A., COPE, G. H.: Compartmentation of enzymes in rabbit parotid salivary gland. A study by enzyme histochemical, tissue fractionation and morphometric techniques. – Histochem. J. *9*, 573–598 (1977).

PRENNA, G., BOTTIRIOLI, G., MAZZINI, G.: Cytofluorimetric quantification of the activity and reaction kinetics of acid phosphatase. – Histochem. J. *9*, 15–30 (1977).

PRENTO, P., LYON, H.: Nucleoprotein staining. An analysis of a standardised trichloracetic acid-Fast Green FCF procedure. – Histochem. J. *5*, 493–501 (1973).

PRUE, J. E., CAVANAGH, K., SCOTT, J. E.: A thermodynamic study of the specific interactions of sodium, potassium, magnesium, and calcium ions with anions in aqueous solutions. – J. chem. Thermodynamics *6*, 93–101 (1974).

PUCHTLER, H., SWEAT, F.: Amidoblack as a stain for hemoglobin. – Arch. Path. *73*, 245–249 (1962).

PUCHTLER, H., SWEAT, F., BATES, R., BROWN, J. H.: On the mechanism of Resorcin-fuchsin staining. – J. Histochem. Cytochem. *9*, 553–559 (1961).

PUCHTLER, H., SWEAT WALDROP, F.: On the mechanism of Verhoeff's Elastica stain: a convenient stain for myelin sheaths. – Histochemistry *62*, 233–247 (1979).

PUCHTLER, H., SWEAT WALDROP, F., MELOAN, S. N.: Effects of acetylation and benzoylation on dye binding: investigations of molecular alterations in models. – Histochemistry *58*, 65–70 (1978).

PUTNAM, F. W.: The Proteins (NEURATH, H., BAILEY, K., eds.), *1*, Part B, Chapter 9. – Academic Press, New York 1953.

QUINTARELLI, G., SCOTT, J. E., DELLOVO, M. C.: The chemical and histochemical properties of Alcian Blue. II. Dye binding of tissue polyanions. – Histochemie *4*, 86–98 (1964).

QUINTARELLI, G., ZITO, R., CIFONELLI, J. A.: On phosphotungstic acid staining. Part I. – J. Histochem. Cytochem. *19*, 641–647 (1971a).

–: On phosphotungstic acid staining. Part II. – J. Histochem. Cytochem. *19*, 648–653 (1971b).

RADMEHR, B., BUTLER, W. F.: Leaching of glycosaminoglycan from cow uterus and vagina into fixative solution. – Histochem. J. *10*, 465–468 (1978).

RATTEE, I. D.: The use of non-ionic level-dyeing assistants with chrome-complex dyes. – J. Soc. Dyers Colour. *69*, 288–295 (1953).

RATTEE, I. D.: The chemistry of dyeing. – Chem. Soc. Revs. (Lond.) *1*, 145–162 (1972).

RATTEE, I. D., BREUER, M. M.: The Physical Chemistry of Dye Adsorption. – Academic Press, London 1974.

REEMPTS, J. V., BORGERS, M.: A simple polychrome stain for conventionally fixed epon-embedded tissues. – Stain Technol. *50*, 19–23 (1975).

Reid, P. E., Culling, C. F. A., Dunn, W. L.: Saponification-induced increase in the periodic acid-Schiff reaction in the gastrointestinal tract mechanism and distribution of the reactive substance. – J. Histochem. Cytochem. *21*, 473–482 (1973).

–: A histochemical method for the identification of 9-0-acyl sialic acids. An investigation of bovine submaxillary gland and intestinal mucins. – J. Histochem. Cytochem. *26*, 187–192 (1978).

Renaud, S.: Superiority of alcoholic over aqueous fixation in the histochemical detection of calcium. – Stain Technol. *34*, 267–271 (1959).

Ritter, J., Wenk, H., Gerth, C. H.: Quantitative Untersuchungen zum Diffusionsproblem enzymatisch reduzierter Pyridinnucleotide; ein Beitrag zum histochemischen Nachweis pyridinnucleotidabhängiger Dehydrogenasen. – Acta. histochem. (Jena) *40*, 51–63 (1971).

Roberts, G. P.: Histochemical determination of sialic acid residues using periodic acid oxidation. – Histochem. J. *9*, 97–102 (1977).

Rodd, E. H.: Chemistry of the Carbon Compounds. – Numerous volumes published at various dates, the cumulative index was in *5*. – Elsevier, Amsterdam 1962.

Roos, R.: The «chessboard» distribution of glycogen in liver. Artifact of fixation or effect of an enzyme? – Histochem. J. *6*, 511–521 (1974).

Roozemond, R. C.: The effects of fixation with formalin and glutaraldehyde on the composition of phospholipids extractable from rat hypothalamus. – J. Histochem. Cytochem. *17*, 482–486 (1969).

–: The staining and chromium binding of rat brain tissue and of lipids in model systems subjected to Bakers acid hematein technique. – J. Histochem. Cytochem. *19*, 244–251 (1971).

Rosenquist, T. H., Slavin, B. G., Bernick, S.: The Pearson silver-gelatin method for light microscopy of 0.5–2 µm plastic sections. – Stain Technol. *46*, 253–257 (1971).

Rosenthal, S. I., Puchtler, H., Sweat, F.: Paper chromatography of Dyes. – Archs. Path. Lab. Med. *80*, 190–196 (1965).

Rost, F. W. D., Bollmann, R., Moss, D. W.: Characterisation of alkaline phosphatase in tissue sections by microspectrofluorimetry. – Histochem. J. *5*, 567–575 (1973).

Runge, W. J., Vernier, R. L., Hartman, J. G.: A staining method for sections of osmium-fixed methacrylate-embedded tissue. – J. biophys. biochem. Cytol. *4*, 327–328 (1958).

Ruitenbeek, W., Scholte, H. R.: Polyacrylamide gel technique for the histochemical demonstration of soluble enzymes. – Histochemistry *50*, 81–89 (1976).

Rutenburg, A. M., Kim, H., Fischbein, J. W., Hanker, J. S., Wasserkrug, H. L., Seligman, A. M.: Histochemical and ultrastructural demonstration of glutamyl transpeptidase activity. – J. Histochem. Cytochem. *17*, 517–526 (1969).

Rys, P., Zollinger, H.: Fundamentals of the Chemistry and Applications of Dyes. – Wiley-Interscience, London 1972.

Sabatini, D. D., Miller, F., Barrnett, R. J.: Aldehyde fixation for morphological and enzyme histochemical studies with the electron microscope. – J. Histochem. Cytochem. *12*, 57–71 (1964).

Salthouse, T. N.: Alcian Blue staining of nuclear chromatin in insect gut after beta-glucuronidase treatment. – Stain Technol. *36*, 63–67 (1961).

Salthouse, T. N., MacDonald, B. G., Agnello, D.: Vinyl sulphone dyes: possible applications in histology. – Stain Technol. *46*, 245–248 (1971).

Sandritter, W., Hartlieb, J.: Quantitative Untersuchungen über den Nukleinsäureverlust des Gewebes bei Fixierung und Einbettung. – Experientia (Basel) *11*, 313–314 (1955).

Sandritter, W., Kiefer, G., Rick, W.: Introduction to Quantitative Cytochemistry (Wied, G. L. ed.), *1*, p. 295. – Academic Press, New York 1966.

Sarkar, A. K.: Fluorescent Whitening Agents. – Merrow, Watford (England) 1971.

Sasaki, M., Fishman, W. H.: In: Electron Microscopy and Cytochemistry, Proc. Second Int. Symp. (Drienerlo) 1973 (Wisse, E., Daems, W. Th., Molenar, I., van Duijn, P., eds.) p. 59. – North-Holland Amsterdam 1974.

Sasaki, M., Takeuchi, T.: Histochemical and electron microscopic observations of glycogen synthesized from glucose-1-phosphatase by phosphorylase and branching enzyme in human muscle. – J. Histochem. Cytochem. *11*, 342–348 (1963).

Sato, T., Shamoto, M.: A simple rapid polychrome stain for epoxy-embedded tissue. – Stain Technol. *48*, 223–227 (1973).

Saunders, D. R., Wilson, J., Rubin, C. E.: Loss of absorbed lipid during fixation and dehydration of jejunal mucosa. – J. Cell Biol. *37*, 183–187 (1968).

Scallen, T. J., Dietert, S. E.: The quantitative retention of cholesterol in mouse liver prepared for electron microscopy by fixation in a digitonin-containing aldehyde solution. – J. Cell Biol. *40*, 802–813 (1963).

SCARPELLI, D. G., PEARSE, A. G. E.: Physical and chemical protection of cell constituents and the precise localisation of enzymes. – J. Histochem. Cytochem. *6,* 369–376 (1958).

SCHWALBACH, G., LICKFELD, K. G., HOFFMEISTER, H.: Differential staining of osmium tetroxide-fixed, Vestopal W-embedded tissue in thin sections. – Stain Technol. *38,* 15–21 (1963).

SCHWARTZ, D. S., LARSH, H. W., BARTELS, P. A.: Enumerative fluorescent vital staining of live and dead pathogenic yeast cells. – Stain Technol. *52,* 203–210 (1977).

SCOTT, J. E.: CEC effects seen in solution, with DNA plus various salts and an organic cation. – Biochem. J. *84,* 270–275 (1962).

–: On the mechanism of the methyl green-pyronin stain for nucleic acids. – Histochemie *9,* 30–47 (1967).

–: In: The Chemical Physiology of Mucopolysaccharides (QUINTARELLI, G. ed.), p. 219. – Little, Brown, Boston 1968.

–: Histochemistry of Alcian Blue. I. Metachromasia of Alcian Blue, Astra blau and other cationic phthalocyanine dyes. – Histochemie *21,* 277–285 (1970).

–: Phosphotungstate: a «universal» (non-specific) precipitant for polar polymers in acid solution. – J. Histochem. Cytochem. *19,* 689–691 (1971).

–: Histochemistry of Alcian Blue. II. The structure of Alcian Blue 8GX. – Histochemie *30,* 215–234 (1972).

–: Affinity, competition and specific interactions in the biochemistry and histochemistry of polyelectrolytes. – Biochem. Soc. Trans. *1,* 787–806 (1973).

–: The Feulgen reaction in polyvinylalcohol or polyethylene glycol solution. «Fixation» by excluded volume. – J. Histochem. Cytochem. *22,* 833–835 (1974).

SCOTT, J. E., DORLING, J.: Differential staining of acid glycosaminoglycans (mucopolysaccharides) by alcian blue in salt solutions. – Histochemie *5,* 221–233 (1965).

–: Periodate oxidation of acid polysaccharides. III. A PAS method for chondroitin sulphates and other glycosamino-glycans. – Histochemie *19,* 295–301 (1969).

SCOTT, J. E., HARBINSON, J.: Periodic oxidation of acid polysaccharides. Inhibition by the electrostatic field of the substrate. – Histochemie *14,* 215–220 (1968).

SCOTT, J. E., TIGWELL, M. J., SAJDERA, S. W.: Loss of basophilic (sulphated) material from sections of cartilage treated with periodate solution. – Histochem. J. *4,* 155–167 (1972).

SCOTT, J. E., WILLETT, I. H.: Binding of cationic dyes to nucleic acids and other biological polyanions. – Nature (Lond.) *209,* 985–987 (1966).

SEIBERT, G. R., BENJAMINSON, M. A., HOFFMAN, H.: A conjugate of cellulase with fluorescein isothiocyanate: a specific stain for cellulose. – Stain Technol. *53,* 103–106 (1978).

SEIDLER, E., KUNDE, D.: Zur Substantivität von Tetrazoliumsalzen in der Histochemie. – Histochemie *17,* 256–262 (1969).

SEKI, M.: Substantive (direkte) Färbung der histologisch fixierten Präparate. – Folia anat. jap. *10,* 635–654 (1932).

–: Zur physikalischen Chemie der histologischen Färbung. V. Anteilnahme der Dispersität der Farben und der Dichte der zu färbenden Strukturen an histologischen Färbungsvorgängen, besonders an der Elastikafärbung. – Folia anat. jap. *11,* 377–390 (1933).

SELIGMAN, A. M.: Unreliability of ultrastructural demonstration of enzymes which depend upon hydrogen peroxide production from action on their substrates. – J. Histochem. Cytochem. *19,* 809 (1971).

SELIGMAN, A. M., WASSERKRUG, H. L., PLAPINGER, R. E., SAITO, T., HANKER, J. S.: Membraneous ultrastructural demonstration of aminopeptidase and γ-glutamyl transpeptidase activities with a new diazonium salt that yields a lipophobic, osmiophilic azo dye. – J. Histochem. Cytochem. *18,* 542–551 (1970).

SELWYN, J. E., STEINFELD, J. I.: Aggregation equilibria of xanthene dyes. – J. phys. Chem. *76,* 762–774 (1972).

SEVIER, A. C., MUNGER, B. L.: The use of oxone to facilitate specific tissue stainability following osmium fixation. – Anat. Rec. *162,* 43–51 (1968).

SHAALA, A. Y.: Ingestion of dyed-opsonised yeast as a simple way of detecting phagocytes in lymphocyte preparations. Cytophilic binding of immunoglobulins by ingesting cells. – J. immun. Meth. *27,* 175–187 (1979).

SHALLA, T. A., CARROLL, T. W., DEZOETEN, G. A.: Penetration of stain in ultrathin sections of tobacco mosaic virus. – Stain Technol. *39,* 257–265 (1964).

SHEA, J. R.: A method for *in situ* cytophotometric estimation of absolute amount of ribonucleic acid using Azur B. – J. Histochem. Cytochem. *18,* 143–152 (1970).

SHIRES, T. K., JOHNSON, M., RICHTER, K. M.: Hematoxylin staining of tissues embedded in epoxy resins. – Stain Technol. *44,* 21–25 (1969).

SHUGAR, D.: Quantitative staining in histo- and cytochemistry. – Progr. Biophys. biophys. Chem. (BUTLER, J. A. V., HUXLEY, H. E., ZIRKLE, R. E., eds.). 153–210. – Pergamon Press, Oxford 1962.

SIBATINI, A.: Feulgen reaction and quantitative cytochemistry of desoxypentose nucleic acid. V. Chemical determination of colour intensity of the reaction *in situ*. – Biochim. biophys. Acta (Amst.) *13*, 66–71 (1954).

SIBATINI, A., NAORA, H.: Feulgen reaction and quantitative cytochemistry of desoxypentose nucleic acid. IV. Microspectrophotometric study of the Feulgen reaction *in situ*. – Biochim. biophys. Acta (Amst.) *12*, 515–521 (1953).

SIERVERS, J.: Basic two-dye stains for epoxy-embedded 0.3–1 mμ sections. – Stain Technol. *46*, 195–199 (1971).

SILVERMAN, L., GLICK, D.: The reactivity and staining of tissue proteins with PTA. – J. Cell Biol. *40*, 761–767 (1969).

SILVERMAN, L., SCHREINER, B., GLICK, D.: Measurements of thickness within sections by quantitative electron microscopy. – J. Cell Biol. *40*, 768–778 (1969).

SIMARD, R.: The bonding of actinomycin D-^{3}H to heterochromatin as studied by quantitative high resolution autoradiography. – J. Cell Biol. *35*, 716–722 (1967).

SIMKISS, K.: Metal ions in cells. – Endeavour, N. Ser. *3*, 2–6 (1979).

SIMSON, J. A. V.: The influence of fixation on the carbohydrate cytochemistry of rat salivary gland secretory granules. – Histochem. J. *9*, 645–657 (1977).

SINGER, M.: Factors which control the staining of tissue sections with acid and basic dyes. – Int. Rev. Cytol. *1*, 211–255 (1952).

SINGER, M., MORRISON, P. R.: The influence of pH, dye and salt concentration on the dye binding of modified and unmodified fibrin. – J. biol. Chem. *175*, 133–145 (1948).

SINGER, S. J., MCLEAN, J. D.: Ferritin-antibody conjugates as stains for electron microscopy. – Lab. Invest. *12*, 1002–1008 (1963).

SINGER, S. J., PAINTER, R. G., TOKUYASU, K. T.: Ferritin-antibody staining of ultrathin-frozen sections. – In: Electron Microscopy and Cytochemistry, Proc. Sec. Int. Symp. (Drienerlo) 1973 (WISSE, E., DAEMS, W. Th., MOLENAR, I., van DUIJN, P., eds.). – North Holland Amsterdam 1974.

SIPPEL, T. O.: The histochemistry of thiols and disulphides. II. Methodology of differential staining. – Histochem. J. *10*, 585–595 (1978).

SJÖSTRAND, F. S.: Electron Microscopy of Cells and Tissues, *1*, Instrumentation and Techniques. – Academic Press, New York 1967.

SMITH, R. E., FISHMAN, W. H.: p-(Acetoxymercuric) aniline diazotate, a reagent for visualising the Naphthol AS-BI product of acid hydrolase action at the level of the light and electron microscope. – J. Histochem. Cytochem. *17*, 1–22 (1969).

SMITHERMAN, M. L., LAZAROW, A., SORENSON, R. L.: The effect of light microscope fixatives on the retention of glycogen in protein matrices and the particulate state of native glycogen. – J. Histochem. Cytochem. *20*, 463–471 (1972).

SNODGRADD, A. B., DORSEY, C. H., BAILEY, G. W. H., DICKSON, L. G.: Conventional histopathological staining methods compatible with epon-embedded, osmicated tissue. – Lab. Invest. *26*, 329–337 (1972).

The society of dyers and colourists: The Colour Index, 3rd edition. – The Society of Dyers and Colourists, Bradford (England) 1971.

SÖDERSTRÖM, K.-O., PARVINEN, M., PARVINEN, M.: Early detection of cell damage by supravital Acridine Orange staining. – Experimentia (Basel) *33*, 265–266 (1977).

SORVARI, T. E.: Histochemical observations on the role of ferric chloride in the high-iron diamine technique for localizing sulphated mucosubstances. – Histochem. J. *4*, 193–204 (1972a).

–: Binding of ferric ions to nuclei and other tissue sites in sections stained for sulphated mucosubstances by the high iron diamine method. – Stain Technol. *47*, 245–248 (1972b).

SORVARI, T. E., LAUREN, P. A.: The effect of various fixative procedures on the digestability of sialomucins. – Histochem. J. *5*, 405–412 (1973).

SORVARI, T. E., STOWARD, P. J.: Some investigations of the mechanisms of the so-called «methylation» reactions used in mucosubstance histochemistry. I. «Methylation» with methyl iodide, diazomethane and various solvents containing either hydrogen chloride or thionyl chloride. – Histochemie *24*, 106–117 (1970a).

–: Some investigations of the so-called «methylation» reactions used in mucosubstance histochemistry. II. Histochemical differentiation of lactone and ester groups in «methylation» of mucosubstances. – Histochemie *24*, 114–119 (1970b).

SPALL, R. D., AINSWORTH, S. K.: A technique for the isolation of single glomeruli for immunohistochemistry. – J. Histochem. Cytochem. *26,* 284–287 (1978).
SPIERS, R. L.: The reaction between some dyes and synthetic hydroxyapatite. I. The uptake of dyes in relation to their structures. – Histochem. J. *2,* 45–66 (1970).
STACEY, K. A., COBB, M., COUSENS, S. F., ALEXANDER, P.: The reactions of the «radiomimetic» alkylating agents with macromolecules *in vitro.* – Ann. N. Y. Acad. Sci. *68,* 682–701 (1957–58).
STAPLE, P. H.: Effect of periodic acid on histochemical aldehyde blockades. – Nature (Lond.) *181,* 288 (1958).
STAPLE, P. H.: Side effects of reaction media in histochemical blocking procedures. – Histochem. J. *1,* 377–381 (1969).
STEINHARDT, J. S., REYNOLDS, J. A.: Multiple Equilibria in Proteins – Academic Press, New York 1969.
STERNBERGER, L. A.: Techniques of Biochemical and Biophysical Morphology, *1,* 67–88 (D. GLICK, R. M. ROSENBAUM, eds.). – Wiley-Interscience, New York 1972.
STONE, F. W., STRATTA, J. J.: 1,2-Epoxide polymers. Ethylene Oxide Polymers. – In.: Encyclopedia of Polymer Science and Technology, *6,* 103–145 (1967).
STOREY, E.: *In vivo* staining by dis and tris azo dyes, especially of bone and elastic tissue. – Stain Technol. 43, 101–109 (1968).
STOWARD, P. J.: Studies in fluorescence histochemistry. 1. The demonstration of sulphomucin. – J. roy. micro. Soc. *87,* 215–235 (1967).
–: A histochemical study of the apparent deamination of proteins by sodium hypochlorite. – Histochemistry *45,* 213–226 (1975).
–: Criteria for the validation of quantitative histochemical enzyme techniques. – In: Trends in Enzyme Histochemistry and Cytochemistry, Ciba Foundation Symposium 73, pp. 11–31. – Excerpta Medica, Amsterdam 1980.
STRAUS, W.: Inhibition of peroxidase by methanol and by methanol-nitroferricyanide for use in immunoperoxidase procedures. – J. Histochem. Cytochem. *19,* 682–688 (1971).
–: Phenylhydrazine as inhibitor of horseraddish peroxidase for use in immunoperoxidase procedures. – J. Histochem. Cytochem. *20,* 949–951 (1972).
SUMNER, A. T., EVANS, H. J., BUCKLAND, R. A.: Mechanisms involved in the banding of chromosomes with Quinacrine and Giemsa. II. The interaction of the dyes with the chromosomal components. – Exp. Cell Res. *81,* 223–236 (1973).
SWEAT, F., MELOAN, S. N., PUCHTLER, H.: A modified One-step trichrome stain for demonstration of fine connective tissue fibres. – Stain Technol. *43,* 227–231 (1968).
SWIFT, J. A.: Comment on Allen and Perrin's l-chloromercuri ferrocene-silver method for the alleged electron histochemical demonstration of sulphydryl groups. – J. Histochem. Cytochem. *23,* 390–391 (1975).
SYLVÉN, B.: Metachromatic dye-substrate interactions. – Quart. J. micr. Sci. *95,* 327–358 (1954).
TAKAMIYA, H., BODEMER, W., VOGT, A.: Masking of protein antigen by modification of amino groups with carbobenzooxychloride (benzyl chloroformate) and demasking by treatment with non-specific protease. – J. Histochem. Cytochem. *26,* 914–920 (1978).
TAKAYA, K.: Luxol Fast Blue MBS and Phloxine: a stain for mitochondria. – Stain Technol. *42,* 207–211 (1967).
TÄLJEDAL, I. B.: Direct fluorophotometric recording of enzyme kinetics in cryostat sections. – Histochemie *21,* 307–313 (1970).
TAMPION, J., WRIGHT, P., HOLT, G.: 4,5,6,7-Tetrachlorofluorescein; a fluorescent vital stain for cell membranes. – Stain Technol. *48,* 41–42 (1973).
TANDLER, C. J.: A method for the selective removal of DNA from tissue sections. – Stain Technol. *49,* 147–152 (1974).
TANFORD, C.: Protein Denaturation. – Adv. Protein Chem, *23,* 121–282 (1968).
TAS, J.: The Alcian Blue and combined Alcian Blue-Safronin O staining of glycosaminoglycans studied in a model system and in mast cells. – Histochem. J. *9,* 205–230 (1977a).
–: Polyacrylamide films as tools for investigating qualitative and quantitative aspects of the staining of glycosamino-glycans with basic dyes. – Histochem. J. *9,* 267–276 (1977b).
TAS, J., OUD, P., JAMES, J.: The Naphthol Yellow S stain for proteins tested in a model system of polyacrylamide films and tested for practical use in histochemistry. – Histochemistry *40,* 231–240 (1974).
TERNER, J. Y.: Histochemical alkylation: a study of methyl iodide and its effect on tissues. – J. Histochem. Cytochem. *12,* 504 (1964).
TERNER, J. Y. CLARK, G.: Alkylation of amines: a study of methylation effects on amines. – J. Histochem. Cytochem. *8,* 184–188 (1960).
TERNER, J. Y., HAYES, E. R.: Histochemistry of plasmalogen. – Stain Technol. *36:* 265–278 (1961).

Thompson, S. W.: Selected Histochemical and Histopathological Methods. – Thomas, Springfield, Ill. 1966.

Thoss, K., Roth, J.: Histochemical lectin affinity technique by means of FITC-labelled serum protein fractions. – Histochemistry *49,* 67–72 (1976).

Timasheff, S. M.: Protein-solvent interactions and protein conformation. – Accounts chem. Res. *3,* 62–68 (1970).

Tock, E. P. C., Pearse, A. G. E.: The preservation of tissue mucins by freeze-drying and vapour fixation. I. Light microscopy. – J. roy. micr. Soc. *84,* 519–537 (1965).

Tock, E. P. C., Woodward, E. M., Pearse, A. G. E.: Preservation of tissue mucins by freeze-drying and vapour fixation. II. Quantitative chemical studies. – Histochemie *7,* 224–229 (1966).

Tomimatsu, Y., Jansen, E. F., Gaffield, W., Olson, A. C.: Physical chemical observations on the α_2-chymotrypsin glutaraldehyde system during formation of an insoluble derivative. – J. Colloid Interface Sci. *36,* 51–64 (1971).

Tooze, J.: Measurement of some cellular changes during the fixation of amphibian erythrocytes with osmium tetroxide solutions. – J. Cell Biol. *22,* 551–563 (1964).

Trump, B. E., Goldblatt, P. J., Stowell, R. E.: Studies on Necrosis of Mouse Liver *in vitro.* Ultrastructural alterations in the mitochondria of hepatic parenchymal cells.-Lab. Invest. *14,* 343–371 (1965).

Tsou, K. C., Goodwin, C. W., Seamond, B., Lynm, D.: Intracristral localisation of succinic dehydrogenase activity with a new osmium-containing tetrazolium salt. – J. Histochem. Cytochem. *16,* 487–489 (1968).

Tucker, F. L.: Bartholomew, J. W.: Variations in the Gram staining results caused by air moisture. – Stain Technol. *37,* 157–160 (1963).

Tyler, H. M., Williams, M. A., Kountz, S. L., Kapros, C., Dempster, W. J.: A biochemical approach to the study of rejection of canine renal transplants. 1. Some quantitative assessments of enzyme activity. – Brit. J. exp. Path. *45,* 219–225 (1964).

Tzitikas, H., Rdzok, H. J., Vatter, A. E.: Silver staining of ultrathin sections of tissues embedded in polyester resin. – Stain Technol. *37,* 293–298 (1962).

Valko, W.: Particle size in wool dyeing. – J. Soc. Dyers Colour. *55,* 173–182 (1939).

Van Dalen, J. P. R., Ahsmann, W. B. A. M., Van Duijn, P.: A method for the determination of the molar extinction coefficients of structure-linked chromophores. – Histochem. J. *2,* 324–342 (1970).

Van Dalen, J. P. R., Van Duijn, P.: Quantitative aspects of the Ninhydrin-Schiff reaction on amino cellulose films and tissue sections. – Histochemie *26,* 180–189 (1971).

Van Duijn, P.: Acrolein-Schiff, a new staining method for proteins. – J. Histochem. Cytochem. *9,* 234–241 (1961).

–: Fundamental aspects of enzyme cytochemistry. – In: Electron Microscopy Cytochemistry, Proc. Sec. Int. Symp. (Drienerlo) 1973 (Wisse, E., Daems, W. Th., Molenar, I., Van Duijn, P., eds), p. 3. – North-Holland, Amsterdam 1974.

–: Prospects for microscopic cytochemistry. – Histochem. J. *8,* 653–676 (1976).

Van Duijn, P., Pascoe, E., Van Der Ploeg, M.: Theoretical and experimental aspects of enzyme determination in a cytochemical model system of polyacrylamide films containing alkaline phosphatase. – J. Histochem. Cytochem. *15,* 631–645 (1967).

Vanha-Perttula, T., Grimley, P. M.: Loss of proteins and other macromolecules during preparation of cell cultures for high resolution autoradiography. – J. Histochem. Cytochem. *18,* 565–573 (1970).

Varriano-Marston, E., Gordon, J., Davis, E. A., Hutchinson, T. E.: Cryomicrotomy applied to the preparation of frozen hydrated muscle tissue for transmission electron microscopy. – J. Microsc. *109,* 193–202 (1977).

Veis, A.: Biological Polyelectrolytes. – Dekker, New York 1970.

Velican, C., Velican, D.: Techniques of Biochemical and Biophysical Morphology (Glick, D., Rosenbaum, R. M., eds.), *1,* 143–190. – Wiley-Interscience, New York 1972.

Venkataraman, K.: The Chemistry of Synthetic Dyes, *1* and *2* (1952), *3* and *4* (1971). – Academic Press, New York.

–: The Analytical Chemistry of Synthetic Dyes. – Wiley, Chichester (England) 1977.

Vermeer, B. J., Van Gent, C. M., De Bruijn, W. C., Booners, T.: The effect of digitonin-containing fixatives on the retention of free cholesterol and cholesterol esters. – Histochem. J. *10,* 287–298 (1978).

Vickerstaff, T.: The Physical Chemistry of Dyeing, 2nd edit. – Oliver and Boyd, London 1954.

Visser, J. W. M., Jongeling, A. A. M., Tanke, H. J.: Intracellular pH-determination by fluorescence measurements. – J. Histochem. Cytochem. *27,* 32–35 (1979).

VOAK, D., WILLIAMS, M. A.: An explanation of the failure of the direct antiglobulin test to detect erythrocyte sensitization in ABO haemolytic disease of the newborn and observations on pinocytosis of IgG Anti-A antibodies by infant (cord) red cells. – Brit. J. Haemat. *20*, 9–23 (1971).

VOGEL, A. I.: Vogel's Textbook of Practical Organic Chemistry including Qualitative Organic Analysis, 4th edition (revised by B. S. Furniss). – Longman, London 1978a.

–: Vogel's Textbook of Quantitative Inorganic Analysis including Elementary Instrumental Analysis, 4th edition (revised by Bassett, J.). – Longman, London 1978b.

WACHSMUTH, E. D., WOODHAMS, R. L.: A quantitative approach for immunofluorescence in biology: the use of antibody multilayers on nuclei. – Histochemistry *38*, 339–350 (1974).

WAGNER, D.: Zusammenhänge zwischen der Löslichkeit und der Assoziation von Farbstoffen in binaren, wäßrigen Systemen (Hydrotropie). – Text. Prax. *24*, 310–314 (1969).

WALKER, J. F.: Formaldehyde, 3rd edit. – Reinhold, New York 1964.

WANG, E., GOLDBERG, A. R.: Binding of deoxyribonuclease to actin: a new way to visualise microfilament bundles in nonmuscle cells. – J. Histochem. Cytochem. *26*, 745–749 (1978).

WARD, B. J., GLOSTER, J. A.: Lipid losses during processing of cardiac muscle for electron microscopy. – J. Microsc. *108*, 41–50 (1976).

WATERS, W. A.: Mechanisms of oxidation by compounds of chromium and manganese. – Quart. Rev. Chem. Soc. *12*, 277–300 (1958).

WATKINS, W. B.: Use of immunocytochemical techniques for the localisation of human placental lactogen. – J. Histochem. Cytochem *26*, 288–292 (1978).

WEATHERFORD, T. W., MANN, W. V.: A method of microscopic evaluation of the effects of demineralisers on complex carbohydrates of rat tissues. – J. Microsc. *99*, 91–100 (1973).

WEDNER, H. J., HOFFER, B. J., BATTENBERG, E., STEINER, A. L., PARKER, C. W., BLOOM, F. E.: A method for detecting intracellular cyclic adenosine monophosphate by immunofluorescence. – J. Histochem. Cytochem. *20*, 293–295 (1972).

WEIGELE, M., DE BERNARDO, S., LEIMGRUBER, W., CLEELAND, R., GRUNBERG, E.: Fluorescent labelling of proteins. A new approach. – Biochem. biophys. Res. Comm. *54*, 899–906 (1973).

WEINEL, P., GEYER, G.: Histochemische Untersuchungen zur Anwendung von Reaktivfarbstoffen der Remazolreihe. – Acta histochem. (Jena) *28*, 119–133 (1967).

WEISZ, P. B., ZOLLINGER, H.: Sorption-diffusion in heterogeneous systems. Part 4, Dyeing rates in organic fibres. – Trans. Faraday Soc. *64*, 1693–1700 (1968).

WENK, H., KRUG, H., FLETCHER, A. M.: Eine mikrospektrophotometrische Methode zur quantitativen Aktivitätsbestimmung von Acetylcholinesterase an Gewebeschnitten. – Acta histochem. (Jena) *45:* 37–60 (1973).

WHEWELL, C. S., STEVENS, C. B., AMIN, S. C.: Observations on the dyeing behaviour of chemically modified wools. – Appl. Polym. Symp. No. 18, 387–408. – Wiley, New York 1971.

WHITE, T.: Chemistry of the vegetable tannins. – In: The Chemistry and Technology of Leather (O'FLAHERTY, F., RODDY, W. T., LOLLAR, R. M., eds.), chapter 18, *2,* Types of Tannages. – Reinhold, New York 1958.

WHUR, P., HERSCOVICS, A., LEBLOND, C.: Radioautographic visualisation of the incorporation of galactose-^{3}H and mannose-^{3}H by rat thyroids in vitro in relation to the stages of thyroglobulin synthesis. – J. Cell Biol. *43*, 289–311 (1969).

WIDHOLM, J. M.: The use of fluorescein diacetate and phenosafronin for determining viability of cultured plant cells. – Stain Technol. *47*, 189–193 (1972).

WIESE, D. A., SIPPEL, T. O.: Microphotometric and microfluorimetric analysis of thiols and disulfides in nuclear proteins. – J. Histochem. Cytochem. *26*, 219 (1978).

WILLIAMS, G., JACKSON, D. S.: Two organic fixatives for acid mucopolysaccharides. – Stain Technol. *31*, 189–191 (1956).

WILLIAMS, M. A.: The assessment of electron microscopic autoradiographs. – Adv. Optical Electron Micr. *3*, 219–272 (1969).

–: Techniques in Protein Biosynthesis (CAMPBELL, P. N., SARGENT, J. R., eds.), *3*, 125. – Academic Press, New York 1973.

–: Autoradiography and Immunocytochemistry. – North-Holland, Amsterdam 1977.

–: Direct deposition autoradiography: can it really be detecting radioactivity? – J. Microsc. *112*, 331–337 (1978).

WILLIAMS, M. A., BABA, W. I.: The localisation of (^{3}H) Aldosterone and (^{3}H) Cortisol within renal tubular cells by electron microscope autoradiography. – J. Endocr. *39*, 543–554 (1967).

Williams, M. A., Gregory, D. W.: The use of bis-diazotised benzidine for preparing ferritin-conjugated antibodies for electron microscopy. – J. roy. micr. Soc. *86*, 397–415 (1966).
Williams, M. A., Meek, G. A.: Studies on thickness variation in ultrathin sections for electron microscopy. – J. roy. micr. Soc. *85*, 337–352 (1966).
Williams, T.: The staining of nervous elements by the Bodian method. II. The effects of some variable factors at the impregnation stage. – Quart. J. micr. Sci. *103*, 297–310 (1962).
Wolman, M.: Problems of fixation in cytology, histology and histochemistry. – Int. Rev. Cytol. *4*, 79–102 (1955).
Wolters, G. H. J., Pasma, A. I., Konijnendijk, W., Bouman, P. R.: Evaluation of the Glyoxal-bis-(2-hydroxyanil)-method for staining of calcium in model gelatin films and pancreatic islets. – Histochemistry *62*, 137–151 (1979).
Yamada, K., Shimizu, S.: The histochemistry of galactose residues of complex carbohydrates as stained by peroxidase-labelled Ricinus Communis agglutinins. – Histochemistry *53*, 143–156 (1977).
Yamane, Y., Davidson, N.: On the complexing of desoxyribonucleic acid (DNA) by mercuric ion. – J. Am. chem. Soc. *83*, 2599–2607 (1961).
Yates, D. E., Healy, T. W.: Mechanism of anion adsorption at the ferric and chromic oxide/water interface. – J. Colloid Interface Sci. *52*, 222–228 (1975).
Yensen, J.: Removal of epoxy resin from histological sections following halogenation. – Stain Technol. *43*, 344–346 (1968).
Yokota, S.: Assay of antihorseraddish peroxidase antibody in an immunocytochemical model system of polyacrylamide gel films. – J. Histochem. Cytochem. *21*, 779–784 (1973).
Yoshika, S., Umeda, T., Kurahashi, Y.: An effective reactivation of alkaline phosphatase in hard tissue completely decalcified for light and electron microscopy. – Histochemie *29*, 296–306 (1972).
Zanker, V.: Physikalisch-chemische Grundlagen der Farbstoff-Substratbeziehungen in der Histochemie. – Acta histochem. (Jena) Supplement *24*, 151–168 (1981).
Zehr, D. R.: Use of hydrogen peroxide-egg albumin to eliminate nonspecific staining in immunoperoxidase techniques. – J. Histochem. Cytochem *26*, 415–416 (1978).
Zollinger, H.: The dye and the substrate: the role of hydrophobic bonding in dyeing processes. – J. Soc. Dyers. Colour *81*, 345–350 (1965).

Subject Index

Author Index